AI-based Forecasting of Solar Photovoltaics Power Generation

Other related titles:

You may also like

- PBPO228 | Motahhir | Digital Technologies for Solar Photovoltaic Systems: From general to rural and remote installations | 2022

- PBPO238 | Shunli Wang, Liu, Yujie Wang, Stroe, Fernandez, and Guerrero | AI for Status Monitoring of Utility Scale Batteries | 2022

- PBPO217 | Milano | Advances in Power System Modelling, Control and Stability Analysis, 2nd Edition | 2022

We also publish a wide range of books on the following topics:
Computing and Networks
Control, Robotics and Sensors
Electrical Regulations
Electromagnetics and Radar
Energy Engineering
Healthcare Technologies
History and Management of Technology
IET Codes and Guidance
Materials, Circuits and Devices
Model Forms
Nanomaterials and Nanotechnologies
Optics, Photonics and Lasers
Production, Design and Manufacturing
Security
Telecommunications
Transportation

All books are available in print via https://shop.theiet.org or as eBooks via our Digital Library https://digital-library.theiet.org.

IET Energy Engineering 268

AI-based Forecasting of Solar Photovoltaics Power Generation

Edited by
Elham Shirazi and Wilfried van Sark

Institution of Engineering and Technology

About the IET

This book is published by the Institution of Engineering and Technology (The IET).

We inspire, inform, and influence the global engineering community to engineer a better world. As a diverse home across engineering and technology, we share knowledge that helps make better sense of the world, accelerate innovation, and solve the global challenges that matter.

The IET is a not-for-profit organisation. The surplus we make from our books is used to support activities and products for the engineering community and promote the positive role of science, engineering, and technology in the world. This includes education resources and outreach, scholarships and awards, events and courses, publications, professional development and mentoring, and advocacy to governments.

To discover more about the IET, please visit https://www.theiet.org/.

About IET books

The IET publishes books across many engineering and technology disciplines. Our authors and editors offer fresh perspectives from universities and industry. Within our subject areas, we have several book series steered by editorial boards made up of leading subject experts.

We peer review each book at the proposal stage to ensure the quality and relevance of our publications.

Get involved

If you are interested in becoming an author, editor, series advisor, or peer reviewer please visit https://www.theiet.org/publishing/publishing-with-iet-books/ or contact author_support@theiet.org.

Discovering our electronic content

All of our books are available online via the IET's Digital Library. Our Digital Library is the home of technical documents, eBooks, conference publications, real-life case studies, and journal articles. To find out more, please visit https://digital-library.theiet.org.

In collaboration with the United Nations and the International Publishers Association, the IET is a Signatory member of the SDG Publishers Compact. The Compact aims to accelerate progress to achieve the Sustainable Development Goals (SDGs) by 2030. Signatories aspire to develop sustainable practices and act as champions of the SDGs during the Decade of Action (2020–30), publishing books and journals that will help inform, develop, and inspire action in that direction.

In line with our sustainable goals, our UK printing partner has FSC accreditation, which is reducing our environmental impact to the planet. We use a print-on-demand model to further reduce our carbon footprint.

British Library Cataloguing in Publication Data

A catalogue record for this product is available from the British Library

ISBN 978-1-83724-019-7 (hardback)
ISBN 978-1-83724-020-3 (PDF)

Typeset in India by MPS Limited

Cover image credit: FernandoAH/E+ via Getty Images

Contents

10 Solar photovoltaic forecasting for energy system integration and control 241

*Silvana Matrone, Amirhossein Heydarian Ardakani,
Emanuele Ogliari, Sonia Leva and Elham Shirazi*

Preface

Over the last few years, photovoltaic (PV) power has seen exponential growth, becoming the most affordable energy source and leading in new capacity added each year. This is a significant step forward in the effort to meet global climate goals. However, as PV power becomes more widespread, accurate forecasting becomes increasingly critical—and it's a difficult challenge to get it right.

The book "AI-based Solar Photovoltaics Power Generation Forecast" guides the reader from a poetic editorial to stringent definitions of AI methods and their evaluation metrics for PV forecasting. It offers a comprehensive exploration of PV power forecasting.

While the sun's daily presence might suggest a simple task, solar forecasting is a complex challenge – "a story of patterns and surprises, of predictable cycles and chaotic skies" as stated in the book – influenced by the solar path, aerosols, and dynamic atmospheric conditions. Due to the chaotic nature of the atmosphere, the use of AI is tempting and obvious – as the future states of the atmosphere can't be predicted precisely.

While AI is meanwhile commonly used also in solar forecasting, the usage is not well documented and referenced. This text fills the gap of a unified reference in this rapidly evolving field by consolidating theoretical foundations, practical methodologies, and model evaluation techniques. It covers a range of approaches, from traditional statistical methods to advanced machine learning and deep learning models, including hybrid and ensemble techniques.

The book also dives into crucial aspects like data preprocessing, model optimization, and the importance of performance metrics. A key focus is the distinction between point forecasting and probabilistic forecasting, highlighting the necessity of quantifying uncertainty for robust, risk-informed decision-making.

As a timely and structured resource, this book aims to be an essential guide for researchers, students, and energy professionals, navigating the complexities of solar PV forecasting to enhance the reliability and efficiency of renewable energy systems.

Jan Remund
Leader IEA-PVPS-Task 16 "Solar Resource for
High Penetration and Large Scale Applications"

About the editors

Elham Shirazi is an assistant professor at the University of Twente, the Netherlands. Her multidisciplinary research focuses on applying AI/ML methods to forecasting, integration, and control of energy systems. She joined the University of Twente in 2021, following postdoctoral research at KU Leuven's and prior work at IMEC's Energy Department in Belgium. She is a member of IEA PVPS and ETIP PV and serves on technical committees for EUPVSEC, IEEE ISGT, and ACM e-Energy.

Wilfried van Sark is a professor in photovoltaics integration at the Copernicus Institute of Sustainable Development, Utrecht University, The Netherlands. He has over 40 years' experience in PV solar energy R&D. His research includes next-generation PV, performance analysis of photovoltaic modules and systems, smart grids with EV and vehicle-to-grid technology, and solar forecasting. He is an associate editor or board member for several related journals and a senior member of IEEE.

Chapter 1

Introduction to solar photovoltaics forecasting

Elham Shirazi[1] and Wilfried van Sark[2]

Each morning, as the sun rises over rooftops and open fields, millions of solar photovoltaic (PV) panels begin their work, transforming sunlight into electricity. From satellites orbiting the Earth to sensors embedded in smart inverters, data is collected, processed, and transformed into predictions and forecasts: *how much power or energy will this PV panel or PV plant generate now, in 1 h, or over the course of a day?* At first glance, it seems simple. After all, the sun rises every day. But solar forecasting is a complex dance of physics, meteorology, models, and data. A cloud moving across the sky or a shift in temperature can make the difference between a balanced (local) electricity grid and a costly imbalance. As solar generation grows rapidly worldwide, the stakes have never been higher.

In many ways, solar forecasting is the art of reading the future, whether it is minutes, hours, days, or months ahead. It is a story of patterns and surprises, of predictable cycles and chaotic skies. Researchers have traded crystal balls of fortune tellers for weather models, neural networks, and sky imagers, but the goal remains the same: to look ahead, even just a few hours or minutes, and plan wisely. This book explores that journey, how solar PV generation can be forecasted accurately, why it matters, and where the field is heading. It is a story not just of technical progress but of adaptation of how energy systems are learning to live with the sun.

> **Solar photovoltaic forecast**
> The process of predicting the (near) future power/energy output of solar photovoltaic systems.

1.1 Introduction

The widespread deployment of PV systems has emerged as a key element in the global shift toward a carbon-neutral and sustainable energy systems. Driven by a combination of supportive regulatory frameworks, government incentive programs,

[1]Faculty of Engineering Technology, University of Twente, The Netherlands
[2]Faculty of Geosciences, Copernicus Institute of Sustainable Development, Utrecht University, The Netherlands

technical developments, and increasing environmental awareness, the adoption of PV technologies has witnessed remarkable growth in recent years [1]. However, the rapid integration of distributed PV systems into existing electricity grid infrastructure introduces new challenges, particularly concerning voltage regulation, reverse power flow, and congestion within the electricity grid [2]. These issues are intensified when PV systems are integrated without sufficient planning, which is especially important for small-sized residential systems. In this context, solar PV power forecasting has become an essential tool for ensuring the reliable and efficient integration of solar PV systems into power systems. Accurate forecasts enable grid operators and energy market participants to better manage the variability and uncertainty associated with solar generation, therefore contributing to grid stability, enhancing market participation, and reducing reliance on costly reserve capacities.

Despite its significance, PV forecasting is challenging due to the intermittent nature of solar irradiance and its dependence on dynamic atmospheric conditions. Major challenges include modeling the nonlinear and time-varying relationship between weather variables and PV output, managing sudden changes in solar irradiance (e.g., due to cloud movement), and effectively using diverse and high-dimensional data sources, such as satellite imagery, sky cameras, numerical weather prediction (NWP) models, and ground-based measurements [3]. Overcoming these challenges requires developing robust PV forecasting tools. Various approaches are taken in PV forecasting, including physics-based and data-driven [4]. Physics-based models are based on physical principles that govern atmospheric processes, while data-driven models ignore these physical principles and work with the data. Artificial intelligence (AI) is a data-driven approach that, particularly through its subset machine learning (ML), offers significant value in PV forecasting due to its ability to accurately predict solar PV generation by extracting complex relationships between different variables. There are of course hybrid models that combine different approaches, often integrating ML with physics-based models to use the strengths of each approach.

1.2   PV forecasting methods

PV power forecasting can be approached using various methods, each with its own strengths and limitations. These methods can be broadly categorized into data-driven, physics-based, or a hybrid of physics-based and data-driven methods. A high-level classification is given in Figure 1.1.

1.2.1   Physics-based methods

Physics-based models for PV forecasting rely on detailed physical principles and mathematical representations of the various phenomena that influence solar power generation. These models use atmospheric physics to predict weather conditions, such as solar irradiance, temperature, humidity, cloud cover, and wind speed, which impact solar radiation. Examples include the Weather Research and

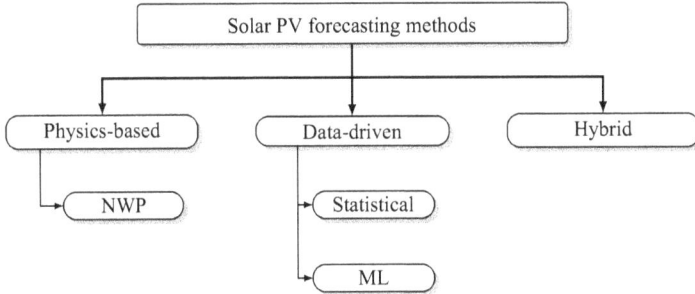

Figure 1.1   High-level classification of solar PV forecasting methods

Forecasting (WRF) [5] and the European Centre for Medium-Range Weather Forecasts (ECMWF) [6] models.

These models incorporate detailed knowledge about the system components, such as solar irradiance, temperature, panel orientation, and geographical location, to predict PV power generation. Physics-based models offer high interpretability and accuracy under known conditions. These models are useful for long-term PV forecasting, particularly for system design. However, their performance may degrade in real-time forecasting, where rapid changes in weather and external factors introduce complexity that is difficult to capture solely through physical equations, which has led to the integration of data-driven and hybrid models for improved forecasting accuracy. As this book is focused on data-driven methods, physics-based approaches will not be further explored.

1.2.2   Data-driven models

Data-driven models can be divided into two main categories, statistical time-series models and machine learning based models (Figure 1.1). Statistical methods, often referred to as classical or traditional forecasting techniques, use historical data patterns to make future predictions without explicitly modeling the physical processes behind PV power generation. These methods assume that future values are dependent on past observations and trends, and they often involve time-series analysis techniques, such as ARIMA (AutoRegressive Integrated Moving Average) [7], SARIMA (Seasonal ARIMA) [8], and Exponential Smoothing (ETS) [9]. Statistical models are generally interpretable, relatively simple to implement, and require less computational power compared to more complex machine learning approaches. However, their accuracy can be limited when dealing with highly nonlinear and variable systems such as PV power generation influenced by dynamic weather conditions. Statistical methods will be discussed in Chapter 3.

ML methods represent a class of data-driven techniques that learn complex patterns from data. In the context of PV power forecasting, ML models can capture the nonlinear dependencies between various input features, such as solar irradiance, temperature, humidity, and historical PV power output, making them highly effective for short- to medium-term forecasting. Unlike traditional statistical methods, ML models do not rely on strong assumptions about data distribution or linearity, offering greater flexibility. However, they typically require extensive training data and careful tuning to avoid overfitting and ensure generalization across different conditions. As seen in Figure 1.2, ML models are classified into supervised, unsupervised, semi-supervised, self-supervised, deep learning, and ensemble. Supervised and unsupervised ML models will be discussed in Chapter 4, while Chapter 5 will discuss methods based on deep learning models. Chapter 6 explores the ensemble models, where multiple models are aggregated or combined to forecast PV output.

1.2.2.1   Supervised learning

Supervised learning is one of the most widely used paradigms in ML, where models are trained on labeled datasets, meaning each input is paired with a corresponding output. The primary goal is to learn a mapping from inputs to outputs, enabling the model to make predictions on new, unseen data. Most common algorithms in supervised learning include Linear Regression [3], Support Vector Machine (SVM), K-Nearest Neighbor (KNN) [10], and Artificial Neural Network (ANN) [11]. The effectiveness of supervised models heavily relies on the quality and quantity of labeled data. With the rise of big data, these models have become increasingly powerful and accurate, especially when combined with feature engineering and tuning techniques. Supervised learning models will be explored in Chapter 4.

1.2.2.2   Unsupervised learning

Unsupervised learning deals with data that has no explicit labels. The main goal is to uncover hidden patterns, structures, or features from the input data without any supervision. Common tasks include clustering, where data points are grouped based on similarity using k-means [12] or hierarchical clustering [13], and dimensionality reduction, where high-dimensional data is compressed into fewer dimensions using techniques like Principal Component Analysis (PCA) [14]. This helps in visualizing, preprocessing, or compressing data while preserving its meaningful structure.Although unsupervised models do not provide direct predictions, they offer valuable insights for feature extraction and data understanding, which can enhance the performance of supervised learning models. Unsupervised models will be explored in Chapter 4.

1.2.2.3   Semi-supervised learning

Semi-supervised learning lies between supervised and unsupervised learning. It uses a small amount of labeled data along with a large amount of unlabeled data during training. This is particularly useful in domains like PV forecasting, where labeled data may be expensive or time-consuming to get, but unlabeled data are abundant. By combining supervised and unsupervised learning techniques, semi-supervised methods can better capture the complex relationships between environmental factors and PV system output, leading to more reliable predictions. Semi-supervised models attempt to learn from the structure of the data, using the labeled examples to guide the learning process from the larger unlabeled dataset [15].

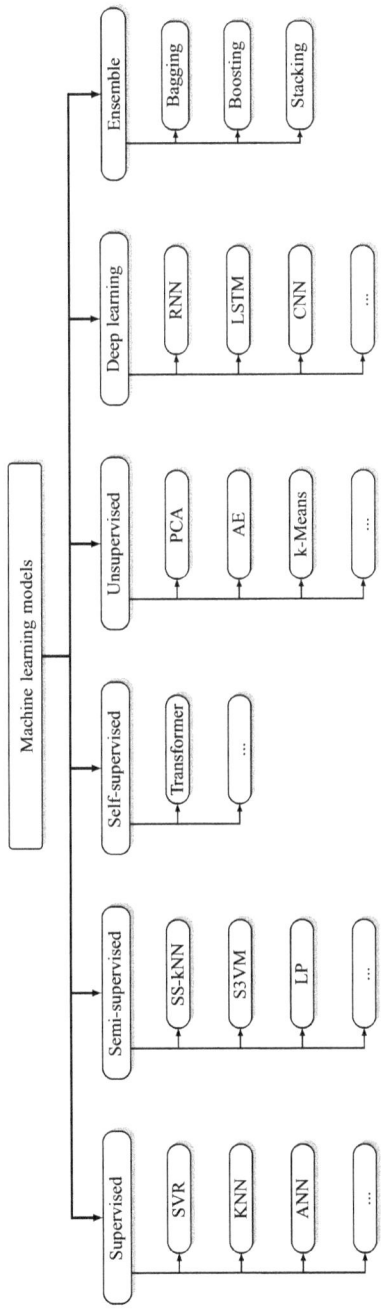

Figure 1.2 Classification of machine learning methods with some example models. Abbreviations and further details can be found in the text.

1.2.2.4   Self-supervised learning

Self-supervised learning does not require any manually labeled data. Instead, it generates labels from the data itself by defining a pretext task, i.e., predicting the next value of PV system output. The model learns useful representations during the pretext task, which can then be fine-tuned for the actual forecasting task. In PV forecasting, self-supervised methods might pretrain models on massive amounts of historical weather or irradiance data, learning patterns that generalize well even with limited actual power output labels. Compared to supervised learning, self-supervised methods offer the advantage of scalability and reduced dependency on labeled datasets. Transformer architecture and attention mechanisms are examples of the self-supervised models [16,17].

   While supervised learning is straightforward and often more accurate with abundant labeled data, semi-supervised and self-supervised methods provide more scalable solutions in scenarios with limited annotations. Semi- and self-supervised methods will not be explored in this book.

1.2.2.5   Deep learning

Deep learning (DL) models consist of multiple layers of neurons, which allow them to learn complex patterns and hierarchical representations from large amounts of data. The strength of deep learning lies in its ability to automatically learn features from raw data, reducing the need for manual feature engineering. These models have shown remarkable performance in fields such as computer vision, speech recognition, and time series forecasting. Key deep learning architectures for PV forecasting include Recurrent Neural Networks (RNNs) and their variants such as Long-Short-Term Memory (LSTM) [18], Gated Recurrent Unit (GRU) [19], and Convolutional Neural Networks (CNNs) [20], which can capture both temporal and spatial dependencies to improve prediction accuracy. DL models will be explored in Chapter 5.

1.2.2.6   Ensemble learning

Ensemble learning combines multiple models to produce a more robust and accurate prediction than any individual model. The main idea is to use the strengths of various models and reduce the risk of overfitting or underfitting. Common ensemble techniques include bagging [21], boosting [22,23], and stacking [24], where different models are combined through a meta-learner. Bagging reduces variance by aggregating predictions from multiple models trained on different subsets of the data [21]. In boosting, different models are trained sequentially to correct the errors of the previous ones, therefore reducing bias and variance [22,23]. Stacking combines predictions from multiple models using a meta-model to improve overall performance [24]. In practice, ensemble methods are highly popular for structured data problems such as classification and regression. These methods are particularly effective in solar PV forecasting due to their ability to handle complex data distributions and improve generalization. Ensemble learning will be explored in Chapter 6.

1.3   Benchmarking forecasting models

Benchmarking in PV power forecasting is a process used to evaluate and compare the performance of forecasting models under standardized conditions. This process typically involves testing new models against established baselines to determine whether they deliver meaningful improvements in predictive accuracy. The persistence model, commonly used as a baseline, assumes that future values will match the most recent observations. While simplistic, it provides a robust baseline for assessing the added value of more sophisticated models.

To ensure objective and comprehensive comparisons, benchmarking employs a set of performance evaluation metrics that quantify forecasting errors from multiple perspectives. These metrics not only standardize the evaluation process but also highlight the specific strengths and weaknesses of each model. As a result, benchmarking facilitates model validation in dynamic and uncertain environments and supports the development of more reliable and accurate forecasting systems.

1.4   Performance evaluation of forecasting models

In the context of solar PV power forecasting, evaluation metrics play a crucial role. These metrics are the quantitative basis for assessing the performance of different models. For classification problems, metrics such as F1 score, recall, and precision are commonly used, while for regression problems, metrics such as root mean squared error (RMSE), mean absolute error (MAE), and the coefficient of determination (R^2) are typically employed. The RMSE quantifies the difference between the predicted value and the actual measured value by calculating the square root of the average of the squared differences. The normalized root mean square error (nRMSE) is the normalized version of the RMSE, which quantifies the average magnitude of prediction errors. The MAE is a straightforward metric that provides insight into the accuracy of predictions by quantifying the average of the absolute differences between the predicted and observed values. Unlike the RMSE, which emphasizes larger errors, the MAE gives equal importance to all errors, making it a simpler and more balanced measure of the overall prediction accuracy. The nMAE is the normalized version of the MAE, which quantifies the average magnitude of prediction errors. Similar to nRMSE, normalization makes nMAE scale-independent, enabling comparison across different datasets. The mean absolute percentage error (MAPE) quantifies the average of the absolute percentage differences between the predicted and observed values. It offers a clear understanding of prediction accuracy by expressing errors as a percentage. The MAPE treats all errors proportionally, relative to the actual observed values. This means the impact of each error is scaled according to the size of the actual value, providing a balanced view of prediction accuracy across different magnitudes. This proportional approach makes the MAPE particularly useful for comparing prediction performance across different datasets or time periods.

The mean bias error (MBE) quantifies the average difference between the predictions and observations, indicating whether the predicted values are systematically overestimating or underestimating the observed values. The coefficient of determination (R^2) measures the proportion of the variance in the observed data that is predictable from the independent variables. It provides insight into the goodness of fit of a model by quantifying how well the predicted values correspond to the actual observed values. An R^2 value of 1 indicates perfect correlation, whereas an R^2 value of 0 suggests that the model does not explain any of the variability in the response data. The R^2 metric provides a normalized measure of model performance, making it useful for comparing the predictive accuracy of different models or evaluating the same model across different datasets. The performance metrics described above will be discussed in more detail in Chapter 7.

1.5   Hyperparameter tuning

Hyperparameter tuning is a critical step in developing accurate forecasting models. Unlike model parameters, which are learned from the data during the training process, hyperparameters are set before training and govern the overall behavior and capacity of the model [25]. Examples of hyperparameters include the learning rate, number of hidden layers in a neural network, tree depth in decision trees, or the number of estimators in ensemble methods. Selecting appropriate hyperparameters can significantly affect the accuracy and generalizability of the forecasting model, especially when dealing with highly nonlinear systems.

Various techniques are used for hyperparameter optimization in PV forecasting, including grid search, random search, and more advanced approaches such as Bayesian optimization [25]. Grid search exhaustively evaluates all possible combinations from a predefined set of hyperparameters, whereas random search samples combinations randomly, often yielding better results with lower computational efforts. Bayesian optimization builds a probabilistic model of the objective function and selects the most promising hyperparameters to evaluate, offering a more efficient and informed search process. These methods help in fine-tuning ML-based models. Hyperparameter tuning and model optimization will also be discussed in Chapter 7.

1.6   Probabilistic vs point forecast

Point forecasting refers to the generation of a single, specific value as the expected output for a given time in the future. This approach is straightforward and widely used due to its simplicity and ease of interpretation. Most forecasting methods focus on producing point forecasts. These forecasts are particularly useful in applications where deterministic values are needed for planning. However, point forecasts do not provide any information about the uncertainty or confidence associated with the prediction, which can be a major limitation, especially in the context of solar energy, where power generation is highly variable due to weather conditions. Nevertheless, forecasting models are developed and optimized using historical data; hence, some idea of typical accuracy can be obtained in relation to different weather types.

In contrast, probabilistic forecasting provides a more comprehensive view by estimating the likelihood of different outcomes, typically in the form of prediction intervals, quantiles, or probability distributions [26]. Ensemble-based weather forecasts from ECMWF generally provide meteograms, indicating the likelihood of cloud cover, precipitation, wind speed, and temperature [27]. The probabilistic forecasting method captures the inherent uncertainty in solar energy generation and allows grid operators to make risk-informed decisions. For instance, knowing the probability of PV output falling within a certain range can help in designing robust energy dispatch strategies or determining reserve margins. Probabilistic models can be developed using techniques, such as quantile regression (QR) [28], Bayesian neural networks (BNN) [29], and ensemble learning approaches [30]. Probabilistic models will be explored in Chapter 8.

1.7 Structure of the book

This book is structured to provide a comprehensive exploration of solar PV power forecasting, including the theoretical foundations, modeling approaches, data considerations, and case studies. It is designed to guide readers through the domain of solar energy forecasting, from traditional statistical methods to machine learning and hybrid approaches, while addressing key challenges such as uncertainty, model evaluation, and hyperparameter tuning. This book includes the following chapters:

Chapter 1 Introduction to solar photovoltaics forecasting
Chapter 2 Data, data collection, and preprocessing for solar photovoltaics forecasting
Chapter 3 Statistical time series for solar photovoltaic forecasting
Chapter 4 Machine learning approaches for PV forecasting
Chapter 5 Deep learning approaches for PV forecasting
Chapter 6 Hybrid and ensemble models for solar energy forecast
Chapter 7 Probabilistic PV forecasting
Chapter 8 Model optimisation, hyperparameter tuning and performance evaluation in machine learning models for solar PV generation forecast
Chapter 9 Sky imager based solar photovoltaic forecast
Chapter 10 Solar photovoltaic forecasting for energy system integration and control

The book begins with this chapter, Chapter 1, which presents an overview of the book and the role of PV forecasting in energy systems, highlighting its importance in the transition toward a carbon-neutral energy future. It introduces the motivations, challenges, and scope of PV power prediction. It reviews the main categories of forecasting methods, including physics-based, statistical, and machine learning approaches and lays the groundwork for subsequent chapters.

Chapter 2 discusses data sources and preprocessing, highlighting the importance of various types of data, satellite imagery, sky images, ground-based measurements, data preprocessing, and feature engineering in enhancing forecast accuracy. Chapter 3 explores statistical time series-based models, covering classical techniques such as ARIMA, SARIMA, and Prophet, along with a discussion of their roles as statistical rather than machine learning models. The chapter also provides insights into their strengths in modeling linear patterns and seasonality in PV data. Chapter 4 introduces machine learning models excluding deep learning models, offering a classification of supervised and unsupervised learning paradigms. It discusses common models such as SVM and ANN. Chapter 5 explores deep learning techniques such as RNN, LSTM, and CNN for PV power forecasting. These models excel at learning complex, non-linear patterns from large and diverse datasets, including weather data and historical power output. The chapter highlights their strengths, practical applications, and recent case studies.

Chapter 6 focuses on hybrid and ensemble methods that combine multiple models to enhance forecasting accuracy and robustness. Ensemble learning helps overcome the limitations of single models. Key approaches are explained with examples of case studies. Chapter 7 addresses hyperparameter tuning and optimization techniques, outlining methods like grid search and random search, and their impact on model performance and robustness. It also focuses on evaluation metrics, explaining the significance of comparing models using metrics such as MAE and RMSE. Chapter 8 contrasts point forecasting with probabilistic forecasting, showing the importance of uncertainty quantification in PV predictions. It introduces techniques such as quantile regression and ensemble learning to capture forecast uncertainty. Chapter 9 discusses sky imagers for solar PV forecast. The book concludes with Chapter 10, which presents solar forecasting for grid integration and energy system control.

When selecting models for solar PV power forecasting, several critical factors must be considered. Data quality and volume are critical, as complex models such as deep learning require large, high-quality datasets to perform effectively. Computational resources also play a significant role, given that neural networks and ensemble methods are computationally intensive and demand considerable processing power. The forecast horizon – whether short-term, medium-term, or long-term – can influence the suitability of different models. Additionally, model interpretability is a key consideration; simpler models such as linear regression and decision trees offer greater interpretability compared to the often black-box nature of deep learning models. By carefully aligning model selection with these considerations and tailoring the approach to the unique characteristics of PV data and stakeholders, the accuracy and reliability of solar PV power forecasts can clearly be enhanced.

1.8   Conclusion

This book serves as a comprehensive and timely contribution to the rapidly evolving field of solar PV forecasting, addressing a critical need for an in-depth and structured resource dedicated exclusively to the application of machine learning and data-driven methods in this domain. Despite the growing number of research articles and technical reports, there has been a lack of a unified reference that consolidates the theoretical foundations, practical methodologies, model evaluations, and current trends in PV forecasting using machine learning. This book fills that gap by providing a cohesive framework that bridges the disciplines of renewable energy systems, artificial intelligence, and data science. As solar PV power becomes increasingly integrated into energy systems, a dedicated book on this subject is incredibly useful, making this book an important resource for researchers, policymakers, energy system operators, and students seeking to improve the reliability and efficiency of solar PV systems and the broader systems into which they are integrated.

References

[1]   Raghoebarsing A., Farkas I., Atsu D., *et al.* The status of implementation of photovoltaic systems and its influencing factors in European countries. *Progress in Photovoltaics: Research and Applications.* 2023;31(2):113–133.

[2] Ahmadi B., Shirazi E. Optimal Allocation of Voltage Regulations to Maximize the Hosting Capacity of Distribution Systems. In: *2023 IEEE 50th Photovoltaic Specialists Conference (PVSC)*; 2023. pp. 1–6.

[3] Shirazi E., Gordon I., Reinders A., *et al.* Sky images for Short-Term Solar Irradiance Forecast: A Comparative Study of Linear Machine Learning Models. *IEEE Journal of Photovoltaics*. 2024;14(4):691–698.

[4] Bazionis I. K., Kousounadis-Knousen M. A., Georgilakis P. S., *et al.* A taxonomy of short-term solar power forecasting: Classifications focused on climatic conditions and input data. *IET Renewable Power Generation*. 2023;17(9):2411–2432.

[5] Skamarock W.C., Klemp J.B., Dudhia J., *et al. A Description of the Advanced Research WRF Version 3.* Boulder, CO: National Center for Atmospheric Research; 2008. NCAR/TN-475+STR.

[6] European Centre for Medium-Range Weather Forecasts. ECMWF Website; 2025. Accessed: 2025-06-27. https://www.ecmwf.int.

[7] Atique S., Noureen S., Roy V., *et al.* Forecasting of total daily solar energy generation using ARIMA: A case study. In: *2019 IEEE 9th Annual Computing and Communication Workshop and Conference (CCWC)*; 2019. pp. 0114–0119.

[8] Vagropoulos S.I., Chouliaras G.I., Kardakos E.G., *et al.* Comparison of SARIMAX, SARIMA, modified SARIMA and ANN-based models for short-term PV generation forecasting. In: *2016 IEEE International Energy Conference (ENERGYCON)*; 2016. pp. 1–6.

[9] De Falco P., Di Noia L.P., Rizzo R. Exponential smoothing model for photovoltaic power forecasting. In: *2021 9th International Conference on Modern Power Systems (MPS)*; 2021. pp. 1–5.

[10] Saxena N., Kumar R., Rao Y. K. S. S., *et al.* Hybrid KNN-SVM machine learning approach for solar power forecasting. *Environmental Challenges*. 2024;14:100838.

[11] Anagnostos D., Schmidt T., Cavadias S., *et al.* A method for detailed, short-term energy yield forecasting of photovoltaic installations. *Renewable Energy*. 2019;130:122–129.

[12] Lu X. Day-ahead photovoltaic power forecasting using hybrid K-Means++ and improved deep neural network. *Measurement*. 2023;220:113208.

[13] Park S., Park Y.B. Photovoltaic power data analysis using hierarchical clustering. In: *2018 International Conference on Information Networking (ICOIN)*; 2018. pp. 727–731.

[14] Kazem H.A., Yousif J.H., Chaichan M.T., *et al.* Long-term power forecasting using FRNN and PCA models for calculating output parameters in solar photovoltaic generation. *Heliyon*. 2022;8(1):e08803.

[15] Jin X., Xiao F., Zhang C., *et al.* Semi-supervised learning based framework for urban level building electricity consumption prediction. *Applied Energy*. 2022;328:120210.

[16] Liu J., Fu Y. Renewable energy forecasting: A self-supervised learning-based transformer variant. *Energy*. 2023;284:128730.

[17]   Tran M., Luis A.D., Liao H., *et al. S3Former: Self-supervised high-resolution transformer for solar PV profiling.* arXiv:240504489v2. 2024

[18]   Sardarabadi A., Ardakani A. H., Matrone S., *et al.* Multi-temporal PV power prediction using long short-term memory and wavelet packet decomposition. *Energy and AI.* 2025;21:100540.

[19]   Gong G., Lou K., Yin J., *et al.* Forecast of photovoltaic power generation based on GRU. In: *Proceedings of the 2022 6th International Conference on Electronic Information Technology and Computer Engineering. EITCE '22.* New York, NY, USA: Association for Computing Machinery; 2023. pp. 283–286.

[20]   Ogliari E., Sakwa M., and Cusa P. Enhanced convolutional neural network for solar radiation nowcasting: All-Sky camera infrared images embedded with exogeneous parameters. *Renewable Energy.* 2024;221:119735.

[21]   Aoyang H., Tongtong L., Zhen W., *et al.* Photovoltaic power forecasting model based on the fusion of bagging algorithm and generalized regression neural networks. In: *2023 International Conference on Power Energy Systems and Applications (ICoPESA);* 2023. pp. 520–523.

[22]   Hanif M. F., Siddique M. U., Si J., *et al.* Enhancing solar forecasting accuracy with sequential deep artificial neural network and hybrid Random Forest and Gradient Boosting Models across varied terrains. *Advanced Theory and Simulations.* 2024;7(7):2301289.

[23]   Boutahir M. K., Farhaoui Y., Azrour M., *et al.* Advancing solar power forecasting: Integrating boosting cascade forest and multi-class-grained scanning for enhanced precision. *Sustainability.* 2024;16(17):7462.

[24]   Natarajan K. P., Singh J. G. Solar power forecasting using stacking ensemble models with Bayesian meta-learning. In: *2023 14th International Conference on Computing Communication and Networking Technologies (ICCCNT);* 2023. pp. 1–6.

[25]   Shirazi E. and Quest H. Model tuning and performance evaluation in machine learning models for PV power forecasting: Case study of a BIPV system in Switzerland. In: *2024 IEEE 52nd Photovoltaic Specialist Conference (PVSC);* 2024. pp. 438–442.

[26]   Visser L. R., AlSkaif T. A., Khurram A., *et al.* Probabilistic solar power forecasting: An economic and technical evaluation of an optimal market bidding strategy. *Applied Energy.* 2024;370:123573.

[27]   European Centre for Medium-Range Weather Forecasts. *ENS Meteograms;* 2025. Accessed: 2025-07-25. https://charts.ecmwf.int/products/opencharts _meteogram?base_time=202507250000&epsgram=classical_10d&lat=51.4 333&lon=-1.0&station_name=Reading.

[28]   Huang H. H. and Huang Y. H. Probabilistic forecasting of regional solar power incorporating weather pattern diversity. *Energy Reports.* 2024;11:1711–1722.

[29]   Alghamdi H., Maduabuchi C., Okoli K., *et al.* Bayesian neural networks for solar power forecasts in advanced thermoelectric systems. *Case Studies in Thermal Engineering.* 2024;61:104940.

[30]   Ahmad T. and Zhou N. Ensemble methods for probabilistic solar power forecasting: A comparative study. In: *2023 IEEE Power & Energy Society General Meeting (PESGM);* 2023. pp. 1–5.

Chapter 2
Data, data collection, and preprocessing for solar photovoltaics forecasting

Lennard Visser[1] and Wilfried van Sark[1]

As a result of the increased worldwide deployment of photovoltaic (PV) solar power, accurate power forecasting is becoming increasingly important to guarantee a reliable electricity grid. The key variable in PV power forecasting is solar irradiance, with ambient temperature as an important second variable. Hence, accurate weather forecasting at appropriate spatiotemporal timescales is necessary. Many forecasting methods are developed as evident in this book, and their accuracies are critically dependent on the input data used for developing and operating these methods. Therefore, methods for retrieving, cleaning, and utilizing data are essential to provide reliable datasets for the development and operation of forecasting methods. This chapter describes various pre- and postprocessing steps that can be applied to increase the value of the input data.

2.1 Introduction

Reliable solar power forecasting models are widely recognized as a key element to support the effective integration of a rapidly increasing installed capacity of solar PV into the electricity grid [1,2]. Accurate forecasts support the timely scheduling of power generators and batteries, thereby enhancing grid stability and reducing the reliance on balancing reserves. Furthermore, the effective implementation of solar power forecasting models can significantly reduce the costs associated with the integration of solar PV.

Today, solar forecasts already play a critical role in many regions, particularly in countries, regions, and/or islands with high PV penetration rates, such as California, Puerto Rico, and Germany. With the fast-growing deployment of PV systems, interest in solar forecasting has increased. Consequently, in the last decade, substantial research efforts have been directed toward improving solar irradiance and PV power forecasts [3]. With the expected continued growth of solar PV systems in the coming years, solar forecasting will play an increasingly vital role in the development of a cost-effective electricity system.

[1]Faculty of Geosciences, Copernicus Institute of Sustainable Development, Utrecht University, The Netherlands

In the literature, a wide variety of predictor variables have been explored and subsequently used to forecast the power output of solar PV systems [4–7]. Since the power output of PV systems is primarily determined by the irradiance received on the plane of the PV array, state-of-the-art solar power forecasting models generally rely on other methods that predict solar irradiance. Other predictor variables that are relevant for determining the power output of a PV system and are therefore commonly considered include ambient temperature, pressure, humidity, wind speed, and wind direction.

The collection, filtering, and processing of solar PV data, as well as weather predictions form the foundation for the successful development and adoption of solar power forecasting models. Without accurate and clean data, solar forecasting models would lack reliability, fail to achieve their full potential, and ultimately fall short of delivering the benefits outlined above. This chapter, therefore, focuses on methods for retrieving, cleaning, and utilizing the data necessary for developing and applying effective solar power forecasting models.

The chapter proceeds as follows. Sections 2.2 and 2.3 dive into typical solar power forecasting set-ups and data resources. Section 2.4 outlines preprocessing steps that can be adopted to increase the value of input data and thus have the ability to improve the reliability of solar power forecasting models. Next, Sections 2.5 and 2.6 present public datasets that feature PV power measurements and explore the methods that can be applied to filter and validate such PV power recordings, respectively. Finally, a summary and an outlook are provided in Section 2.7.

2.2 Forecasts and applications

2.2.1 Solar forecasting

A typical process for predicting the power output of PV systems is illustrated in Figure 2.1. The forecasting scheme consists of three distinct stages: first, data is collected, filtered, and (pre-)processed; next, a forecasting model is employed to predict the solar power output by (post-)processing the prepared data; and subsequently, the forecast is used for its intended application.

The accuracy of solar forecasts is widely recognized to be influenced by the spatial and temporal resolution considered, as well as the forecast horizon of interest. These factors are contingent upon the specific application of the forecasting model, which, in turn, dictates the most suitable approach. Consequently, the selection of an optimal solar forecasting framework is inherently tied to its intended application. In the following sections, various applications will be further explored.

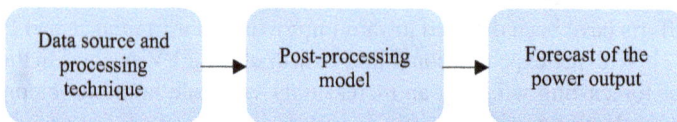

Figure 2.1   General overview of a typical solar power forecast scheme, adopted from [8]

The specific requirements for an operational solar power forecasting model are determined by its intended application, which also dictates the most effective forecasting approach. In the following subsections, we explore three main categories that define the requirements, i.e., models used for participation in energy markets, for the operation of microgrids, and for ensuring compliance with (local) policies and regulations.

2.2.2 Energy market participation

For utilities, aggregators, and other participants in energy markets, forecasting requirements are dictated by the organizational structure of the market. In deregulated electricity markets, the trading of electricity is commonly organized in the day-ahead, intraday, and balancing markets [9–11] (see Figure 2.2 for an overview). It is important to recognize that market structures may vary between regions and/or countries. These variations often pertain to aspects of the balancing markets or organizational details, such as gate closure times, settlement periods for day-ahead and intraday markets, and auction mechanisms (e.g., pay-as-bid or market-clearing). The transmission system operator facilitates the operation of these markets, also ensuring the stability and reliability of the electricity grid. Market participants engage in these markets by submitting volume and price bids, thereby committing to sell or purchase electricity at specified times.

2.2.2.1 Day-ahead market

Currently, the majority of electricity is traded in the day-ahead market [10]. In most European countries, the day-ahead market operates with a gate closure time around noon, e.g., at 12:00 local time in Belgium, Denmark, France, Germany, and the Netherlands [12]. This requires that power purchase and sale bids for the following day be submitted in advance, which requires forecasts with a lead time of 12 h

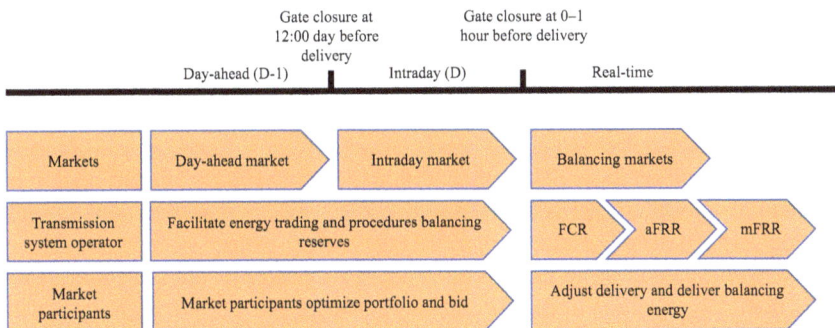

Figure 2.2   *An overview of the operation of energy markets and the role of the local transmission system operator and market participants, adopted from [11]. FCR: frequency containment reserves, aFRR and mFRR: automatic and manual frequency restoration reserves.*

and a time horizon of 24 h (that is, 12–36 h ahead). Furthermore, electricity is commonly traded at a hourly or shorter resolution. As a result, solar forecasts used for day-ahead market bidding must provide hourly (or subhourly) resolutions [13]. Additionally, since forecasts must be generated daily for the subsequent trading day, the forecasting approach must support a daily update rate.

2.2.2.2   Intraday market

The intraday market complements the day-ahead market by enabling market participants to trade closer to the time of delivery. Specifically, intraday markets typically open at 15:00 on the day prior to delivery and allow power purchase and sale bids to be placed up to 0–60 min before actual delivery. These markets are characterized by a combination of auction-based and continuous trading mechanisms. Subsequently, the lead-time and horizon depend on the user's intention. Depending on the region or country of interest, bids generally feature a temporal resolution of 15, 30, or 60 min. In practice, forecasts used for intraday market participation should be updated at least hourly [12]. The shorter forecast horizon of the intraday market enables participants to refine their day-ahead bids using updated and more accurate forecasts for the same day. In recent years, the volume of electricity traded in intraday markets has surged, driven by the growing integration of variable renewable energy sources (VRESs) [13]. As renewable energy capacity continues to expand, further growth in both the number of trades and the volume of electricity traded in intraday markets is anticipated.

2.2.2.3   Balancing markets

Real-time deviations between electricity supply and demand are managed through the deployment of balancing energy or reserve power. Balancing markets, administered by the TSO, consist of several submarkets designed to maintain real-time supply–demand equilibrium. These include frequency containment reserves (FCR), automatic frequency restoration reserves (aFRR), manual frequency restoration reserves (mFRR), and, in some cases, replacement reserves (RR) [9].

Balancing markets are crucial for solar plant operators, who are contractually obligated to deliver power as scheduled in their day-ahead or intraday market bids. Failure to meet these commitments results in penalties, which are based on the deviation between scheduled and actual power, as well as the cost of activating reserve power [14]. Consequently, forecasting errors can be costly and reduce the value of electricity produced, motivating operators to improve their forecasting models. To mitigate this risk, solar plant operators may deploy battery systems that can be charged or discharged when there is an excess or shortage of the scheduled power production. Alternatively, under specific circumstances (such as a negative penalty or excess production), solar plant operators may also decide to curtail a part of the power production to maximize net revenues. To this end, solar plant operators require very short-term forecasts, also referred to as nowcasting, with a 15–30 min horizon and 0 min lead time to minimize discrepancies between forecasted and actual power output. The temporal resolution of such forecasts typically aligns with the balancing period set by the TSO, such as the 15 min interval used in the Netherlands [14], and is often provided at a 1-min resolution and update rate.

2.2.3   Microgrid management

Solar power forecasting also plays a critical role in the operation and management of microgrids and isolated power systems, including off-grid installations, standalone systems, and small island networks. In these environments, where traditional energy markets are typically absent, forecasting requirements are primarily determined by technical constraint sets that may be affected by the characteristics of backup power generators and, in particular, their start-up times and ramping capabilities. Accurate solar forecasts enable improved grid stability and power quality in microgrid settings with high photovoltaic penetration rates [15–17]. Furthermore, reliable forecasting contributes to overall system optimization by reducing energy costs, minimizing CO_2 emissions, and decreasing the required (back-up) capacity of, e.g., battery storage systems [5,18].

2.2.4   Policy and regulation

Policy frameworks and regulatory requirements constitute a third category driving the need for operational solar power forecasts. Multiple countries have implemented regulations designed to mitigate grid disturbances caused by rapid power fluctuations from photovoltaic systems. These regulations typically set limits on the maximum allowable ramp rates to ensure grid stability. For instance, Germany restricts power ramp-up rates to 10% of contracted capacity per minute. Puerto Rico applies this same 10% limit to both ramp-up and ramp-down events. Denmark permits maximum ramp rates of 100 kW/s (6 MW/min) in either direction, whereas Hawaii allows up to 2 MW/min and Ireland permits ramp-up rates of 30 MW/min [19,20].

Accurate solar power forecasts are essential for predicting these ramp events and enabling PV plant operators to comply with regulatory constraints cost-effectively. Consequently, these applications require forecasts with short horizons, typically up to 30 min ahead with minimal lead time (i.e., 0 min). Various power ramp rate control strategies utilizing short-term PV forecasts have been proposed [21,22], while alternative approaches employing battery storage systems to mitigate ramp rate violations are discussed in [19,20,23,24].

2.3   Data sources

2.3.1   Data collection

State-of-the-art solar PV power forecasting relies primarily on predictions of meteorological variables, with solar irradiance being the most critical parameter. Several distinct approaches are employed to obtain information about current conditions and future values of the relevant weather variables. These methodologies can be categorized into four principal techniques: numerical weather predictions (NWPs), satellite imaging, all-sky imaging, and sensor networks. The selection of the most appropriate

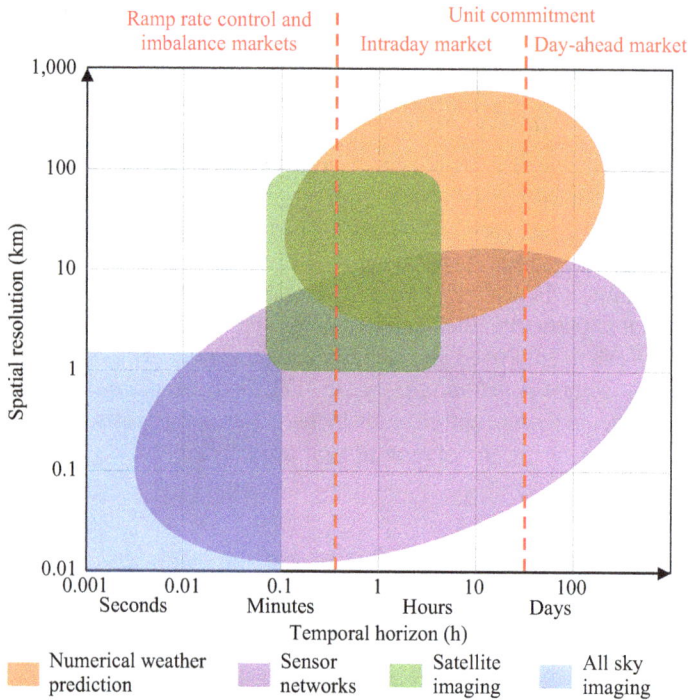

Figure 2.3   Classification of primary information sources for PV power forecasting based on application domains, spatial resolution characteristics, and temporal forecast horizons

technique depends on the required forecast horizon, which is directly determined by the intended application as discussed in Section 2.2. Figure 2.3 provides an overview of these techniques and their respective applications. The following sections examine each of these four approaches in detail.

2.3.2   Numerical weather predictions

Solar power forecasts using NWPs outperform alternative techniques for forecast horizons exceeding 4–6 h [5], making them the preferred information source for intraday and day-ahead market trading applications.

Approximately 15 weather services operate global NWP models [25], including the integrated forecast system (IFS) by the European Centre for Medium-Range Weather Forecasts (ECMWF). These models describe the atmospheric temporal development by solving prognostic equations governing atmospheric and oceanic processes [26]. NWP models rely on fundamental physical laws including conservation of momentum (Navier–Stokes equations), mass continuity, energy conservation (first law of thermodynamics), the ideal gas law, and transport equations to describe spatiotemporal changes in wind, pressure, density, and temperature.

Each simulation initializes with recent meteorological observations of three-dimensional wind fields, temperature, humidity, and two-dimensional surface pressure, as well as boundary variables such as snow cover and sea surface temperature [26,27]. To make the most efficient use of the available meteorological observations, data assimilation systems use the forecast model to process the observations and produce a more complete picture of the state of the atmosphere through reanalysis. These calculations are performed in four dimensions, ensuring that the analysis is consistent in space and time. This process involves running simulations for previous time steps, generating short-term forecasts for current conditions, and correcting forecasted states using recent observations to fill data gaps and integrate satellite measurements consistently.

As a next step, the physical equations are numerically solved on a grid, involving parameterizations to describe processes that are not explicitly resolved by the model. The grid presents the spatial resolution of the NWP model, which is usually developed on global or mesoscale domains. A detailed description of the basic features introduced above is given in [26].

In general, higher resolution improves forecast accuracy but increases computational demands, creating trade-offs between global coverage and spatiotemporal resolution. Current global NWP models feature 10–50 km spatial resolution with 3–6 h temporal output, while mesoscale models achieve 1–20 km resolution with 1 h output despite internal calculations using 30 s time steps. Nevertheless, NWP accuracy has steadily improved through enhanced physical process representation, advanced data assimilation algorithms, expanded observational data integration, and increased computing power—trends expected to continue, yielding higher prediction accuracy and resolution [27].

2.3.3   Satellite imaging

Solar power forecasts using satellite imaging demonstrate good performance for forecast horizons of up to 6 h ahead. Satellite-derived information thus serves as a primary source for solar power forecast models deployed in intraday market trading.

Satellite imaging utilizes geostationary satellite observations at the top of the atmosphere to determine the atmospheric state and retrieve information on cloud properties, upwelling radiances, and surface/atmospheric albedos from which other variables are inferred. To derive surface irradiance, four distinct methodological approaches are employed: empirical/statistical, empirical/physical, physical, and subjective methods [25]. Empirical/statistical approaches derive irradiance based on observed relationships between satellite- and ground-based observations. Empirical/physical approaches use retrieval schemes to estimate temporal atmospheric properties critical to radiative transfer processes, from which surface irradiance can be estimated directly. Since both empirical/statistical and empirical/physical approaches rely on establishing intermediate relationships, these are often referred to as semiempirical models. Physical models typically first determine cloud optical properties from satellite radiances and subsequently apply radiative transfer modeling to establish surface irradiance. Generally, semiempirical approaches estimate only the global horizontal irradiance (GHI), whereas the physical approaches provide both

GHI and direct normal irradiance (DNI) estimates. When only GHI information is available, decomposition models can be deployed to determine DNI (see Section 2.4).

Satellite imaging-based forecasting models rely on establishing cloud motion vectors (CMVs) to simulate future atmospheric cloud states, assuming that cloud structures remain temporally stable over the forecast horizon (up to 6 h ahead). The forecasting process involves establishing independent grid points or fields, estimating cloudiness through cloud indices, deriving CMVs by analyzing sequential satellite images [28], and extrapolating the CMV field to predict future cloud spatial coverage. Satellite-based irradiance models subsequently estimate surface irradiance from these cloud simulations, with optional spatial smoothing applied to minimize prediction errors.

Current geostationary satellites provide near-global coverage with high spatiotemporal resolution at latitudes within ±60°, offering temporal resolution 15 min or better and spatial resolution up to 1 km. These capabilities are delivered through advanced satellite systems including Meteosat Second and Third Generation satellites, Geostationary Operational Environmental Satellite, and Geostationary Meteorological Satellite [25].

2.3.4   All-sky imaging

Solar power forecasts using all-sky imaging typically outperform alternative techniques for forecast horizons up to several minutes ahead. Thus, making these approaches particularly valuable for nowcasting applications, ramp rate detection, and real-time supply management [29–31].

All-sky imaging employs ground-based all-sky imagers (ASIs), also known as total sky imagers, whole sky imagers, or cloud cameras, to capture sequential sky images at predetermined frequencies for estimating cloud coverage and tracking cloud movement dynamics. Most approaches follow a similar process: sky images are captured, cloud detection algorithms identify clouded pixels, and cloud motion vectors (CMVs) are extracted from image sequences to construct CMV fields. These fields enable estimation of future cloud geolocations, with varying levels of detail depending on available equipment. For example, ceilometers can determine cloud height while dual or larger ASI systems can construct 3D cloud imagery [32]. The shadow field is then projected onto the surface to infer local irradiance components, with an optional final step converting GHI predictions to PV power production using conversion models (see Section 2.4).

2.3.5   Sensor networks

The fourth overarching technique for collecting information for solar power forecasting involves sensor networks. Unlike the three approaches previously discussed (NWP, satellite imaging, and all-sky imaging), sensor networks do not constitute a unified methodology with specific equations, singular applications, or clear preferences for particular temporal or spatial resolutions. Instead, sensor networks encompass a diverse collection of standalone approaches that support solar power forecasting through distributed information gathering, often incorporating big data analytics. Additionally, in forecasting, sensor networks may also rely on PV power output recordings from surrounding PV systems. As a result, sensor network applications exhibit considerable variability of (available) variables, lead times, forecast horizons, and both temporal and spatial resolutions.

An example of a sensor network concerns a network of 202 rooftop PV systems that was operated from 2013 to 2017, covering a 35 by 54 km area in the Netherlands [33]. Real-time power output data from these systems were used, among others, to forecast production at downwind locations with 30-min horizons and 1-min time steps [34].

A second sensor network consisting of 200 pyranometers was installed on Oahu Island, Hawaii, to collect solar irradiance data in 2010 and 2011 [35]. Among others, the data collected in the network was used to forecast the solar power output of PV systems up to 10 s ahead [36].

Finally, a smaller network consisting of approximately 10 ASIs is under construction for the city of Utrecht, the Netherlands, which also includes an ASI at the location of the Royal Netherlands Meteorological Institute (KNMI) [37].

2.4 Preprocessing input data

2.4.1 The value of preprocessing

The collection, selection, and preprocessing of predictor variables are critical steps in developing high-performance solar power forecasting models. In contrast to electricity price and wind power forecasting, where predictor variable evaluation receives considerable attention, the literature dedicates limited effort to assessing predictor variables for solar irradiance and power forecasting, particularly with regard to the evaluation of individual variable contributions. This research gap is notably evident, given that power curves establishing predictor–target relationships have gained significant attention in wind forecasting [38–40], while analogous studies remain largely absent in solar forecasting applications. An example of such an exception is provided in [41].

Similarly, limited attention is devoted to preprocessing predictor variables and creating additional expert variables that carry potential values for improving model performance [42,43], despite the widespread availability of open-source models that can readily provide these enhanced variables. The different irradiance components and in-plane variables can be efficiently derived using well-established decomposition and transposition models available through libraries such as *pvlib* [44]. The importance of incorporating these expert variables into solar power forecasting frameworks for both single-point and probabilistic models was demonstrated in [42], with transposition models proving particularly valuable for improving forecast accuracy. In addition, Visser *et al.* [42] found that the estimated PV power output, constructed using a model chain and a number of (predicted) weather variables, proved to be the single most valuable variable in solar power forecasting.

2.4.2   Preprocessing steps PV power forecasting

A first set of expert variables can be created using a decomposition (or separation) model. Such models decompose the GHI into its components, DNI and diffuse horizontal irradiance (DHI). Some well-known decomposition models available in [44] include the DISC [45], DIRINT [46], and Erbs [47].

Subsequently, transposition models can be applied to obtain the in-plane counterparts of the irradiance variables, which describe the irradiance received in the plane of the PV array. These transposition models output the global, direct, diffuse, ground diffuse, and sky diffuse irradiance received on the plane of the PV array. Open-source transposition models include isotropic [48], Hay–Davies [49], King [50], Perez [51], and Reindl [44,52].

Next, given the in-plane irradiance values, a PV and inverter model can be used to estimate the DC and AC PV power output for a predefined system. Assumptions or recording of additional weather variables, including ambient temperature and wind speed, are required to operate these models. Widely known publicly available PV and inverter models in [44] include, respectively, PVWatts [53], CEC [44], and Sandia [54] and PVWatts [53], Driesse [55], and Sandia [54].

2.5   PV datasets

The availability of open-access or public datasets containing PV power measurements remains limited, posing significant challenges to the advancement of related scientific fields, such as solar forecasting (as discussed in this book), model predictive control, and home energy management systems [56]. This scarcity impedes progress for three key reasons. First, the lack of publicly available data restricts researchers and stakeholders from conducting studies in relevant domains. Second, it undermines the ability to compare methodologies and results across studies. Third, it compromises (academic) reproducibility, as studies relying on proprietary datasets cannot be independently verified or replicated [11].

Table 2.1 provides an overview of open-access PV power datasets and their main characteristics. This overview reveals not only the limited number of datasets but also the shortcomings. Moreover, most of the identified datasets either feature coarse temporal resolutions (≥ 15 min) or span short durations (< 3 months). In addition, power measurements are often confined to a small number of locations, whereas it is known that the accuracy of forecasting models depends on the regional climate characteristics [57].

Moving forward, greater availability of comprehensive datasets could accelerate advancements in the scientific fields that rely on such data and subsequently improve the robustness of future research. Thus, increased attention is needed to stimulate open data practices, including the availability of open-access PV power datasets. Ideally, such datasets include long-term power measurements from solar PV systems across diverse climatic regions. The PVoutput.org [58] and PEARL-PV [59] initiatives provide rich datasets, but these are based on voluntary contributions from individuals and the academic community, respectively.

Table 2.1 Overview of public datasets

Reference	Name	Type of system	Number of systems	Duration	Resolution	Location
[60,61]	Pecan Street	Residential rooftop PV	>300	from Feb '11	From 1 s	US
[62]	Ausgrid	Residential rooftop PV	300	Jul '10–Jun '13	30 min	Australia
[63]	Australia	Residential rooftop PV	1287	Sep '16–Mar '17	10-min	Australia
[64]	Japan	Rooftop PV	1	Jan '15–Apr '18	1 s	Japan
[65]	UCSD microgrid	Rooftop PV	22	Jan '15–Feb '20	15-min	San Diego, US
[33,66]	Utrecht	Residential rooftop PV	175	Jan '14–Dec '17	1-min	Utrecht, The Netherlands
[58]	PVoutput.org	Rooftop and standalone PV systems	>10,000	daily updated	5 min	Global
[59]	PEARL-PV	PV performance and solar irradiance	20	Various	Various	Global

Table 2.2   Overview of filters for quality controlling PV power measurements

Filter	Description	Requirements	Reference
Data availability	Sets a daily minimum to the number of reported values	Single PV	[33]
Night (sundown values)	Limiting PV power measurements to zero at night-time	Single PV	[33,67]
Lower limit	Lower limitation for PV power measurements	Single PV	[33,67]
Upper limit	Upper limitation for PV power measurements	Single PV	[33]
Linear values	Identify linear trends (up/down) and constant data	Single PV	[33]
Persistence	Checks the variability of measurements	Single PV	[33,67]
Daily energy ratio	Identifies days with a very low amount of	Single PV	[67]
Cloud enhancement	Cloud enhancement events must also be reported by other systems	Multiple PV	[67]
Spurious data	Benchmark measurements to reported data	Multiple PV	[67]
Daily energy ratio	Benchmarks daily energy generated	Multiple PV	[33,67]

2.6   PV data cleaning

2.6.1   Filtering PV power measurements

The use of high-quality data forms a prerequisite for all research studies. To this end, it is essential to validate the quality of data. The validation of PV power measurements is a highly underexposed topic in the literature today [11]. First, only a few studies report the application of data quality checks when reporting results that involve power measurements of PV systems. Additionally, only two studies were found in the literature that aimed to establish a standardized protocol for the quality control of PV power measurements, i.e., Killinger *et al.* [67] and Visser *et al.* [33].

This section highlights the state-of-the-art of filtering techniques that are used to validate PV power measurements. These techniques are subdivided into single and across filters. Single filters are techniques that can be applied to control the quality of the power measurements of PV systems, without having reference systems available. Such filters are typically based on considering the physical limits of the potential power output of PV systems. In contrast, across filter techniques rely on the data that is collected from nearby PV systems during the same period to execute intersystem comparisons. An overview of the categories and filters is given in Table 2.2.

2.6.2   Single filters

2.6.2.1   Data availability filter

The data availability filter assesses the availability of data measurements over a period of time, such as a single day. This filter simply ensures the quality of PV power data, as well as its usability by posing a minimum data availability rate. This ensures a minimal daily data availability. To this end, a threshold is used. In Visser *et al.* [33], a complete day of power measurements was removed from a dataset in case the availability rate of daily power measurements, i.e., number of registered power values to the number of (daily) instances, fall below 50% of the number of timestamps for the solar zenith angle $\theta_z < \mathbf{85°}$.

2.6.2.2   Night or sundown filter

The night or sundown filter sets all power output values during night-time to zero. Applying this filter requires selecting night-time values daily. Killinger *et al.* [67] define night-time values by timestamps where the solar zenith angle $\theta_z > 95°$, indicating power outputs recorded outside daylight hours (including diffuse irradiance at dawn and dusk). Visser *et al.* [68] alternatively define night-time values as timestamps, where the expected power output of a photovoltaic (PV) system under clear-sky conditions is zero or negative, based on system characteristics such as tilt, azimuth, and DC and AC capacity. To identify these values, the power output of a PV system (p_m) is simulated under clear-sky conditions (p_{cs}), as outlined in (2.1). Additionally, to correct erroneous values near sunrise and sunset, Visser *et al.* [68] propose filtering out negative power output values within a 30-min window around these times, as shown in (2.2). Such negative values are attributed to PV system orientation, low GHI during sunrise and sunset, or shading from surrounding obstacles, such as buildings or trees.

$$p_{cs}(t) \leq 0 \Longrightarrow p_m(t) = 0, \tag{2.1}$$

$$p_m(t) < 0 , \forall \ t \leq (\text{sunrise} + 30 \text{ min}) \lor t \geq (\text{sunset} - 30 \text{ min}) \Longrightarrow p_m(t) = 0 \tag{2.2}$$

2.6.2.3   Lower limit filter

The lower limit filter implies a minimum threshold to validate recorded power measurements. In [67,68], a logical minimum limit of 0 is used, leading to setting values as NaN (not a number).

$$p_m(t) < 0 \Longrightarrow p_m(t) = \text{NaN} \tag{2.3}$$

2.6.2.4   Upper limit filter

An upper limit filter is applied to identify unrealistically high values in solar PV power output measurements. These values can arise either when measurements exceed the technical power limits of solar PV systems or when they surpass physical limits on irradiance that may be received at the plane of the PV array [33]. The former can be addressed by applying an upper limit subfilter, which validates power measurements against the system's technical constraints, specifically by considering the PV system's AC capacity (S_{AC}), as shown in (2.4). Here, υ is a constant, based on findings that the inverter capacity can be exceeded for short periods up to 10 min [69]. In Visser *et al.* [33], this constant is set to 1.025.

A second subfilter is applied to validate power measurements against physical limits, as unrealistically high values may still occur. This addresses cases where

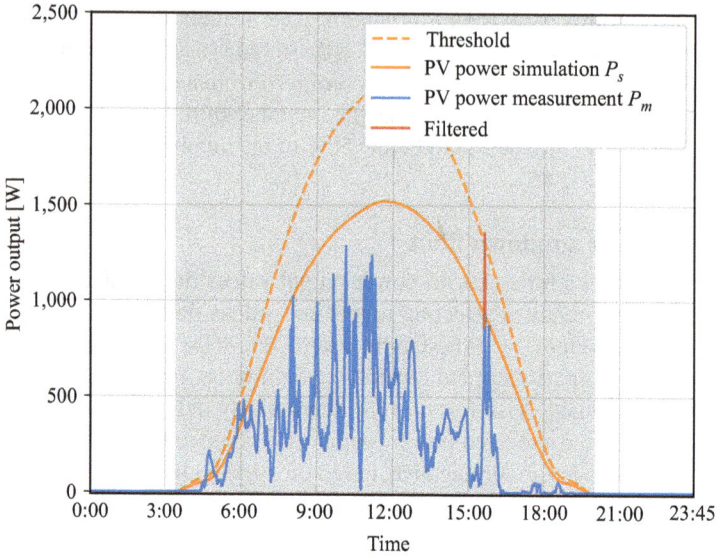

Figure 2.4   An example of the operation of the upper limit filter, adopted from
[33]. The figure presents the threshold value, simulated (P_s) and
measured (P_m) power output of a single PV system for July 6, 2014.
Potential erroneous values are marked (red) and set to NaN by
applying the upper limit filter.

PV power output measurements exceed expected outputs under clear-sky conditions [33,67]. For high-resolution power measurements, however, cloud enhancement effects may occur, which should be accounted for [69]. Additionally, since the clear-sky model's GHI estimations carry inherent uncertainty (due to simplifications such as those related to atmospheric turbidity), a threshold value (k_{PV}) can be applied to the system DC capacity (S_{DC}). For small values of p_{cs}, corresponding to high solar zenith angles, the upper limit may still be exceeded [67] because the clear-sky model and simulated power output are less reliable at high zenith angles ($\theta_z > \mathbf{80°}$). To address this, Visser *et al.* [33] propose adjusting the upper limit filter for such high zenith angles, as shown in (2.5)–(2.7). An example of the application of this subfilter is presented in Figure 2.4.

$$p_m(t) > \upsilon * S_{AC} \implies p_m(t) = \text{NaN} \tag{2.4}$$

$$p_m(t) \geq p_{cs}(t) * k_{PV} \dots \forall t \text{ where } \theta_z < 80° \qquad \implies p_m(t) = \text{NaN} \tag{2.5}$$

$$\left(p_m(t) \geq p_{cs}(t) * k_{PV} \right) \wedge \left(p_m(t) \geq 0.125 * S_{DC} \right) \dots \forall t \text{ where } \theta_z \geq 80° \wedge < 85° \implies p_m(t) = \text{NaN} \tag{2.6}$$

$$\left(p_m(t) \geq p_{cs}(t) * k_{PV} \right) \wedge \left(p_m(t) \geq 0.075 * S_{DC} \right) \dots \forall t \text{ where } \theta_z \geq 85° \qquad \implies p_m(t) = \text{NaN} \tag{2.7}$$

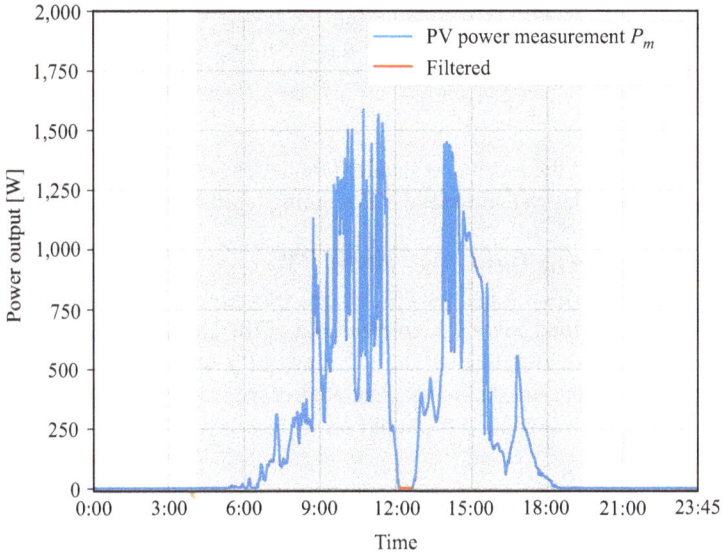

Figure 2.5   An example of the application of the linear data filter, adopted from [33]. The figure presents the measured (P_m) power output of a single PV system for August 7, 2014. Values around noon are marked (red) as these values are detected and registered as erroneous values by the filter.

2.6.2.5   Linear values filter

The linear data filter is designed to identify periods in which the recorded values display either constant or linearly increasing or decreasing trends, as shown in (2.8). This filter, introduced by Visser *et al.* [33], is designed to detect instances of temporary data logger failures, where such patterns might occur.

$$\frac{\partial p_m(t)}{\partial t} = C \ , \forall \ t = \{t, t-1, \ldots, t-K\} \quad \wedge \theta_z < 85° \implies p_m(t) = NaN \qquad (2.8)$$

where C is a constant, equal to zero for a period that features constant power values, or negative (positive) for a linear decrease (increase) of recordings. Equation (2.8) ensures that power measurements are set to NaN when this linear pattern persists for at least K consecutive time steps, which thus covers a predefined period (e.g., 20-min as used in [33]). The condition $\theta_z < 85°$ limits the operation of the filter to times when the solar zenith angle is below $85°$, excluding low-sun periods. Figure 2.5 illustrates an example of this filter's application.

2.6.2.6   Persistence filter

The persistence filter identifies days characterized by either consistently stable or highly variable power values, resulting in minimal or extreme fluctuations in PV power output measurements, respectively. This filter was introduced by Journée and Bertrand [70] to detect spurious GHI data. In later studies, it was adapted by Killinger *et al.* [67] and Visser *et al.* [33] for filtering PV power measurements, specifically targeting anomalous data. Spurious data is flagged based on the following condition:

$$\frac{1}{8}\mu\left(\frac{p_{\mathrm{m}}/S_{\mathrm{DC}}}{E_{\mathrm{ext}}}\right) \le \sigma\left(\frac{p_{\mathrm{m}}/S_{\mathrm{DC}}}{E_{\mathrm{ext}}}\right) \le \tau, \tag{2.9}$$

where μ and σ denote the mean and standard deviation, respectively; E_{ext} represents the extraterrestrial solar radiation; and τ is a defined threshold constant.

2.6.2.7   Daily energy ratio filter, single PV

The daily energy ratio filter for single PV systems was originally developed to identify days with extensive cloud cover, i.e., overcast days [70]. However, it can also be used to detect potential system failures or the covering of PV modules by snow [67]. This filter operates by comparing the daily generated energy (volume) to the potential or theoretical PV power production (volume) for a clear-sky day:

$$r_{\mathrm{de}}(d) = \frac{\sum_{t=1}^{T} p_{\mathrm{m}}(t)}{\sum_{t=1}^{T} p_{\mathrm{cs}}(t)} \tag{2.10}$$

based on empirical findings, a threshold value of 0.05 was proposed [67] to filter daily reported power measurements, as follows:

$$r_{\mathrm{de}}(d) > 0.05 \tag{2.11}$$

2.6.3   Across filters

2.6.3.1   Cloud enhancement filter

The cloud enhancement filter, introduced in [67], is designed to validate cloud enhancement events in PV system measurements. To this end, potential cloud enhancement events are first identified per system by considering the measured power output as well as the power output for clear-sky conditions. Next, to validate the recordings, it is assumed that a true cloud enhancement event is also observed at a neighboring PV system. In addition, this filter considers both the intensity and duration of these events: cloud enhancement events are expected to increase the PV system power output by up to 30% above standard test conditions and should typically last up to 140 s.

2.6.3.2   Spurious data filter

The spurious data filter detects erroneous recordings by comparing statistical data across neighboring PV systems, operating on the assumption that adjacent systems should exhibit similar statistical patterns. This filter specifically targets erratic fluctuations in data recordings, as demonstrated in [67]. The filter flags data based on the following condition:

$$-\sigma_{ij}^2 < \mu(\sigma_i^2) + t\sqrt{\sigma_i^2} \tag{2.12}$$

where σ_{ij}^2 denotes the variance at site j and time i, $\mu(\sigma_i^2)$ is the mean variance across all sites at time i, and t represents a threshold value. In [67], t is set to 3 standard deviations.

2.6.3.3   Daily energy ratio filter, multi-PV

The daily energy ratio filter (across PV systems) leverages data from neighboring PV systems to benchmark daily energy (volume) production values [33,67]. Analogous to the spurious data filter, this approach assumes that neighboring PV systems should produce similar daily energy outputs (if the systems are identical). In this filter, the daily energy ratio (r_{de}) (2.13) for each system is compared to the mean daily energy ratio of all PV systems, $\bar{r}_{de}$ (2.14). This comparison allows the filter to flag and remove days (d) for a single PV system, where significantly deviating production values are observed to the observed mean of surrounding PV systems.

$$r_{de}(d) = \frac{\sum_{t=1min}^{T=24hr} p_{m}(t)}{\sum_{t=1min}^{T=24hr} p_{cs}(t)} \tag{2.13}$$

$$-\lambda \bar{r}_{de}(d) \leq r_{de,i}(d) \leq \lambda \bar{r}_{de}(d) \tag{2.14}$$

where λ is a constant used as a tolerance factor.

2.7   Summary and outlook

This chapter provided a comprehensive overview of data collection, filtering, and processing methods essential for developing reliable solar power forecasting models. The accuracy of forecasting systems fundamentally depends on input data quality and preprocessing techniques employed. The chapter examined three primary applications of solar power forecasting, each with distinct temporal and spatial requirements: energy market participation, microgrid operations, and regulatory compliance.

Four principal information sources support these applications across different forecast horizons. Weather predictions from NWP systems, satellite data, local meteorological and power data (sensor networks), and ASI-derived irradiance data are described. Furthermore, a detailed description of methods that can be used to increase the value of the retrieved (input) information through preprocessing steps, thereby enhancing model performance. In addition, an overview of publicly available solar PV datasets and a standardized procedure for cleaning PV power measurements are given.

Looking forward, the solar forecasting community would greatly benefit from increased availability and widespread usage of public datasets. Comprehensive PV power datasets spanning diverse climatic regions and extended time periods would accelerate model development and enable more robust validation across different environments. This chapter presented an overview of a number of public datasets, considering PV power measurements for different regions covering various time horizons and time spans. However, the number of public datasets is very limited, both in terms of the number of datasets available and the key characteristics they feature, including global coverage representing different climates (e.g., according to the Köppen–Geiger climate classification [71]), period, time resolution, and spatial resolution. Therefore, the availability of a more extensive set of public datasets would highly benefit research in the scientific domain(s) as it, among others, increases accessibility and enables for more robust model development and testing.

The widespread adoption of standardized data filtering procedures represents another critical advancement opportunity for the domain of solar power forecasting. Implementing consistent quality control protocols would enhance research reproducibility, as proposed in [33,72], enable meaningful interstudy comparisons, and establish a level playing field for evaluating forecasting methodologies across different geographical locations and research groups. Such standardization efforts would strengthen the scientific foundation of solar power forecasting and support its continued evolution as a mature research discipline.

References

[1]   Yang D, Wang W, Gueymard C A, *et al.* A review of solar forecasting, its dependence on atmospheric sciences and implications for grid integration: Towards carbon neutrality. *Renewable and Sustainable Energy Reviews.* 2022;161:112348.

[2]   Ahmed A and Khalid M. A review on the selected applications of forecasting models in renewable power systems. *Renewable and Sustainable Energy Reviews.* 2019;100:9–21.

[3]   Yang D, Kleissl J, Gueymard C A, *et al.* History and trends in solar irradiance and PV power forecasting: A preliminary assessment and review using text mining. *Solar Energy.* 2018;168:60–101.

[4]   Ahmed R, Sreeram V, Mishra Y, *et al.* A review and evaluation of the state-of-the-art in PV solar power forecasting: Techniques and optimization. *Renewable and Sustainable Energy Reviews.* 2020;124:109792.

[5]   Antonanzas J, Osorio N, Escobar R, *et al.* Review of photovoltaic power forecasting. *Solar Energy.* 2016;136:78–111.

[6]   Barbieri F, Rajakaruna S, and Ghosh A. Very short-term photovoltaic power forecasting with cloud modeling: A review. *Renewable and Sustainable Energy Reviews.* 2017;75:242–263.

[7]   AlSkaif T A, Dev S, Visser L R, *et al.* A systematic analysis of meteorological variables for PV output power estimation. *Renewable Energy.* 2020;153: 12–22.

[8]   Visser L R, Lorenz E, Heinemann D, *et al.* 1.11 - Solar power forecasts. In: Letcher TM, editor. *Comprehensive Renewable Energy* 2nd edn. Oxford: Elsevier; 2022. pp. 213–233.

[9]   Lampropoulos I, Alskaif T A, Blom J, *et al.* A framework for the provision of flexibility services at the transmission and distribution levels through aggregator companies. *Sustainable Energy, Grids and Networks.* 2019;17:100187.

[10]   Tijdink A, Hoffman M, Vrolijk R, *et al. Annual Market Update 2021: Electricity market insights.* TenneT; 2022. Available from: https://tennet-d rupal.s3.eu-central-1.amazonaws.com/default/2022-07/Annual_Market_Up date_2021_0.pdf.

[11]   Visser L R. *Forecasting to Welcome the Solar Era: Supporting the Integration of Solar Photovoltaics in the Electricity Grid.* The Netherlands: Utrecht University; 2024.

[12]   EPEX SPOT. *The European Power Exchange*; 2022. Available from: https: //www.epexspot.com/en.

[13]   Silva-Rodriguez L, Sanjab A, Fumagalli E, *et al.* Short term wholesale electricity market designs: A review of identified challenges and promising solutions. *Renewable and Sustainable Energy Reviews.* 2022;160:112228.

[14]   TenneT. *Market Types*; 2022. Available from: https://www.tennet.eu/market -types.

[15]   Diagne M, David M, Lauret P, *et al.* Review of solar irradiance forecast- ing methods and a proposition for small-scale insular grids. *Renewable and Sustainable Energy Reviews.* 2013;27:65–76.

[16]   Rikos E, Tselepis S, Hoyer-Klick C, *et al.* Stability and power quality issues in microgrids under weather disturbances. *IEEE Journal of Selected Topics in Applied Earth Observations and Remote Sensing.* 2008;1(3):170–179.

[17]   Simoglou C K, Kardakos E G, Bakirtzis E A, *et al.* An advanced model for the efficient and reliable short-term operation of insular electricity networks with high renewable energy sources penetration. *Renewable and Sustainable Energy Reviews.* 2014;38:415–427.

[18]   Yamamoto S, Park J S, Takata M, *et al.* Basic study on the prediction of solar irradiation and its application to photovoltaic-diesel hybrid generation system. *Solar energy materials and solar cells.* 2003;75(3–4):577–584.

[19]   Martins J, Spataru S, Sera D, *et al.* Comparative study of ramp-rate control algorithms for PV with energy storage systems. *Energies.* 2019;12(7):1342.

[20]   Patarroyo-Montenegro J F, Vasquez-Plaza J D, Rodriguez-Martinez O F, *et al.* Comparative and cost analysis of a novel predictive power ramp rate control method: A case study in a PV power plant in Puerto Rico. *Applied Sciences.* 2021;11(13):5766.

[21]   Chen X, Du Y, Wen H, *et al.* Forecasting-based power ramp-rate con- trol strategies for utility-scale PV systems. *IEEE Transactions on Industrial Electronics.* 2018;66(3):1862–1871.

[22]   Zhang C, Du Y, Chen X, *et al.* Cloud motion tracking system using low-cost sky imager for PV power ramp-rate control. In: *2018 IEEE International Con- ference on Industrial Electronics for Sustainable Energy Systems (IESES).* IEEE; 2018. pp. 493–498.

[23]   Visser L R, Schuurmans E M B, AlSkaif T A, *et al.* Regulation strategies for mitigating voltage fluctuations induced by photovoltaic solar systems in an urban low voltage grid. *International Journal of Electrical Power & Energy Systems.* 2022;137:107695.

[24]   Brinkel N B G, Gerritsma M K, AlSkaif T A, *et al.* Impact of rapid PV fluc- tuations on power quality in the low-voltage grid and mitigation strategies using electric vehicles. *International Journal of Electrical Power & Energy Systems.* 2020;118:105741.

[25]   Sengupta M, Habte A, Wilbert S, *et al. Best Practices Handbook for the Collection and Use of Solar Resource Data for Solar Energy Applications: Fourth Edition.* Golden, CO (United States): National Renewable Energy Lab.(NREL), 2024. https://docs.nrel.gov/docs/fy24osti/88300.pdf.

[26]   Kalnay E. *Atmospheric Modeling, Data Assimilation and Predictability.* Cambridge: Cambridge University Press; 2003.

[27]   Bauer P, Thorpe A, and Brunet G. The quiet revolution of numerical weather prediction. *Nature.* 2015;525(7567):47–55.

[28]   Wang P, van Westrhenen R, Meirink J F, *et al.* Surface solar radiation forecasts by advecting cloud physical properties derived from Meteosat Second Generation observations. *Solar Energy.* 2019;177:47–58.

[29]   Lin F, Zhang Y, and Wang J. Recent advances in intra-hour solar forecasting: A review of ground-based sky image methods. *International Journal of Forecasting.* 2023;39:244–265.

[30]   Logothetis S A, Salamalikis V, Nouri B, *et al.* Solar irradiance ramp forecasting based on all-sky imagers. *Energies.* 2022;15(17):6191.

[31]   Logothetis S A, Salamalikis V, Wilbert S, *et al.* Benchmarking of solar irradiance nowcast performance derived from all-sky imagers. *Renewable Energy.* 2022;199:246–261.

[32]   Peng Z, Yu D, Huang D, *et al.* 3D cloud detection and tracking system for solar forecast using multiple sky imagers. *Solar Energy.* 2015;118:496–519.

[33]   Visser L R, Elsinga B, AlSkaif T A, *et al.* Open-source quality control routine and multi-year power generation data of 175 PV systems. *Journal of Renewable and Sustainable Energy.* 2022;14(4):043501.

[34]   Elsinga B and Van Sark W G J H M. Short-term peer-to-peer solar forecasting in a network of photovoltaic systems. *Applied Energy.* 2017;206:464–1483.

[35]   Sengupta M and Andreas A. Oahu solar measurement grid (1-year archive): 1-second solar irradiance; Oahu, Hawaii (data). Golden, CO (United States): National Renewable Energy Lab.(NREL); 2010.

[36]   Amaro e Silva R and Brito M. Spatio-temporal PV forecasting sensitivity to modules' tilt and orientation. *Applied Energy.* 2019;255:113807.

[37]   Van Sark W. Personal communication; 2025.

[38]   Jeon J and Taylor J W. Using conditional kernel density estimation for wind power density forecasting. *Journal of the American Statistical Association.* 2012;107(497):66–79.

[39]   Xu M, Pinson P, Lu Z, *et al.* Adaptive robust polynomial regression for power curve modeling with application to wind power forecasting. *Wind Energy.* 2016;19(12):2321–2336.

[40]   Yang D and van der Meer D. Post-processing in solar forecasting: Ten overarching thinking tools. *Renewable and Sustainable Energy Reviews.* 2021;140:110735.

[41]   Yang D, Xia X, and Mayer M J. A tutorial review of the solar power curve: Regressions, model chains, and their hybridization and probabilistic extensions. *Advances in Atmospheric Sciences.* 2024;41(6):1023–1067.

[42]   Visser L R, AlSkaif T A, Hu J, *et al.* On the value of expert knowledge in estimation and forecasting of solar photovoltaic power generation. *Solar Energy.* 2023;251:86–105.

[43]   Mayer M J. Benefits of physical and machine learning hybridization for photovoltaic power forecasting. *Renewable and Sustainable Energy Reviews.* 2022;168:112772.

[44]   Holmgren W F, Hansen C W, and Mikofski M A. pvlib python: A python package for modeling solar energy systems. *Journal of Open Source Software.* 2018;3(29):884.

[45]   Maxwell E L. *A quasi-physical model for converting hourly global horizontal to direct normal insolation.* Golden, CO (USA): Solar Energy Research Inst.; 1987.

[46]   Ineichen P, Perez R, Seal R, *et al.* Dynamic global-to-direct irradiance conversion models. *Ashrae Transactions.* 1992;98(1):354–369.

[47]   Erbs D, Klein S, Duffie J. Estimation of the diffuse radiation fraction for hourly, daily and monthly-average global radiation. *Solar Energy.* 1982;28(4):293–302.

[48]   Loutzenhiser P, Manz H, Felsmann C, *et al.* Empirical validation of models to compute solar irradiance on inclined surfaces for building energy simulation. *Solar Energy.* 2007;81(2):254–267.

[49]   Davies J A and Hay J E. Calculation of the solar radiation incident on a horizontal surface. In: *Proc. First Can. Solar Radiation Data Workshop.* Canada: Ministry of Supply and services; 1980. pp. 32–58.

[50]   King D L, Kratochvil J A, and Boyson W E. *Photovoltaic array performance model.* United States: Department of Energy; 2004.

[51]   Perez R, Ineichen P, Seals R, *et al.* Modeling daylight availability and irradiance components from direct and global irradiance. *Solar Energy.* 1990;44(5):271–289.

[52]   Reindl D T, Beckman W A, Duffie J A. Evaluation of hourly tilted surface radiation models. *Solar Energy.* 1990;45(1):9–17.

[53]   Dobos A P. PVWatts version 5 manual. Golden, CO (United States): National Renewable Energy Lab. (NREL); 2014.

[54]   Boyson W E, Galbraith G M, King D L, *et al. Performance model for grid-connected photovoltaic inverters.* Albuquerque, NM, and Livermore, CA: Sandia National Laboratories (SNL); 2007.

[55]   Driesse A, Jain P, and Harrison S. Beyond the curves: Modeling the electrical efficiency of photovoltaic inverters. In: *2008 33rd IEEE Photovoltaic Specialists Conference.* IEEE; 2008. pp. 1–6.

[56]   Van Sark W. Photovoltaics performance monitoring is essential in a 100% renewables-based society. *Joule.* 2024;7(7):1388–13936.

[57]   Lauret P, Alonso-Suárez R, e Silva R A, *et al.* The added value of combining solar irradiance data and forecasts: A probabilistic benchmarking exercise. *Renewable Energy.* 2024;237:121574.

[58]   PVoutput org. *Monitoring live solar photovoltaic (PV) and energy consumption data*; 2025. https://pvoutput.org/.

[59]   PEARL-PV. *COST ACTION PEARL-PV data sets*; 2025. https://ckan.pearl pv-cost.eu.

[60]   Street P. *Dataport*, Pecan Street Inc; 2018. https://dataport.cloud/data/intera ctive.

[61]   Smith C A. *The Pecan Street Project: developing the electric utility system of the future*. Citeseer; 2009.

[62]   Ratnam E L, Weller S R, Kellett C M, *et al.* Residential load and rooftop PV generation: an Australian distribution network dataset. *International Journal of Sustainable Energy*. 2017;36(8):787–806.

[63]   Bright J M, Killinger S, and Engerer N A. Data article: Distributed PV power data for three cities in Australia. *Journal of Renewable and Sustainable Energy*. 2019;11(3):035504.

[64]   Vink K, Ankyu E, and Koyama M. Multiyear microgrid data from a research building in Tsukuba, Japan. *Scientific Data*. 2019;6(1):1–9.

[65]   Silwal S, Mullican C, Chen Y A, *et al.* Open-source multi-year power generation, consumption, and storage data in a microgrid. *Journal of Renewable and Sustainable Energy*. 2021;13(2):025301.

[66]   Visser L, Elsinga B, AlSkaif T, *et al. Open-source quality control routine and multi-year power generation data of 175 PV systems.* Zenodo; 2022. Available from: https://doi.org/10.5281/zenodo.6906504.

[67]   Killinger S, Engerer N, Müller B. QCPV: A quality control algorithm for distributed photovoltaic array power output. *Solar Energy*. 2017;143:120–131.

[68]   Visser L R, Elsinga B, AlSkaif T A, *et al. Open-source quality control routine and multi-year power generation data of 175 PV systems*. 2022; Available from: https://doi.org/10.5281/zenodo.6906504.

[69]   Kreuwel F P M, Knap W H, Visser L R, *et al.* Analysis of high frequency photovoltaic solar energy fluctuations. *Solar Energy*. 2020;206:381–389.

[70]   Journée M, Bertrand C. Quality control of solar radiation data within the RMIB solar measurements network. *Solar Energy*. 2011;85(1):72–86.

[71]   Kottek M, Grieser J, Beck C, *et al.* World map of the Köppen-Geiger climate classification updated. *Meteorologische Zeitschrift*. 2006;15(3):259–263.

[72]   Yang D. A guideline to solar forecasting research practice: Reproducible, operational, probabilistic or physically-based, ensemble, and skill (ROPES). *Journal of Renewable and Sustainable Energy*. 2019;11(2):022701.

Chapter 3
Statistical time series for solar photovoltaic forecasting

Ioannis K. Bazionis[1], Athina P. Georgilaki[1] and Pavlos S. Georgilakis[1]

Time series analysis is a fundamental statistical methodology concerned with the examination of data points collected or recorded at successive and typically uniform time intervals. The principal aim of this analysis is to detect, model, and interpret underlying patterns within the data—such as long-term trends, seasonal fluctuations, cyclical behaviors, and irregular components. By capturing these structures, time series analysis facilitates the forecasting of future values, a function that holds critical importance across a range of domains, including finance, economics, meteorology, and energy systems. A defining characteristic of time series data is its temporal dependency. Unlike cross-sectional data, in which individual observations are assumed to be independent, time series observations are often autocorrelated—meaning each value is potentially influenced by its preceding values. This inherent dependency introduces analytical complexities, rendering many traditional statistical techniques inadequate. Consequently, time series analysis employs specialized approaches—such as autoregressive models, moving averages, and their extensions—that explicitly account for such dependencies, thereby enhancing modeling accuracy and predictive power.

A comprehensive time series analysis generally begins with the decomposition of the data into four principal components: trend, seasonality, cyclical variation, and noise. The trend represents the long-term progression of the series, indicating whether it increases, decreases, or remains stable over time. Seasonality involves patterns that recur at regular intervals, such as daily, monthly, or yearly cycles. Examples include increased retail sales during the holiday season or higher electricity usage in the summer due to air conditioning. Cyclical behavior, while analogous to seasonality, does not follow a strictly regular interval and is often driven by macroeconomic or environmental forces. Finally, noise encompasses the random, unsystematic variations that do not align with the other components and are typically treated as stochastic errors.

An emerging and significant application of time series analysis is in the domain of solar photovoltaic (PV) power forecasting. As solar energy assumes a growing

[1]School of Electrical and Computer Engineering, Department of Electric Power, National Technical University of Athens (NTUA), Greece

role within the global energy portfolio, the need for accurate and reliable forecasting methodologies becomes increasingly pronounced. Effective solar forecasting is essential for achieving a balance between energy supply and demand, optimizing the operation of energy storage systems, and ensuring the stability and reliability of modern power grids [1]. Furthermore, improved forecasting reduces dependence on fossil fuel-based backup generation, thereby advancing the transition toward more sustainable and resilient energy infrastructures.

Forecasting solar power output presents distinct challenges due to the inherent variability and intermittency of solar energy generation. Key influencing factors include solar irradiance, ambient temperature, cloud cover, and the geographical location of the solar panels [2]. These variables introduce substantial fluctuations both intraday and across seasons, thereby necessitating robust analytical frameworks capable of capturing complex temporal patterns [3]. In this context, time series analysis serves as an indispensable tool for modeling solar energy data and enhancing the predictability of renewable power systems. In this chapter, the pivotal factors affecting the time series solar power forecasting are explained and analyzed. Furthermore, time-series forecasting methodologies are presented, highlighting their advantages and disadvantages, considering the different forecasting cases.

3.1   Univariate and multivariate time series

Time series forecasting in PV power generation can be approached through either univariate or multivariate frameworks. Each framework offers distinct modeling perspectives depending on the type of available data, as well as the goal of the forecasting process. Understanding the structural differences and classification of these types of time series is crucial for selecting the appropriate approach for accurate predictions.

3.1.1   Univariate time series

Univariate time series consist of sequential observations of a single variable recorded over time. Considering PV power forecasting, such observations may involve generated power or irradiance recorded at specific intervals. Such models rely on the historical values of the target variable to identify temporal patterns, cyclical trends, seasonality, and autocorrelation, as well as project future values.

Univariate modeling is often applied without including external factors. Such models are generally simpler, easier to interpret, and have lower computational cost. However, they may lack the ability to account for variations caused by external influences, such as weather conditions, which can significantly impact PV output. Despite their limitations, univariate time series can be valuable for short-term forecasting, especially in cases where data availability is limited (Figure 3.1).

3.1.2   Multivariate time series

A multivariate time series consists of multiple variables, where each variable can influence the behavior of the target variable. In PV systems, this often includes not

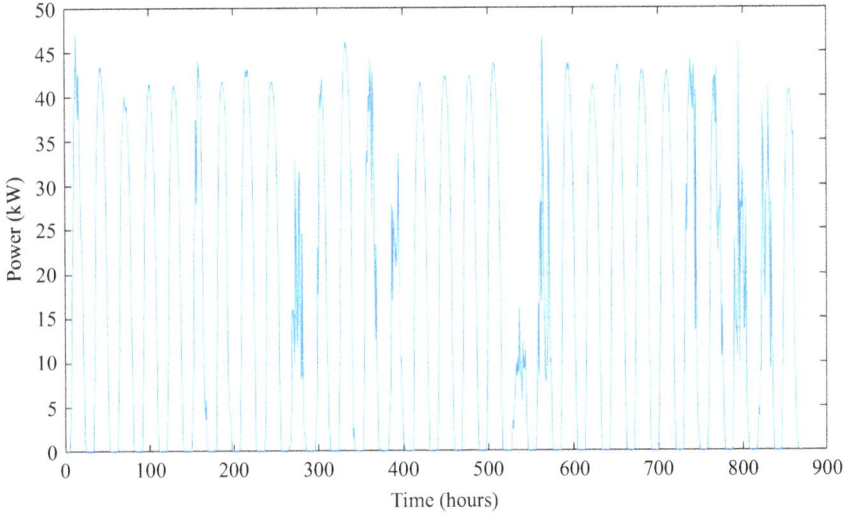

Figure 3.1   Example of a univariate time series with only historical PV time series considered as input [4]

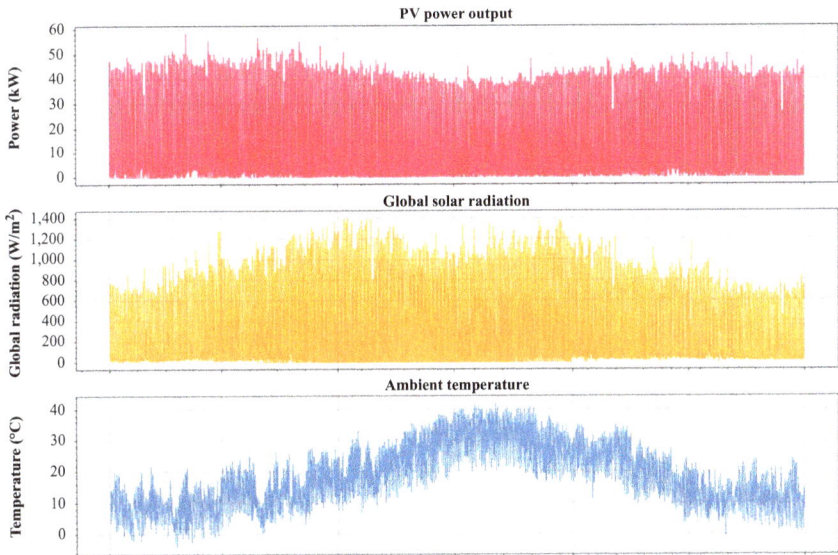

Figure 3.2   Example of a multivariate time series with historical PV time series considered as input, along with other meteorological factors [4]

only historical PV values but also meteorological parameters, such as solar irradiance, ambient temperature, wind speed, humidity, and cloud cover. By capturing these interdependencies, multivariate models can capture the complex physical and environmental dynamics of PV power generation (Figure 3.2).

3.1.3   Classifications into univariate and multivariate time series

The classification of time series data into univariate or multivariate types is a result of various parameters, such as forecasting goals, available data, and technical specifications. Univariate approaches are often preferable for real-time monitoring, particularly when historical data patterns are stable and periodic or when data availability is limited. In contrast, multivariate approaches are better suited for complex scenarios, large-scale PV facilities, or situations where external parameters play a significant role. These models are particularly advantageous in handling nonstationary behavior induced by changing weather patterns, system aging, or varying load profiles.

There are cases where hybrid strategies that incorporate both univariate and multivariate components are needed to leverage the advantages of both components. Such approaches allow models to maintain interpretability while also capturing exogenous influences that enhance forecast robustness.

3.2   Key factors in time series forecasting in PV power generation

The most critical factors in PV power forecasting using time series methods include the quality and resolution of input data, the degree of seasonality and non-stationarity in the power output, and the relevance of external variables, such as temperature, irradiance, and cloud cover. Model-related aspects, such as the choice of algorithm, the structure and complexity of the model, and the approach to hyperparameter tuning, also play a decisive role.

As forecasting applications become more widespread in operational environments, there is a growing need for models that are not only precise but also computationally efficient and resilient to changing patterns in solar energy production. Recognizing and addressing these factors are essential for advancing forecasting methods and ensuring the reliable integration of solar power into future energy systems.

3.2.1   Data quality

Data quality is a fundamental element in the development of accurate and reliable forecasting models for PV power generation. High-quality data ensures that the statistical and machine learning models used for forecasting are trained on inputs that accurately represent the behavior of the PV system, as well as the environmental conditions that affect the power generation process. In contrast, poor-quality data can introduce noise, false learning patterns, and negatively affect the forecasting accuracy. Data quality encompasses several aspects, including completeness, consistency, accuracy, and reliability. In time series forecasting, where patterns and dependencies are learned from sequential observations, even minor anomalies or inconsistencies can affect the model, amplifying forecasting errors and reducing trust in the predictive outputs.

Common issues that compromise data quality in PV systems include missing values, outliers, and measurement errors. Missing values may result from technical parameters, such as sensor failures, data transmission errors, or maintenance periods. If not appropriately addressed, missing data can interrupt the temporal continuity of the series and bias model training. Moreover, systematic measurement errors from faulty equipment can create persistent distortions in the dataset. These data quality issues not only affect model training but also impair real-time inference, making robust data preprocessing critical for the whole process.

3.2.2 Non-stationarity

Non-stationarity refers to a condition in which the statistical properties of a time series, such as its mean, variance, and autocovariance, change over time. In PV power generation, non-stationarity is inherent to the data due to the dynamic nature of solar irradiance and weather-dependent generation. Non-stationarity poses a challenge because many time series models assume that the underlying data is stationary, meaning its statistical properties do not change over time.

Non-stationarity can appear in various cases, the main one for PV power generation being the cyclical and seasonal variations, such as daily solar cycles and seasonal changes in day length and sun angle. These variations introduce repeating yet nonconstant fluctuations in the data, producing temporal structures that require special attention during preprocessing and modeling. Another source of non-stationarity in PV time series is deterministic trends, which can arise from external, long-term factors such as panel degradation, changes in system configuration, or gradual shifts in climate patterns. Over time, these influences cause the baseline level of power output to either increase or decrease, resulting in a time-dependent mean.

Moreover, high-frequency fluctuations due to localized weather phenomena may create volatile periods with increased variance, especially during transitional times of day (e.g., sunrise and sunset). These characteristics make nonstationary data particularly difficult to model without proper transformation or decomposition.

To mitigate this issue, techniques such as differencing and seasonal decomposition can be applied. These methods help transform nonstationary data into a stationary form, making it more suitable for accurate model fitting and improving overall forecasting performance.

3.2.3 Model complexity

Complex models, while capable of capturing intricate patterns in data, may overfit the training data. Complexity reflects the model's capacity to represent patterns in the data, such as temporal dependencies, nonlinear relationships, and interactions between multiple influencing factors. While more complex models generally offer a greater ability to capture fine-grained dynamics, their development can be highly challenging.

Overfitting occurs when a model becomes too tailored to the specific data it was trained on, reducing its ability to generalize to new, unseen data. This can lead to poor forecasting performance when the model is applied in real-world scenarios. However, underfitting occurs when a model is too simplistic, failing to capture essential patterns in the data. This often leads to poor forecasting performance across all conditions. The balance between underfitting and overfitting is central to understanding the trade-offs associated with model complexity.

To balance model complexity and accuracy, cross-validation techniques and model simplification strategies are employed. These approaches help ensure that the model remains robust and generalizable, providing reliable forecasts without being overly complex.

3.2.4 External parameters

PV power generation is influenced by multiple external factors that introduce variability and complexity into forecasting efforts. Weather conditions, shading, and system degradation are among the key factors that can cause deviations in power output. These factors may not be fully captured by traditional time series models, leading to inaccuracies in forecasts. Incorporating such exogenous variables helps bridge the gap between temporal trends and real-world phenomena, offering a more robust basis for prediction.

Among external factors, weather conditions play a particularly critical role in shaping PV output. Variables such as cloud cover, ambient temperature, and humidity directly affect the amount of solar radiation available and the efficiency of energy conversion. Sudden shifts in these conditions can lead to sharp fluctuations in power generation, underscoring the importance of including real-time and forecasted weather data in modeling architectures. By capturing the dynamic interaction between environmental variables and temporal dependencies, forecasting models become better equipped to anticipate and adapt to short-term variations in solar energy production.

3.3 Temporal dependency modeling

Modeling temporal dependencies in time series analysis is crucial for accurately forecasting PV generation, as it captures the inherent time-based patterns and dependencies in solar power output. Effective modeling of these temporal dynamics is essential for predicting the energy produced by PV systems, considering factors such as seasonality, trends, cyclic behaviors, and external influences.

Seasonality has a significant impact on PV generation due to the variations in solar radiation caused by changes in the sun's angle, daylight hours, and weather conditions throughout the year [5]. These seasonal variations are important because they directly affect the amount of solar energy generated by the PV panels. For instance, during the summer, longer daylight hours and a higher sun angle result in increased energy production, whereas winter months may see reduced output. On a typical day, solar power output follows a bell-shaped curve, peaking around mid-day when the sun is at its highest point. Over the course of a year, the generation pattern often follows a sinusoidal curve, reflecting seasonal variations in solar energy availability.

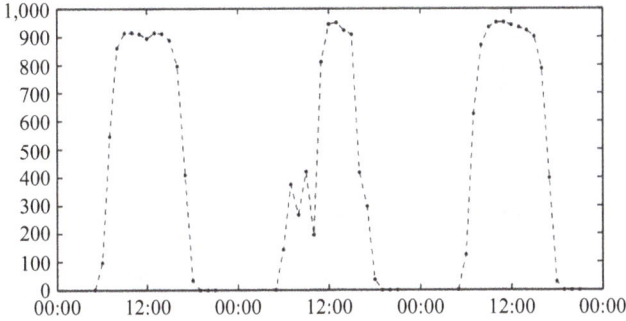

Figure 3.3   Typical daily cyclical patterns of power production [6]

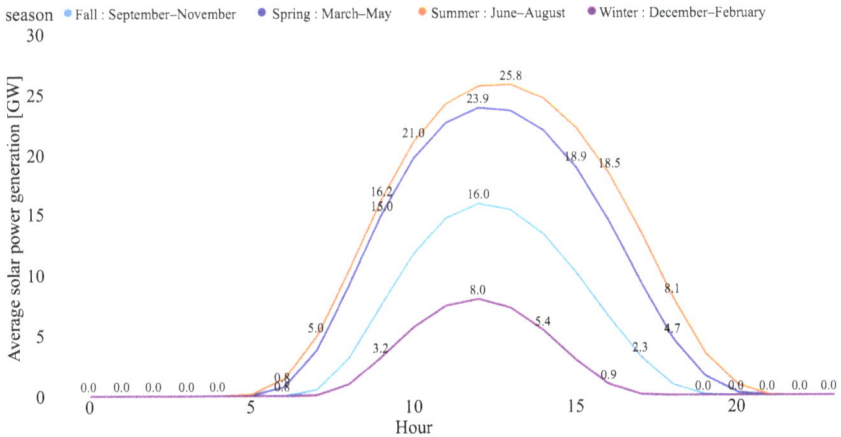

Figure 3.4   Seasonal variation in PV output profiles of the typical PV system [7]

Accurately modeling these seasonal patterns is crucial for reliable solar power forecasts, ensuring that predictions account for expected fluctuations in energy generation across different seasons (Figures 3.3 and 3.4).

Trends in PV generation may develop over time due to factors such as changes in equipment efficiency, solar panel degradation, or modifications in installation angles. As solar panels age, they may experience a gradual decline in efficiency, leading to a long-term decrease in power output. At a broader scale, however, technological advancements can lead to increased efficiency in newer installations, resulting in upward trends in generation. Recognizing and modeling these distinct types of trends is crucial for understanding both the long-term performance of specific solar installations and the evolving landscape of PV power generation.

3.3.1   Seasonal and trend decomposition

Seasonal and trend decomposition (STL) is a technique used to break down a time series into its trend, seasonal, and cyclical components (Figure 3.5).

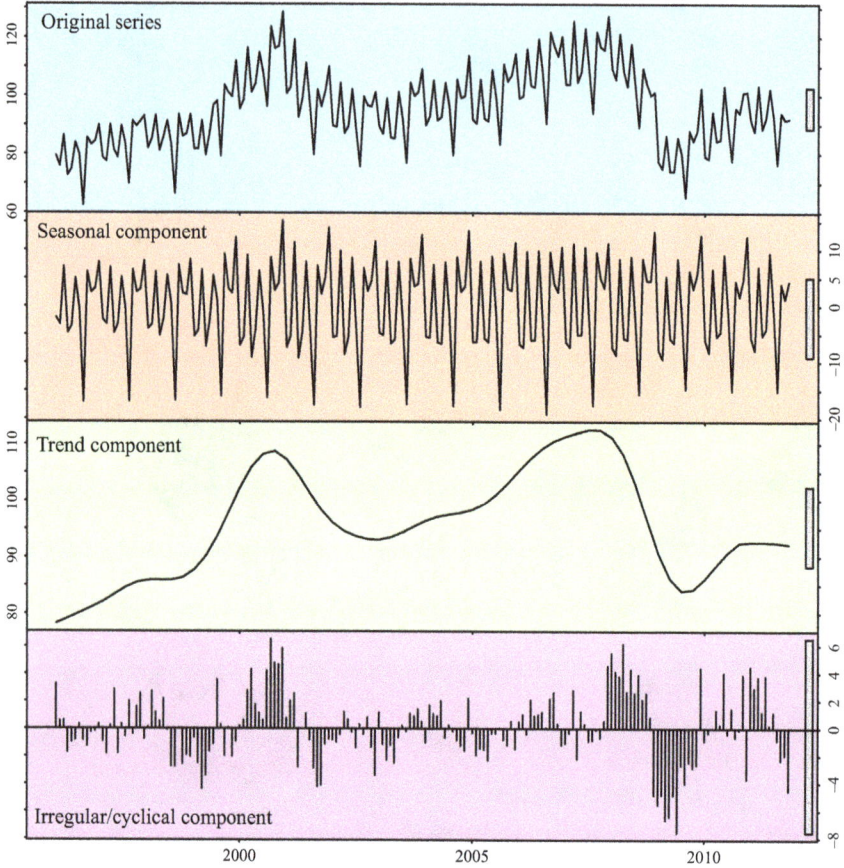

Figure 3.5   Time series components [8]

This process involves applying a smoothing function to the data and iteratively removing the fitted curve to isolate the trend and seasonality elements. The residuals, or the differences between the original data and the fitted curve, represent noise or irregularities in the data. One of the smoothing functions commonly used in the STL method is LOESS, a nonparametric smoothing technique that applies local regression to fit a smooth curve to the data. LOESS is particularly effective at smoothing noisy data and is more flexible than many other smoothing methods [9].

Within the framework of STL decomposition, the LOESS method is utilized to fit individual smooth curves to the seasonal, trend, and residual components of the time series. These curves are then combined to create an overall fit to the data, which can be leveraged for forecasting. To perform STL decomposition, it is essential to first determine the period of the seasonal component, which typically corresponds to the number of time units (such as months or years) in one complete cycle of the seasonal pattern [10]. Once the seasonal component's period has been set, the STL technique proceeds as follows:

- The time series data are separated into seasons, each with a set number of time units.
- Each season's data are fitted with a LOESS smooth curve independently. This generates a sequence of smooth curves, one for each data season.
- Smooth curves are then joined to give a global fit to the data. The seasonal component of the time series is represented by this global fit.
- To create a detrended series, the seasonal component is eliminated from the original time series data. This detrended series only includes the data's trend and residual components.
- The detrended series is then fitted with a LOESS smooth curve. This smooth curve represents the time series' trend component.
- The residual component is then calculated by subtracting the trend component from the detrended series. These are the data that cannot be explained by seasonal or trend components.

After extracting the seasonal, trend, and residual components from the time series data, each can be analyzed separately to gain deeper insights into the underlying patterns and trends. The trend component, for instance, can highlight significant long-term movements in the data, while the seasonal component reveals patterns that occur at regular intervals. The residual component, by contrast, captures random noise or volatility that is not explained by the trend or seasonality. Analyzing the residuals can be useful for assessing the reliability of the trend and seasonal estimates and identifying potential outliers or anomalies in the data.

The STL approach can be used to forecast time series data after the components have been removed and examined. This is accomplished by extrapolating the smooth curves fitted to the data. The seasonal and trend curves are projected to future time points and then combined to produce a time series data forecast. The equation for STL decomposition is as follows:

$$y_t = S_t + T_t + R_t \tag{3.1}$$

where y_t is the original time series data, S_t is the seasonal component, T_t is the trend component, and R_t is the residual component.

Seasonal decomposition with LOESS is accomplished mathematically by fitting a smooth curve to the data using locally weighted regression. This is often accomplished using a weighted least squares method, with the weights determined based on the data's local features. The equation for locally weighted regression is as follows:

$$y = f(x) + \epsilon \tag{3.2}$$

where y is the dependent variable (i.e., the time series), $f(x)$ is the locally weighted regression curve, and ϵ is the error term representing the difference between the predicted value of y and the actual value.

The locally weighted regression curve can be estimated as follows:

$$\min \sum_{i \in X} w_i \left(y_i - \theta^T x_i \right)^2 \tag{3.3}$$

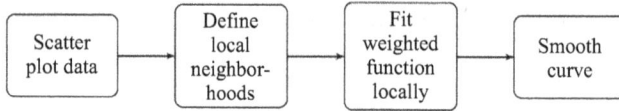

Figure 3.6   LOESS procedure [11]

where x is the independent variable (i.e., the time of the data point), and w is a non-negative weight associated with the training point. The points in the training set that are close to x are given a higher "preference" than the points that are far away from x. As a result, the value of w_i is large for x lying close to the query point x. The coefficients associated with a certain training point x are typically chosen as follows:

$$w_i = \exp -\frac{(x_i - x)^T (x_i - x)}{2\tau^2} \tag{3.4}$$

where τ is the bandwidth parameter and controls the rate at which weight falls with distance from x. The parameters can be calculated by the closed-form solution:

$$\theta = \left(X^T W X\right)^{-1} \left(X^T W Y\right) \tag{3.5}$$

where $X = \{x_1, x_2, \ldots, x_n\}$, $Y = \{y_1, y_2, \ldots, y_n\}$, and $W = \{w_1, w_2, \ldots, w_n\}$

After fitting a smooth curve to the data, its characteristics can be used to distinguish the individual components of the time series. For instance, applying a linear or nonlinear regression model to the curve and analyzing the general trend over time can help estimate the trend component. Seasonality can be identified by observing recurring patterns at regular intervals, such as monthly variations in sales figures. The noise component is detected by examining random or unpredictable fluctuations in the data that do not align with any discernible pattern. The LOESS procedure is presented in Figure 3.6.

Decomposing time series data into seasonal and trend components can enhance the understanding and analysis of the data. This approach helps in recognizing significant patterns and trends that inform decision-making. Techniques such as the STL method are useful for breaking down data into its seasonal, trend, and residual elements, thereby providing deeper insights into the underlying patterns.

3.4   Methodologies for PV power forecasting

Understanding and accurately modeling temporal dependencies are fundamental to effective PV power forecasting. PV power output is inherently time-dependent, shaped by periodic solar cycles, seasonal patterns, and rapidly changing weather conditions. These temporal characteristics usually include complex autocorrelation structures that, if properly addressed, can significantly improve forecasting performance. To capture these dynamics, various methodologies have been developed, ranging from classical linear models such as autoregressive integrated moving average (ARIMA) and exponential smoothing state space (ETS) models to more flexible approaches such as seasonal decomposition of time series.

Furthermore, with the increasing capabilities of computational resources, machine learning and deep learning models are used increasingly for temporal modeling. Long short-term memory (LSTM) networks and convolutional neural networks (CNN) are capable of capturing long-range dependencies and nonlinear temporal patterns in PV power data. Machine learning and deep learning models will be further analyzed in Chapter 5.

Hybrid approaches that integrate physical modeling with data-driven techniques are also emerging, offering improved generalizability and robustness across varying climatic and geographical conditions. This chapter systematically explores these methodologies, emphasizing their theoretical foundations, practical implementations, and comparative effectiveness in modeling temporal dependencies for PV power forecasting.

3.4.1   Autoregressive integrated moving average

The ARMA model is a combination of the Autoregressive (AR) model and the Moving Average (MA) model, which can forecast future values of a time series by performing linear regression on the previously observed values of that series [1]. A typical ARMA(p,q) model can be described as follows:

$$X_t = \sum_{i=1}^{p} \varphi_i X_{t-1} + \alpha_t - \sum_{j=1}^{q} \theta_j a_{t-j} \tag{3.6}$$

where X_t represents the time series value for time t, p and q are the orders of the AR and MA models, respectively, φ_i is the autoregressive parameter, θ_j is the moving average parameter, and a_t is the normal white noise.

The model in (3.6) can either contain the autoregressive part AR(p), the moving average part MA(q), or both parts ARMA(p,q). The AR component handles the previously observed power values, while the MA component addresses previous model errors to enhance the final prediction [1].

The primary drawback of this model is its reliance on a large amount of high-quality stationary historical data to achieve accurate results. However, since the data is often nonstationary, additional steps are required to ensure reliable predictions. To address the issue of nonstationary data in the ARMA model, a form of differencing transformation is employed. The original time series is differenced to achieve the stationarity of the data [5]. As a result, the model is called Auto Regressive Integrated Moving Average (ARIMA(p,d,q)) when the integration part is implemented.

The $I(d)$ part is responsible for the number of different times of the original time series [12]. The general equation of successive differences at the dth difference of the forecasted prediction X_t is

$$\Delta^d X_t = (1 - B)^d X_t \tag{3.7}$$

where d is the difference order and is usually 1 or 2, and B is the backshift operator and is defined as follows:

$$B X_t = X_{t-1} \tag{3.8}$$

When the backshift operator is applied to a value X_t of the time series, it can shift the data back one period. For example, following the previous equation, the successive difference at one time lag is as follows:

$$\Delta^1 X_t = (1 - B)X_t = X_t - X_{t-1} \tag{3.9}$$

3.4.2   Exponential smoothing state space model

The ETS model is a robust approach for time series forecasting that captures level, trend, and seasonality components in data. It leverages three primary components: error (E), trend (T), and seasonality (S). These components can be additive (A), multiplicative (M), or even non-computed (N), allowing the model to handle various types of time series patterns [13]. Some of the most commonly used ETS variations are the Simple Exponential Smoothing (ETS(A,N,N)), which addresses data without trends or seasonality, Holt's Linear Trend Model (ETS(A,A,N)), which considers data with trends but without the seasonal component, and the Holt-Winters Seasonal Model (ETS(A,A,A)), which captures both additive trend and seasonality. The above models are formulated as presented below:

Simple Exponential Smoothing (ETS(A,N,N))

$$y_t = l_{t-1} + e_t \tag{3.10}$$

$$l_t = ay_t + (1 - a)l_{t-1} \tag{3.11}$$

where y_t is the actual observed value of the time series at time t, l_t represents the level or baseline value of the series at time t, which is the smoothed estimate of the data at time t, adjusted for trend and seasonality, and α is the smoothing parameter for the level component of the series and determines how much weight is given to the most recent observations when updating the level estimate.

Holt's Linear Trend Model (ETS(A,A,N))

$$y_t = l_{t-1} + b_{t-1} + e_t \tag{3.12}$$

$$l_t = ay_t + (1 - a)(l_{t-1} + b_{t-1}) \tag{3.13}$$

$$b_t = \beta(l_t - l_{t-1}) + (1 - \beta)b_{t-1} \tag{3.14}$$

where b_t is the estimate of the trend at time t, which captures the long-term direction or slope of the series, and β is the smoothing parameter for the trend component, which controls how quickly the trend component is updated based on new data.

Holt-Winters Seasonal Model (ETS(A,A,A))

$$y_t = l_{t-1} + b_{t-1} + s_{t-m} + e_t \tag{3.15}$$

$$l_t = a(y_t - s_{t-m}) + (1 - a)(l_{t-1} + b_{t-1}) \tag{3.16}$$

$$b_t = \beta(l_t - l_{t-1}) + (1 - \beta)b_{t-1} \tag{3.17}$$

$$s_t = \gamma(y_t - l_t - b_t) + (1 - \gamma)s_{t-m} \tag{3.18}$$

where s_t represents the estimate of the seasonal effect at time t, which captures the repeating cyclical pattern at a given time, and γ is the smoothing parameter for the seasonal component, which determines how much weight is given to recent seasonal patterns when updating the seasonal estimate.

ETS models are highly flexible and interpretable. They provide clear decompositions of the time series into level, trend, and seasonal components, making it easier to understand underlying patterns. These models are also computationally efficient and relatively simple to implement, making them a popular choice for many forecasting tasks [14]. However, ETS models assume linear relationships, which might not always be appropriate, especially in complex real-world scenarios. They also have limited capacity to incorporate external factors, such as weather data, compared to more advanced machine learning models.

To implement an ETS model, one typically starts with data preparation, ensuring that the time series is clean and any missing values are handled appropriately [15]. The model is then selected based on the data's characteristics, such as the presence of trends and seasonality. Smoothing parameters (α, β, γ) are estimated using optimization techniques, and the model is fitted to the training data.

ETS models are particularly effective for PV generation forecasting due to their ability to capture and model seasonal patterns and trends in the data. For example, PV generation data typically exhibits strong daily and seasonal cycles, which can be well-represented by an ETS model with appropriate trend and seasonal components. However, for enhanced accuracy, particularly when external variables significantly impact the forecast, hybrid models or machine learning approaches may be considered.

3.4.3   Prophet

Prophet, developed by Facebook, is an open-source tool designed for forecasting time series data. It is particularly adept at handling data with strong seasonal effects and historical patterns, making it a robust choice for many forecasting applications. Prophet is designed to be user-friendly, requiring minimal configuration and delivering high-quality forecasts with interpretable components [16].

One of the key strengths of Prophet is its flexibility in modeling various types of time series data. It decomposes the time series into three main components: trend, seasonality, and holidays (or special events), as presented in Figure 3.7. The trend component models the long-term progression of the data, while the seasonality component captures periodic patterns that occur at regular intervals, such as daily, weekly, or yearly cycles. The holidays component allows the model to account for the impact of specific dates or events that can cause deviations from the regular patterns, such as public holidays or special promotions. Prophet uses an additive model, where the effects of these components are summed to generate the forecast [17]. This approach makes the model highly interpretable, as each component can be individually analyzed to understand its contribution to the overall forecast. For instance, users can easily visualize how seasonality affects the data at different times of the year or how certain holidays impact the forecast.

The math powering Facebook Prophet entails identifying and modeling these components using a number of mathematical and statistical methodologies: (i) linear regression is used to model the trend component of a time series; (ii) Fourier series are used to model the seasonality component of a time series; and (iii) additive models are used to describe the holidays component of a time series [18].

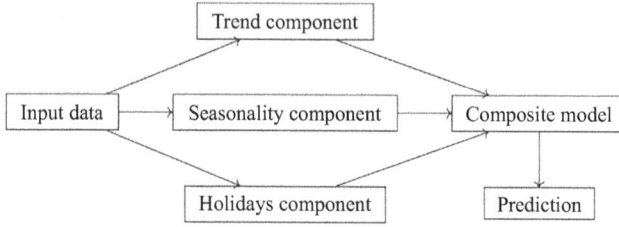

Figure 3.7   Representative flowchart of the Facebook Prophet [11]

A linear or nonlinear regression model is often used to discover and model any underlying patterns or trends in the data for the trend component of the time series. A linear regression model's equation can be given as follows:

$$y = \beta_0 + \beta_1 + \epsilon \tag{3.19}$$

where y is the dependent variable (i.e., the time series), x is the independent variable (i.e., the time of the data point), β_0 and β_1 are the coefficients of the model that are determined by the regression, and ϵ is the error term representing the difference between the predicted value of y and the actual value.

A Fourier series, a mathematical tool for describing periodic functions, is generally used to model the seasonality component of a time series. A Fourier series can be represented as the sum of sines and cosines:

$$f(t) = a_0 + \sum_{k=1}^{n} \left(a_k \cos \frac{2\pi kt}{T} + b_k \sin \frac{2\pi kt}{T} \right) \tag{3.20}$$

where $f(t)$ is the time series, t is the time of the data point, T is the period of the time series, n is the number of terms in the series, a_0 is the average value of the time series, and a_k and b_k are the coefficients of the cosines and sines that are determined by the Fourier analysis, respectively.

The holidays component of the time series is often modeled using a generative additive model (GAM), a statistical model that allows for the insertion of additional elements that may influence the time series. GAM is a flexible and nonparametric regression model that may be used to match complex data patterns. This is accomplished by representing the relationship between the response variable (the time series data) and the predictor variables (the factors influencing the data) as a sum of smooth functions. The GAM used by the Prophet method can be expressed as follows:

$$y(t) = g_1(t) + g_2(t) + \cdots + g_k(t) + \epsilon(t) \tag{3.21}$$

where $y(t)$ is the response variable (the time series data), $g_i(t)$ is the smooth function that models the ith predictor variable, and $\epsilon(t)$ is the error term that represents the random noise or variation in the data.

Smooth functions $g_i(t)$ can be described using a number of basic functions that can capture the data's complexity, such as splines or Fourier series. Maximum likelihood estimation or other optimization approaches can be used to estimate the parameters of these basis functions.

Prophet is also distinguished by its capacity to handle missing data and outliers effectively. Unlike traditional time series models that often require uninterrupted data, Prophet can work effectively with irregular time series data. It includes features to identify and adjust for outliers, ensuring that they do not disproportionately affect the forecast.

Designed for both scalability and user-friendliness, Prophet provides an intuitive interface that simplifies the forecasting process and often requires just a few lines of code. This makes it accessible even to users with limited experience in time series analysis. For more experienced users, Prophet allows for the adjustment of various parameters to tailor the model to the specific nuances of their data.

In PV generation forecasting, Prophet's capabilities are particularly advantageous. Given that PV generation data frequently displays strong seasonal variations due to changes in sunlight throughout the year, Prophet's seasonality component can effectively capture these patterns. Additionally, external factors such as weather conditions and special events that may impact energy production can be incorporated into the model through its holidays component.

The interpretability of Prophet's forecasts is another asset, providing valuable insights into the underlying factors driving the predictions. This feature helps stakeholders such as energy producers and grid operators to plan maintenance, manage energy storage, and optimize grid operations based on anticipated changes in PV output.

3.4.4   Hybrid methodologies for time-series PV forecasting

Hybrid models combine different modeling techniques to use their individual strengths and mitigate their weaknesses, resulting in more accurate and robust forecasts. Chapter 6 will explore hybrid and ensemble AI-based PV forecasting in detail. In the context of time series analysis, hybrid models can address the complexities and nuances of data that might be challenging for a single model to capture comprehensively.

Hybrid models offer several key benefits. Enhanced accuracy is achieved by combining models to capture a broader range of patterns and relationships in the data, integrating both linear and nonlinear relationships, short-term fluctuations, and long-term trends. Robustness is another advantage, as hybrid models are less susceptible to the weaknesses of individual models; if one model fails to capture certain aspects of the data, another in the hybrid setup can compensate, resulting in more stable and reliable forecasts. Additionally, flexibility is a significant benefit, allowing hybrid models to be tailored to specific data characteristics and forecasting requirements. This flexibility enables the inclusion of domain-specific knowledge and the adjustment of model components to better fit the data. In the context of PV generation forecasting, hybrid models can address the complex nature of the data, which is influenced by various factors, such as weather conditions, seasonal cycles, and long-term trends. Reviewing the hybrid models presented above, various conclusions emerge.

Considering the models presented above, some examples of hybrid time series forecasts with ML models are presented below:

(i)   An ARIMA–LSTM hybrid model can combine the strengths of ARIMA and LSTM models. ARIMA is effective at capturing linear relationships and short-term dependencies, while LSTMs excel at learning complex, long-term patterns and nonlinear relationships. By combining these models, the hybrid can handle both linear trends and complex temporal dependencies. The ARIMA–LSTM hybrid can effectively model PV generation data by capturing the linear trends and seasonality using ARIMA, while LSTM handles the nonlinear relationships and long-term dependencies introduced by factors such as changing weather patterns. In [19], a hybrid ARIMA–LSTM model was developed to assess the siting of PV power generation in the case of China. Different geographical cases were analyzed, while indicators affecting the accuracy of the forecasts were also reviewed.

(ii)   SARIMA–Prophet Hybrid: A SARIMA–Prophet hybrid model can utilize the SARIMA model's ability to capture seasonality and trends in the data, as well as the Prophet's flexibility in modeling holidays and handling missing data. A hybrid of these two methodologies could leverage SARIMA's precise seasonality adjustments and Prophet's ease of handling external factors and irregular time series. In [20], SARIMA-based forecasting models were effectively applied in four different PV plant sites in Greece. Those models showed improved forecasting accuracy thanks to incorporating external solar radiation data. In [21], the Prophet's forecasting capabilities were evaluated for the case of Oman. Both under clear and cloudy sky conditions, Prophet outperformed traditional comparative models in terms of modeling seasonality and trend effects with uncertain outliers. By applying the SARIMA–Prophet hybrid model, the SARIMA is used to model the seasonality and trends, combined with Prophet's ability to incorporate holiday effects and handle missing data, providing a comprehensive approach to PV forecasting. This hybrid can be particularly useful in managing periods with irregular data or significant external events affecting energy production.

3.5   Conclusion

In conclusion, time series forecasting in PV power generation requires a comprehensive understanding of the underlying temporal patterns and the ability to model both regular and irregular variations in the data. Accurate forecasting supports better energy management, grid stability, and efficient planning, particularly given the fluctuating nature of solar power production. Various modeling strategies aim to capture key time series components, such as trends, seasonality, and irregular fluctuations, enabling more reliable predictions of future energy output. These methods often focus on extracting meaningful signals from complex, noisy datasets while adapting to the cyclical and evolving behavior of PV systems.

Moreover, the effectiveness of forecasting depends not only on the ability to represent temporal patterns but also on how well the models can generalize across different scenarios and environmental conditions. This involves addressing challenges such as data non-stationarity, missing values, and sudden changes due to weather variability. Combining multiple modeling strategies or adopting adaptable frameworks can enhance both the flexibility and accuracy of forecasting tools. Overall, the thoughtful integration of temporal structure, data-driven patterns, and system-specific characteristics is central to advancing PV power forecasting and meeting the growing demands of renewable energy systems.

References

[1]   Manz D., Walling R., Miller N., LaRose B., D'Aquila R., Daryanian B. 'The grid of the future: Ten trends that will shape the grid over the next decade'. *IEEE Power Energy Magazine*. 2014;**12**(3):26–36.

[2]   Bazionis I., Kousounadis-Knousen M., Georgilakis P., Shirazi E., Soudris D., Catthoor F. 'A taxonomy of short-term solar power forecasting: Classifications focused on climatic conditions and input data'. *IET Renewable Power Generation*. 2023;**17**(9):2411–2432.

[3]   Shirazi E., Gordon I., Reinders A., Catthoor F. 'Sky images for short-term solar irradiance forecast: a comparative study of linear machine learning models'. *IEEE Journal of Photovoltaics*. 2024;**14**(4):691–698.

[4]   Ali M., Rabehi A., Souahlia A., *et al.* 'Enhancing PV power forecasting through feature selection and artificial neural networks: a case study'. *Scientific Reports*. 2025;**15**, 22574.

[5]   Strzalka A., Alam N., Duminil E., Coors V., Eicker U. 'Large scale integration of photovoltaics in cities'. *Applied Energy*. 2012;**93**:413–421.

[6]   Pedro H., Coimbra C. 'Assessment of forecasting techniques for solar power production with no exogenous inputs'. *Solar Energy*, 2012;**86**:2017–2028.

[7]   Allison M., Akakabota E., Pillai G. 'Future load profiles under scenarios of increasing renewable generation and electric transport'. *5th International Conference on Renewable Energy: Generation and Applications (ICREGA)*; Al Ain, United Arab Emirates, 2018.

[8]   Libesa 'Using decomposition to improve time series prediction'. 2014. https://quantdare.com/decomposition-to-improve-time-series-prediction/

[9]   Boland J., David M., Lauret P. 'Short term solar radiation forecasting: island versus continental sites'. *Energy*, 2016;**113**:186–192.

[10] Dong Z., Yang D., Reindl T., Walsh W. 'Short-term solar irradiance forecasting using exponential smoothing state space model'. *Energy*. 2013;**55**:1104–1113.

[11] Stefenon S., Seman L., Mariani V., Coelho, L. 'Aggregating Prophet and seasonal trend decomposition for time series forecasting of Italian electricity spot prices'. *Energies*, 2023;**16**(3):1371.

[12] De Felice M., Soares M., Alessandri A., Troccoli A. 'Scoping the potential usefulness of seasonal climate forecasts for solar power management'. *Renewable Energy*. 2019;**142**:215–223.

[13] Box G. E. P., Jenkins G. M., Reinsel G. C. *Time series analysis: Forecasting and Control*. New York: John Wiley & Sons, Inc.; 2008.

[14] Wan C., Zhao J., Song Y., Xu Z., Lin J., Hu Z. 'Photovoltaic and solar power forecasting for smart grid energy management'. *CSEE Journal of Power and Energy Systems*. 2015;**1**(4):38–46.

[15] Diagne M., David M., Lauret P., Boland J., Schmutz N. 'Review of solar irradiance forecasting methods and a proposition for small-scale insular grids'. *Renewable and Sustainable Energy Reviews*. 2013;**27**:65–76.

[16] Liu C.-H., Gu J.-C., Yang M.-T. 'A simplified LSTM neural networks for one day-ahead solar power forecasting'. *IEEE Access*. 2021;**9**:17174–17195.

[17] Hossain M., Mahmood H. 'Short-term photovoltaic power forecasting using an LSTM neural network and synthetic weather forecast'. *IEEE Access*. 2020;**8**:172524–172533.

[18] Shohan M. J. A., Faruque M. O., Foo S. Y. 'Forecasting of electric load using a hybrid LSTM-neural Prophet model'. *Energies*. 2022;**15**(6):2158.

[19] Chen G., Qi X., Wang Y., Du W. 'ARIMA-LSTM Model-based siting study of photovoltaic power generation technology'. *2024 4th Asia-Pacific Conference on Communications Technology and Computer Science (ACCTCS)*; Shenyang, China, 2024.

[20] Vagropoulos S., Chouliaras G., Kardakos E., Simoglou C., Bakirtzis A. 'Comparison of SARIMAX, SARIMA, modified SARIMA and ANN-based models for short-term PV generation forecasting'. *2016 IEEE International Energy Conference (ENERGYCON)*; Leuven, Belgium, 2016.

[21] Baloch M., Honnurvali M., Kabbani A., Ahmed T., Chauhdary S., Saeed M. 'Solar energy forecasting framework using Prophet based machine learning model: an opportunity to explore solar energy potential in Muscat Oman'. *Energies*. 2025;**18**(1);205.

Chapter 4
Machine learning approaches for PV forecasting

*Markos A. Kousounadis-Knousen[1], Athina P. Georgilaki[1]
and Pavlos S. Georgilakis[1]*

Machine learning (ML) is described as the data-driven branch of artificial intelligence (AI), using historical data to learn relationships between variables. ML models are characterized by their learning process, parameters, and hyperparameters. The learning process typically involves either iterative optimization, known as training, or a rule-based construction. During the learning process, model parameters are adjusted to capture hidden patterns and underlying relationships in historical data, enabling ML models to generalize to unseen data and perform complex tasks, such as classification and forecasting. In contrast, hyperparameters are predefined settings that shape the ML model's structure and influence the learning process.

The main advantages that make ML approaches attractive are their strong inference capabilities for deriving complex nonlinear relationships in nonstationary data. Furthermore, ML models exhibit high levels of interpretability and flexibility. This capability helps mitigate the challenges associated with PV power forecasting, as PV power generation is a highly intermittent and volatile stochastic variable, with nonlinear dependencies on multiple meteorological, geographical, and degradation factors. ML approaches are capable of modeling these nonlinear dependencies, provided that the historical data are of high quality or have been appropriately preprocessed.

Over the past two decades, the number of published works on ML-based PV power forecasting has increased exponentially [1]. The availability of meteorological data and monitoring data of PV systems has enabled the use of ML for PV power forecasting. Most ML approaches are developed for generating predictions, particularly in the short-term, i.e., from intraday and day-ahead horizons up to several days ahead [2]. Maintaining stability in power systems with high PV integration requires rapid decision-making through effective energy management and control, in which ML-based PV power forecasting plays a crucial role. Furthermore, the significant increase in PV integration into local distribution-level power systems creates a plethora of case studies with different characteristics that are difficult to model physically. ML-based PV power forecasting offers flexibility for retraining and application across various target case studies.

[1]School of Electrical and Computer Engineering, National Technical University of Athens (NTUA), Greece

There are two main categories of ML approaches: supervised learning and unsupervised learning. This chapter focuses only on shallow ML models. Approaches based on deep learning will be discussed in Chapter 5. A graphical tree of the most common supervised and unsupervised learning approaches employed for PV power forecasting is presented in Figure 4.1. The main characteristics of each approach, as well as implementation details for PV power forecasting, are described in the following sections of this chapter.

4.1   Supervised learning

Supervised learning is a type of ML that involves learning a mapping between input and output data. The input data, also known as explanatory or independent variables, consist of low-level features that impact a corresponding high-level output, commonly referred to as the label or dependent variable. Supervised learning relies on labeled datasets, in which each set of explanatory variables is associated with a known ground-truth label, forming input–output pairs. These pairs are used in supervised learning to adjust the parameters of an ML model and optimize a given objective function. Apart from learning a mapping function between the input and the output data of the labeled dataset, the goal in supervised learning is to increase the generalization ability, i.e., the ability to generate accurate outputs for unseen input data.

Depending on the output type, supervised learning is typically divided into two main categories: regression and classification. Regression involves continuous numerical outputs, in contrast to classification, where the outputs are discrete-dependent variables that represent categories or classes. Since the output of PV systems can be fully represented as a continuous numerical variable, PV power forecasting is treated as a regression task. Commonly employed regression-based supervised learning techniques for PV power forecasting are linear regression (LR), artificial neural networks (ANNs), support vector regression (SVR), and regression trees (RTs).

4.1.1   Regression models

Regression analysis aims to model the effects of K explanatory variables $x_1, x_2, \ldots, x_K$ on a dependent stochastic variable y, such as PV power generation. The stochastic variable is decomposed into a systematic component f, dependent on the explanatory variables, and a stochastic component ε_r [3]:

$$y = f(x_1, x_2, \ldots, x_K) + \varepsilon_r \tag{4.1}$$

Provided n sample sets of explanatory variables $x_1, x_2, \ldots, x_K$ and their corresponding dependent variables y, regression models try to isolate the systematic component and approximate it. The choice of a suitable regression model depends on the complexity of the relationship between the dependent variable and the explanatory variables, the number of explanatory variables, as well as the data type.

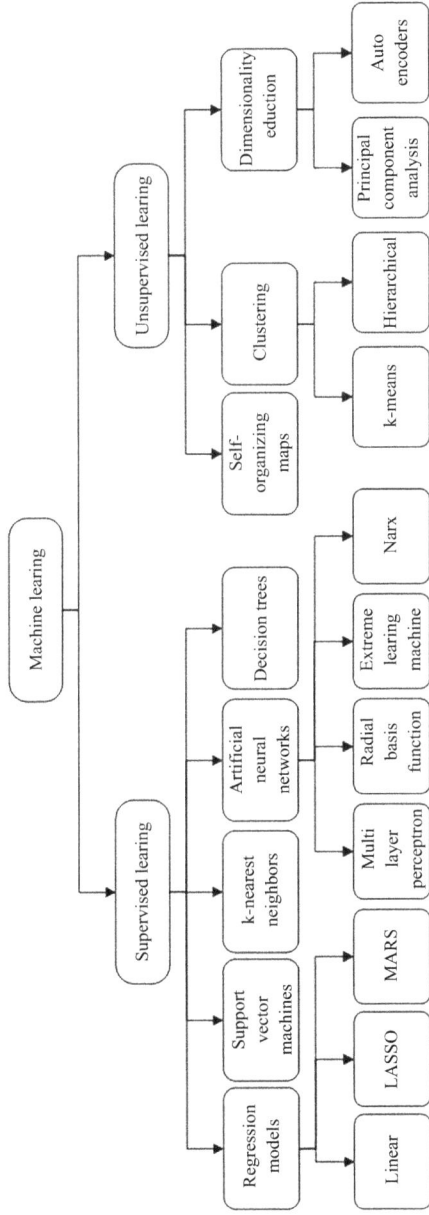

Figure 4.1   Graphical tree of the most common supervised and unsupervised learning approaches employed for PV power forecasting

LR is the most used regression technique, owing to its simplicity and interpretability [1]. LR assumes that the systematic component f is a linear combination of the explanatory variables $x_1, x_2, \ldots, x_K$:

$$f(x_1, x_2, \ldots, x_K) = \beta_0 + \beta_1 x_1 + \cdots + \beta_K x_K \tag{4.2}$$

If K is set to 1, y is dependent on a single explanatory variable and LR is referred to as simple or univariate; otherwise, LR is referred to as multiple or multivariate. In PV power forecasting, multivariate LR has exhibited better performance compared to univariate LR [4].

Coefficients $\beta_0, \beta_1, \ldots, \beta_K$ are estimated using gradient descent algorithms or fitting methods, such as least squares (LS):

$$LS(\beta_0, \beta_1, \ldots, \beta_K) = \min \sum_{i=1}^{n} (y - f(x_1, x_2, \ldots, x_K))^2 \tag{4.3}$$

The flowchart of an LR model is shown in Figure 4.2. This LR model is based on gradient descent and iteratively adjusts its coefficients until it reaches the desired accuracy or the maximum iterations.

Gradient descent algorithms are well-suited options for optimizing convex objective functions. However, traditional gradient descent is prone to local optima entrapment and becomes slow on large datasets, as it updates coefficients only once per iteration using all n sample sets. An alternative to traditional gradient descent is

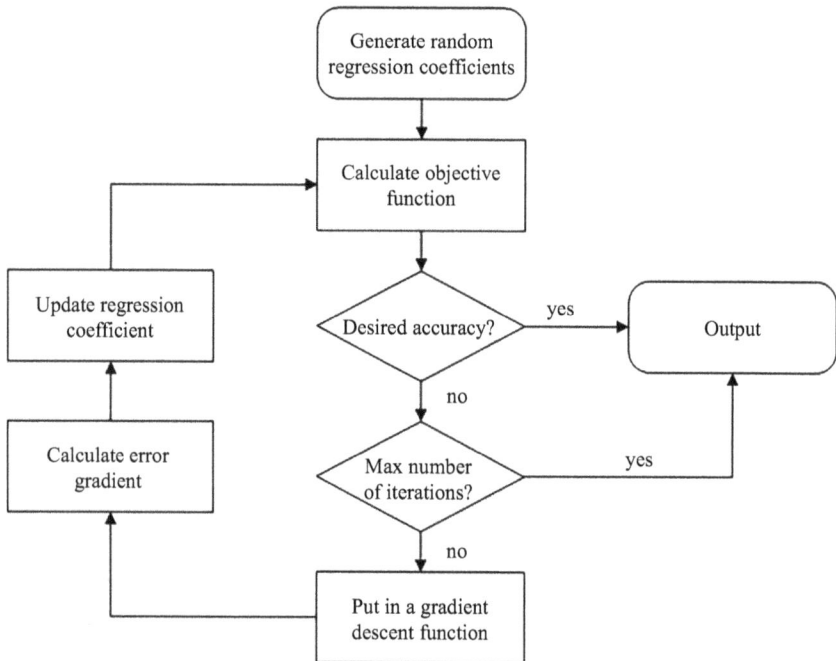

Figure 4.2   Flowchart of the linear regression model

stochastic gradient descent (SGD), which updates the coefficients multiple times per iteration, after evaluating each individual sample set [5]. Unlike traditional gradient descent, SGD can escape local optima by introducing a noisy gradient evaluated from a single sample set or a minibatch of sample sets [6]. While SGD is efficient and easy to implement, it requires careful tuning of multiple hyperparameters and is sensitive to feature scaling.

Another fitting algorithm is the random sample consensus (RANSAC) regressor. RANSAC defines a hypothesis using randomly selected sample sets and evaluates the hypothesis using the remaining sample sets in the dataset [7]. If the number of sample sets satisfying the defined hypothesis is less than a specified threshold, then a new hypothesis must be defined. This iterative process continues until the total number of satisfying sample sets exceeds a threshold, or a maximum number of iterations is reached [8]. The flowchart of the RANSAC algorithm is illustrated in Figure 4.3. In Figure 4.3, S is the smallest number of sample sets needed to define a hypothesis, D is the total number of sample sets, K is the number of sample sets threshold to assert if a model fits well, T is the threshold used to identify if a sample set fits the hypothesis, and I is the maximum number of iterations. The maximum number of iterations should be set such that it can guarantee that at least one hypothesis has been validated with probability p. There are many hyperparameters in RANSAC that make this algorithm sensitive to hyperparameter tuning. To tackle this issue, some of the hyperparameters can be calculated based on the values of others. For example, the maximum number of iterations I can be calculated based on the number of sample sets S, the validation probability p, and the inlier ratio r.

$$I = \frac{\log(1-p)}{\log(1-r^S)} \tag{4.4}$$

LR models often suffer from overfitting, particularly when the number of available sample sets is limited. To overcome the problem of overfitting, (4.3) is extended with the addition of a regularization term:

$$\begin{aligned} LS(\beta_0, \beta_1, \ldots, \beta_K) = \min \sum_{i=1}^{n} (y - f(x_1, x_2, \ldots, x_K))^2 \\ + \lambda_r p_r(\beta_0, \beta_1, \ldots, \beta_K) \end{aligned} \tag{4.5}$$

where p_r denotes the imposed regularization penalty function to smooth the fitting curve, and $\lambda_r \geq 0$ controls the smoothing level [3]. Tuning λ_r is critical in regularization-based regression. When $\lambda_r = 0$, the penalty term has no effect and regression is identical to LR. However, increasing λ_r to infinite will shrink the regression coefficients toward zero. Regularization-based regression generally follows the same flowchart as LR (Figure 4.2). A common approach to tune λ_r is Bayesian regression, in which λ_r is considered a random variable and is estimated from the sample sets, assuming a Gaussian distribution [3]. However, it should be considered that this procedure could be very time-consuming because of the stochastically tuned repeated simulations that are necessary.

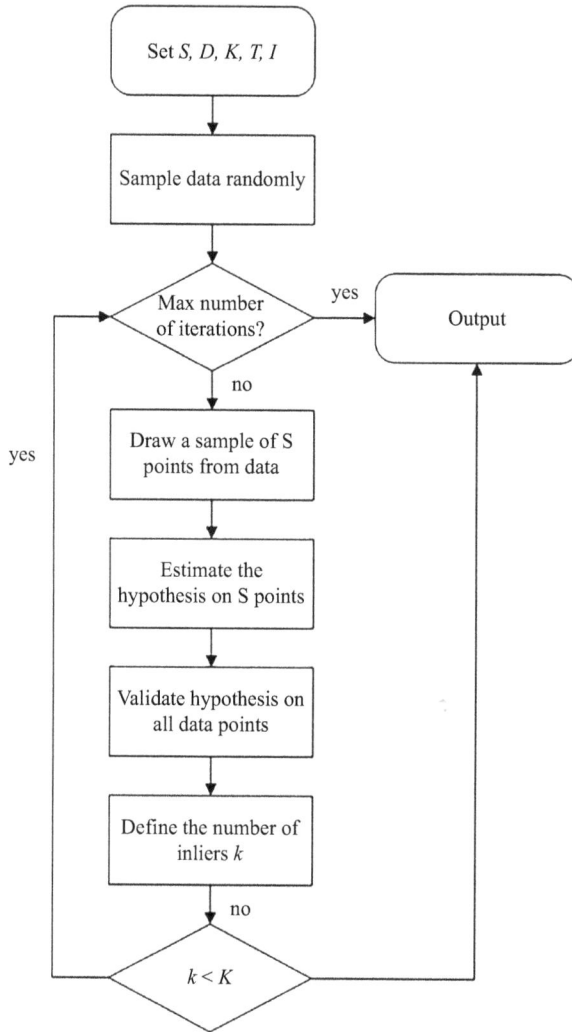

Figure 4.3   Flowchart of the RANSAC algorithm

A popular regularization-based regression model is ridge regression, introduced by Hoerl and Kennard in [9]. The regularization penalty function in ridge regression is L2 regularization, which is the sum of squared regression coefficients:

$$p_r\left(\beta_0, \beta_1, \ldots, \beta_K\right) = \sum_{j=1}^{K} \beta_j^2 \tag{4.6}$$

The quadratic penalty of L2 regularization in ridge regression results in a significant impact on large coefficient values and a small impact for coefficient values close to zero, thus preventing them from shrinking toward zero.

Another popular regularization-based regression model is the least absolute shrinkage and selection operator (LASSO) [10]. LASSO uses L1 regularization as the penalty function, which is given as the sum of absolute coefficients $\beta_0, \beta_1, \ldots, \beta_K$:

$$p_r(\beta_0, \beta_1, \ldots, \beta_K) = \sum_{j=1}^{K} |\beta_j| \qquad (4.7)$$

Compared to ridge regression, LASSO tends to shrink smaller coefficients toward zero, while keeping the larger coefficients less affected by the regularization penalty term. Setting some coefficients to zero can be beneficial in some cases, as it will remove the impact of irrelevant explanatory variables and reduce the overall number of features. In this way, LASSO simultaneously performs regularization and feature selection.

Regularization-based LR has been successfully applied for PV power forecasting [11]. However, the interest in regularization-based LR models is currently declining, as bigger, less biased PV-related datasets are more easily obtainable in recent years [1]. Improved regularization-based LR techniques are mainly employed as part of hybrid forecasting models [12].

An alternative to simple LR models is piecewise LR. With piecewise LR, the nonlinear relationship between PV power generation and the explanatory variables related to it is approximated by multiple local LR models with varying spline functions. Multivariate adaptive regression splines (MARS) is a piecewise LR model that divides the sample sets into groups and applies local LR for each group [13]. MARS has become an attractive alternative for PV power forecasting, as it can approximate nonlinear relationships while maintaining high levels of interpretability, scalability, and computational efficiency [1,14,15].

With the MARS algorithm, the systematic component of (4.1) is approximated as follows:

$$f(x_1, x_2, \ldots, x_K) = \beta_0 + \sum_{i=1}^{M} \beta_i BF_i(x_1, x_2, \ldots, x_K) \qquad (4.8)$$

where BF_i denotes the basis function i, and M is the total number of nonconstant basis functions. Each basis function takes the form of a hinge function or a product of two or more hinge functions. A hinge function is defined as follows:

$$h(x) = \begin{cases} x - c, \ x > c \\ 0, \ \text{otherwise} \end{cases} \qquad (4.9)$$

Basically, the sample sets are partitioned into M different pieces, and an LR model is created for each piece in the form of a basis function. The MARS algorithm involves two main steps: the forward pass and the backward pass. During the forward pass, new basis functions are iteratively added to the model to increase its accuracy. The backward pass is a pruning process in which the basis functions with the least contribution are removed to prevent overfitting.

Regression models have been extensively tested in the field of PV power forecasting [16]. LR is considered a simplistic method incapable of accurately modeling the nonlinearities of a complex task, such as PV power forecasting; thus, it is usually

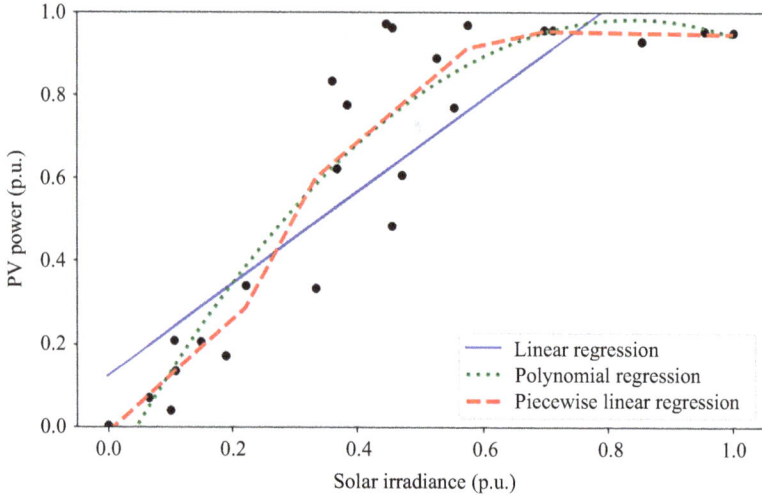

*Figure 4.4   Linear, piecewise linear, and polynomial regression lines for 24
random samples of a PV dataset. Measurements are taken from a 1.2
MW PV power plant located in Western Macedonia, Greece (inverter
clipping included).*

employed for simple, coarse-grained PV power forecasting tasks or benchmarking
purposes [1]. The MARS algorithm achieves a good trade-off between modeling
accuracy and computational efficiency; however, it still provides a piecewise linear
approximation rather than an actual nonlinear model. However, nonlinear modeling
of the relationships between PV power and its explanatory variables can be achieved
using polynomial regression. Examples of linear, piecewise linear, and polynomial
regression lines computed for PV power forecasting are shown in Figure 4.4. The data
samples are randomly selected from a PV power dataset, with solar irradiance serving as the explanatory variable. In this case, piecewise LR contains four knots, and
the polynomial regression line corresponds to a second degree polynomial function.
As the knots and the polynomial degree increase, there is a higher risk of overfitting. Complex polynomial regression models often suffer from overfitting, while
simpler polynomial regression models usually underperform compared to other ML
approaches, such as ANNs. Thus, polynomial regression is rarely proposed for PV
power forecasting, with its use limited to benchmarking purposes.

4.1.2   Artificial neural networks

ANNs are computational models inspired by the function and structure of biological nervous systems [17]. The basic processing units of ANNs, called neurons, are
interconnected by synaptic weights and typically include a bias. A group of neurons
organized in parallel constitutes a layer. ANNs consist of three types of layers: the
input layer, one or more hidden layers, and the output layer. The neurons of the input
layer are passive, in the sense that they serve as entry points for the input data and do
not perform any computations. All processing occurs within the hidden and output

layers. For regression tasks, each neuron of the input layer represents an explanatory variable, whereas the output layer neurons represent the dependent variable. In PV power forecasting, the neurons of the input layer usually correspond to meteorological parameters and the neurons of the output layer represent the predicted PV power. Typically, the output layer contains a single neuron that represents the point forecast of PV power for a given set of input variables. A wide variety of ANN types exists, each with its own structure, computational processes, and characteristics, suitable for PV power forecasting tasks at specific temporal and spatial scales. ANNs are particularly effective in modeling complex, nonlinear relationships between variables. The nonlinear relationship between the input and the output is modeled by the neurons of the hidden layer and the weight synapses. It has been shown that any mapping function can be sufficiently extracted by an ANN with up to only two hidden layers [18]. Therefore, ANNs are among the most popular supervised ML methods for predicting PV power [1].

Each of the hidden and output layers in an ANN incorporates an activation function, which plays a key role in capturing the nonlinear relationships between the explanatory and dependent variables. The output of each neuron is passed through the activation function of its layer, which determines the neuron's final output. Common activation functions of ANNs are described in Table 4.1 and illustrated in Figure 4.5. The selection of an appropriate activation function for each layer primarily depends on the nature of the nonlinearities present in the relationship between the explanatory and dependent variables.

The trainable parameters of ANNs are their weight and bias values. During training, the weights and biases are appropriately adjusted to learn a mapping function between the explanatory and dependent variables by optimizing a given objective function. In regression tasks such as PV power forecasting, objective functions typically involve the minimization of a loss function, such as the mean squared error (MSE), the root mean squared error (RMSE), the mean absolute error (MAE), and the mean absolute percentage error (MAPE), described by (4.10)–(4.13):

$$MSE = \frac{1}{n} \sum_{i=1}^{n} (y_i - \widehat{y}_i)^2 \tag{4.10}$$

Table 4.1 *Description and formula of common activation functions of ANNs*

Activation function	Formula	Notes
Sigmoid	$f_a^{sig}(x) = \frac{1}{1+e^{-x}}$	Nonlinear, continuous, differentiable Range: [0, 1]
Hyperbolic tangent sigmoid (tanh)	$f_a^{tanh}(x) = \frac{e^x - e^{-x}}{e^x + e^{-x}}$	Nonlinear, continuous, differentiable Range: [−1, 1]
Gaussian radial basis function (RBF)	$f_a^{RBF}(x) = e^{-\frac{(x-\mu)^2}{2\sigma^2}}$	Nonlinear, continuous, differentiable Range: [0, 1], μ: mean, σ: standard deviation
Rectified linear unit (ReLU)	$f_a^{ReLU}(x) = \max(0, x)$	Linear, continuous Range: [0, +∞]

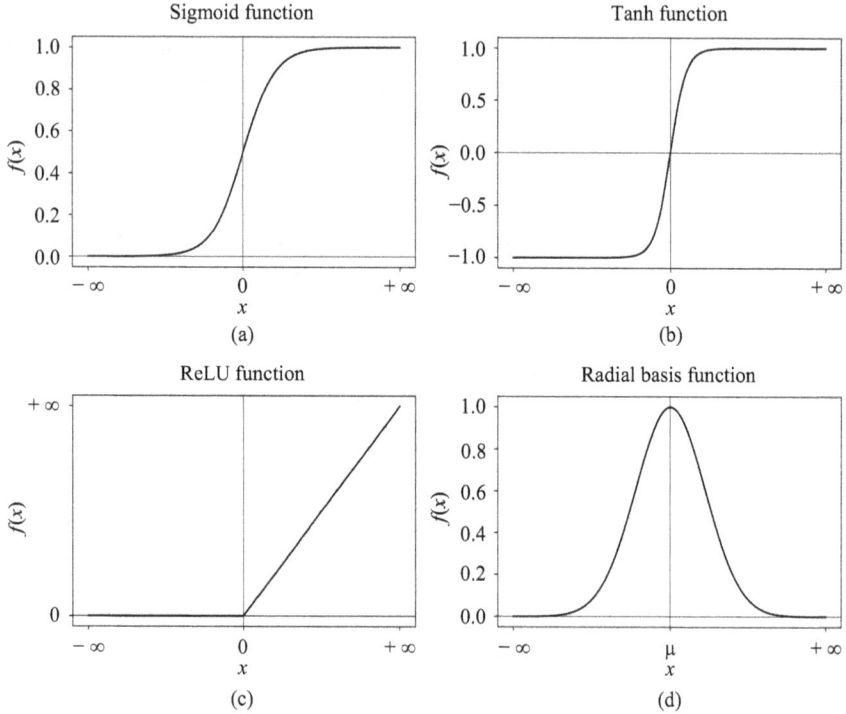

Figure 4.5   Graphs of common activation functions for ANNs. (a) Sigmoid. (b) Hyperbolic tangent sigmoid. (c) Rectified linear unit. (d) Gaussian radial basis function.

$$\text{RMSE} = \sqrt{\frac{1}{n}\sum_{i=1}^{n}(y_i - \widehat{y}_i)^2} \tag{4.11}$$

$$\text{MAE} = \frac{1}{n}\sum_{i=1}^{n}|y_i - \widehat{y}_i| \tag{4.12}$$

$$\text{MAPE} = \frac{1}{n}\sum_{i=1}^{n}\frac{|y_i - \widehat{y}_i|}{y_i} \tag{4.13}$$

where y_i and $\widehat{y}_i$ are the real and predicted values, respectively, for the sample set i of the dependent variable y, i.e., PV power generation.

ANNs are classified into two main categories, based on the direction of the information flow: feed forward ANNs (FFNNs) and recurrent ANNs (RNNs). In FFNNs, the information propagates unidirectionally from the input layer to the output layer through the hidden layers, while RNNs contain loops where information is fed back to the same or previous layers. Preservation of previous hidden states introduces depth in the temporal dimension, decreasing the interpretability and computational efficiency of RNNs. As a result, RNNs are typically classified as part of deep learning.

Fully connected ANNs (FCNNs) are a common class of FFNNs. In FCNNs, each neuron of layer L is connected to every neuron of layer $L+1$ with a simple weight synapse, forming a dense network of connections between the layers. This section focuses only on shallow FCNNs (one or two hidden layers), as other types of FFNNs typically fall into the category of deep learning and will be analyzed in the next chapters. Furthermore, as multiple variations of FCNNs exist, this section is constrained to the FCNN types that are frequently applied for PV power forecasting.

4.1.2.1   Multilayer perceptron

One of the most widely used FCNNs is the multilayer perceptron (MLP). MLPs have been widely employed for PV power forecasting due to their simplicity, flexibility, and interpretability. Examples of MLP applications for predicting solar power can be found in [19, 20]. The structure of an MLP with a single hidden layer and one neuron in the output layer is depicted in Figure 4.6(a). Figure 4.6(b) depicts the block diagram of the neurons in the hidden and output layers. Each active neuron receives weighted information from the neurons of the previous layer and a bias input. In other words, to compute the input of each layer, MLPs use dot products of the output of the previous layer and the weight synapses, followed by the addition of the bias. The input is then passed through the activation function f_a – typically a sigmoidal activation function or other monotonic functions such as ReLU – to produce the output of the neuron. The output y of the MLP of Figure 4.6(a) is calculated as follows:

$$ y = f_{a,\text{out}} \left(b_{\text{out}} + \sum_{i=1}^{N} w_i f_{a,\text{hid}} \left(b_i + \sum_{j=1}^{K} w_{ji} x_j \right) \right) \qquad (4.14) $$

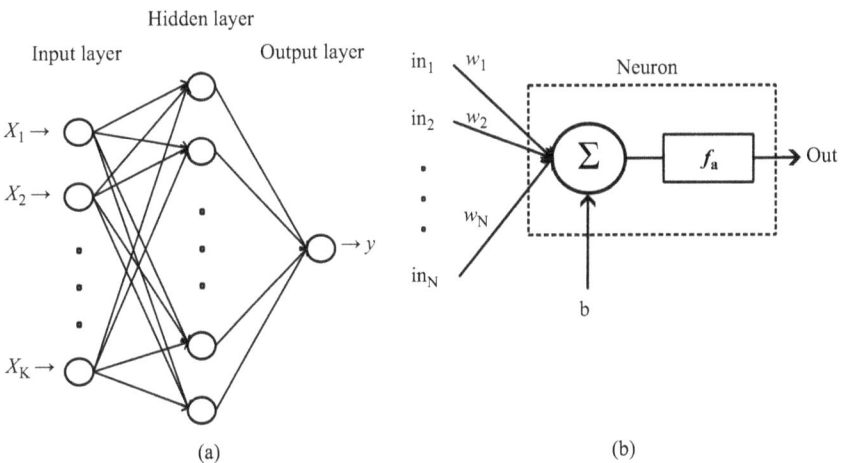

(a)   (b)

Figure 4.6   (a) *MLP with one hidden layer and a single neuron in the output layer.*
(b) *Block diagram of the neurons in the hidden and output layers of an MLP.*

where $f_{a,\text{out}}$ is the activation function of the output layer, $f_{a,\text{hid}}$ is the activation function of the hidden layer, b_{out} is the bias of the neuron in the output layer, b_i is the bias of neuron i of the hidden layer, w_i is the weight synapse connecting neuron i of the hidden layer to the neuron in the output layer, and w_{ji} is the weight synapse connecting neuron j of the input layer to neuron i of the hidden layer.

There are several methods for updating the weights of an MLP during training, the most popular being the gradient descent backpropagation algorithms [17]. Gradient descent backpropagation algorithms are based on two main principles. First, the weights are updated such that the derivative of the loss function with respect to the weight values (gradient) remains negative, i.e., the loss function decreases. Second, the gradients are propagated backward from layer to layer, starting from the output layer, until all weights are updated. In each iteration it, the weight $w_{ji}^{L,L+1}$ connecting neuron j of layer L to neuron i of layer $L + 1$ is updated as follows:

$$w_{ji}^{L,L+1}(it + 1) = w_{ji}^{L,L+1}(it) + \Delta w_{ji}^{L,L+1}(it + 1) \tag{4.15}$$

$$\Delta w_{ji}^{L,L+1}(it + 1) = lr \cdot \delta_j^L \cdot out_i^{L+1} + m \cdot \Delta w_{ji}^{L,L+1}(it) \tag{4.16}$$

where lr is the learning rate, m is the momentum, out_i^{L+1} is the output of neuron i of layer $L + 1$ and δ_j^L is the delta coefficient for neuron j of layer L. The delta coefficient is related to the gradient, and its calculation formula depends on the loss function, the activation functions, as well as the layer where the neuron of interest is located. The learning rate controls the amount of effect the gradients have on the weight updates, while the momentum utilizes information from previous iterations to avoid getting stuck in local minima [17].

Common gradient descent algorithms employed for MLPs are the adaptive gradient optimizer (Adagrad), the adaptive moment estimation (Adam), and RMSprop. Unlike classic gradient descent algorithms, which keep the learning rate constant, Adagrad adaptively updates the learning rate in each iteration independently for each weight. For each weight, the learning rate in Adagrad is inversely proportional to the sum of all the squared gradients recorded for that weight up to that point; thus, the learning rate adaptively decreases as training proceeds, increasing the convergence speed. However, Adagrad tends to underperform when dealing with nonconvex objective functions with multiple local minima, due to the rapid learning rate decrease caused by gradient explosion. RMSprop and Adam deal with this issue by normalizing the gradients with the moving average of the squared gradients, calculated for only a few past iterations. Adam further boosts its convergence capabilities by normalizing the momentum with the moving average of the gradients.

Figure 4.7 illustrates an example of the final weight values of an MLP trained on the data samples of Figure 4.4. Note that biases are not included in Figure 4.7. In addition to solar irradiance, the temperature of the PV modules is added as an explanatory variable; thus, the MLP has two neurons in the input layer, corresponding to each of the explanatory variables. The MLP consists of a single hidden layer containing four neurons, and a single neuron in the output layer corresponding to the PV power output. Thus, the MLP has 15 parameters (weight synapses and biases) to optimize during training. Both layers use the sigmoid function as the activation function. The MLP was trained on minimizing the MSE, using batch gradient descent (BGD) and

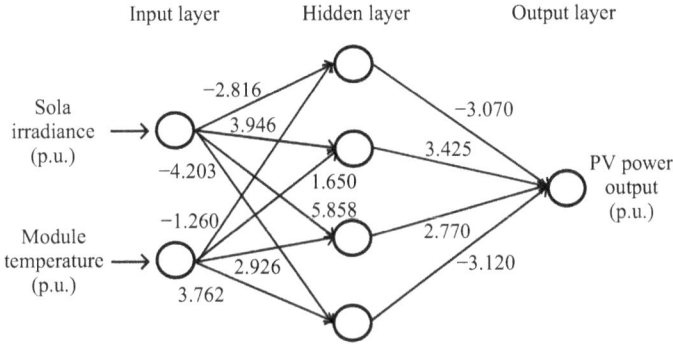

*Figure 4.7   Weight values of MLP trained for 24 random samples of a PV dataset.
Measurements are taken from a 1.2 MW PV power plant located in
Western Macedonia, Greece.*

the Adam algorithm in 10,000 iterations. Successful training significantly depends
on the initialization of the weights and biases. In the example of Figure 4.7, the MLP
manages to fit the training samples with an MSE of only 1.31%.

The regression problem in Figure 4.7 can be considered a relatively simple task,
as the MLP is trained to predict the PV power output (dependent variable) using
explanatory variables that correspond to the same timestep. Thus, the regression problem in Figure 4.7 is merely a modeling problem rather than a forecasting task. In
forecasting tasks, the explanatory variables corresponding to the forecasting horizon
usually include noise, which increases forecasting uncertainty and optimization difficulty. In general, gradient descent backpropagation algorithms underperform when
dealing with complex nonconvex objective spaces [21]. SGD and improved gradient descent algorithms, such as Adam, increase the probability of escaping a local
optimum; however, they still tend to converge prematurely when encountering multiple local optima. Thus, when dealing with more challenging PV power forecasting
tasks with fine-grained spatiotemporal resolutions or longer forecasting horizons, it
is common to enhance MLP performance either by hybridizing it with other types
of optimization algorithms, such as evolutionary and metaheuristic algorithms [22],
or by deepening its architecture through additional hidden layers and incorporating
techniques such as dropout and batch normalization.

4.1.2.2   Radial basis function neural networks

Radial basis function neural networks (RBFNNs) offer a distinct ANN architecture
that has been effectively applied in PV power forecasting. A schematic example of
an RBFNN with a single neuron in the output layer is presented in Figure 4.8. Unlike
MLPs, RBFNNs typically consist of a single hidden layer where each neuron employs
an RBF as its activation. Most commonly, a Gaussian RBF is used, which may also be
multivariate. In RBFNNs, the input of each neuron i of the hidden layer is determined
by the Euclidean distance between the input vector **x** and the weight vector connecting

Hidden layer   Output layer

$d_{\text{Eucl.}}(x, c_1)$ ⟶   RBF$_1$

$d_{\text{Eucl.}}(x, c_2)$ ⟶   RBF$_2$

⟶ y

$d_{\text{Eucl.}}(x, c_{N-1})$ ⟶   RBF$_{N-1}$

$d_{\text{Eucl.}}(x, c_N)$ ⟶   RBF$_N$

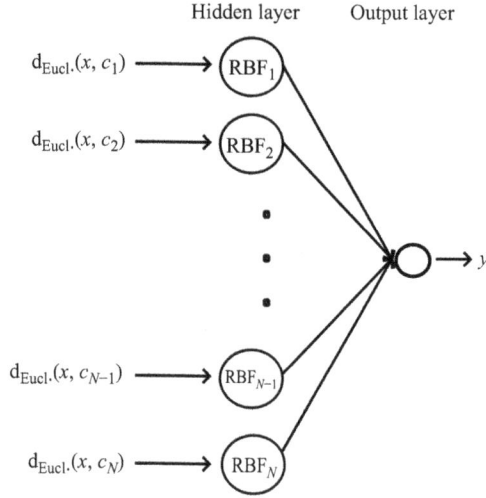

Figure 4.8   RBFNN with N neurons in the hidden layer and a single neuron in the output layer

the input to that neuron, which can be viewed as the center vector $\mathbf{c}_i$ of the RBF corresponding to neuron i of the hidden layer:

$$d_{\text{Eucl.}}(x, c_i) = \sqrt{\sum_{j=1}^{K} (x_j - c_{ij})^2} \tag{4.17}$$

RBFNNs typically involve two learning steps: Determining the centers of the RBF neurons in the hidden layer and training the weights connecting the neurons of the hidden layer to the neurons of the output layer. The centers can be selected randomly or by using unsupervised learning methods such as clustering on the input data. Once the centers are established, the output weights can be learned using back-propagation gradient descent or other approaches such as LR. As the output layer of RBFNNs is typically linear, the output y of an RBFNN is calculated as follows:

$$y = b_{\text{out}} + \sum_{i=1}^{N} \left(w_i e^{-\frac{d_{\text{Eucl.}}(x,c_i)}{2\sigma^2}} \right) \tag{4.18}$$

The distance-based approach of (4.17) makes RBF neurons more locally sensitive, which in turn makes RBFNNs particularly well-suited for capturing novel patterns in time series data such as PV power generation. As illustrated in Figure 4.5(d), RBF neurons achieve maximum activation when their centers match the input values. Due to this property, if each RBF neuron is centered around an input instance of the training set, instances that lie far from all RBF centers can be identified as novel. However, this feature – while beneficial for novelty detection – also limits the ability of RBFNNs to extrapolate beyond the training data, despite their theoretical capacity as universal approximators [23].

Compared to MLPs, RBFNNs are simpler and more computationally efficient. Their locally sensitive RBF neurons enable faster training, as the output weights can often be determined analytically, eliminating the need for iterative optimization algorithms typically required in MLPs. Hyperparameter tuning is generally limited to selecting the number of hidden neurons. Notably, the structure of RBFNNs allows for new neurons to be added during training, increasing their adaptive learning capabilities. However, the performance of RBFNNs heavily depends on the initial placement of the center points of the RBFs. There is no guarantee that the chosen centers will yield an optimal function mapping. Furthermore, determining the spread of the RBFs can also be challenging, as too wide RBFs have difficulty in accurately mapping the output to the input, while too narrow RBFs increase the risk of overfitting. An example of RBFNN employment for day-ahead PV power forecasting can be found in [24].

A popular variant of the RBFNN was proposed in [25], commonly referred to as extreme learning machine (ELM). ELMs share the same principles as RBFNNs; however, they are different in several aspects. First, the activation function of the hidden layer can be any nonlinear function. Second, the input is connected to the neurons of the hidden layer by fixed-weight synapses that are randomly initialized. Finally, the weights connecting the neurons of the hidden layer to the output are calculated using the Moore–Penrose pseudo-inverse of the output matrix formed by the neurons of the hidden layer. Implementation examples of PV power forecasting with ELMs can be found in [26,27].

4.1.2.3   NARX neural networks

A common choice for time-series forecasting tasks is the Non-linear Auto-Regressive eXogenous (NARX) model. The NARX model lies somewhere between FCNNs and RNNs. As illustrated in Figure 4.9, the core network structure of the NARX model is identical to that of an FCNN, except for the inclusion of an output feedback loop. Unlike most RNNs, which unfold their hidden states deeply into time, the feedback involved in the NARX model is limited. Since high levels of interpretability and computational efficiency are maintained, the NARX model is considered a shallow network.

The NARX model allows each sample set of explanatory input variables to be augmented with a set of generated outputs with different feedback delays. At each timestep t, the output y of the NARX model of Figure 4.9 is calculated as follows:

$$y(t) = f_{a,\text{out}}\left(b_{\text{out}} + \sum_{i=1}^{N}\left(w_i f_{a,\text{hid}}\left(b_i + \sum_{j=1}^{K} w_{ji}x_j(t) + \sum_{k=1}^{l} w_{(k+K)i}y(t-d_k)\right)\right)\right) \quad (4.19)$$

The optimal choice of the number of feedback outputs and the corresponding feedback delays depends on the target time-series characteristics and the forecasting horizon. Feedback delays greater than the forecasting horizon have no meaning, as the true value of the target time series will already be available at the time of issuing the forecast.

The computational process conducted at the neurons of each layer, as well as the overall training procedure, is identical to that of the MLP. However, in contrast to

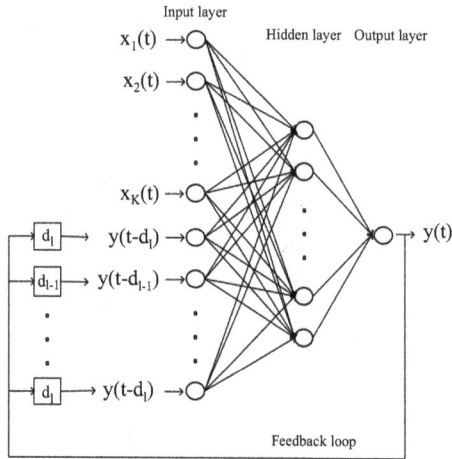

Figure 4.9   NARX model with one hidden layer and a single neuron in the output layer

classic MLPs, the NARX model uses the output of previous timesteps. This makes it a suitable choice for predicting time series with short-term temporal interdependencies. The PV power generation time series usually exhibits strong diurnal patterns and intra-hour temporal autocorrelations. Thus, it can be advantageous to include a feedback loop into the classic MLP architecture when forecasting PV power to introduce a temporal dimension in the forecasting framework without compromising computational efficiency and interpretability. Furthermore, the NARX model is particularly effective in forecasting PV energy yield, as its autoregressive structure fits the accumulative nature of energy [28]. However, the NARX model tends to overfit to the feedbacked outputs, particularly for time series with strong temporal autocorrelations. As a result, the performance of the NARX model is highly sensitive to the quality of the initial forecasts, since early large prediction errors can be propagated to the subsequent predictions. This challenge can potentially be mitigated through several strategies, such as integrating postprocessing techniques like Kalman filters to correct forecast trajectories, employing ensemble models to enhance robustness, fine-tuning feedback delays to reduce over-reliance on recent outputs, and incorporating attention mechanisms to allow the model to dynamically focus on the most relevant past states.

4.1.3   Support vector machines

Support vector machines (SVMs) were originally proposed in [29] to solve classification problems; however, they have also been used to solve regression problems, such as PV power forecasting. Solving regression problems with SVMs is commonly referred to as SVR. In SVR, the vectors formed by the explanatory variables are mapped into a higher-dimensional feature space to transform the nonlinear regression problem into one that can be addressed through linear function approximation.

The mapping of the explanatory variables' vectors in the higher-dimensional feature space is performed by kernel functions. Kernel functions measure the similarity between two instances x and x' of the explanatory variables' vectors directly in the higher-dimensional feature space, instead of explicitly transforming each vector and then performing the computations. Therefore, kernel mapping maintains high levels of computational efficiency. The most commonly used kernel functions in SVR and their parameters are described in Table 4.2 [22,30,31]. Kernel functions play a vital role in the performance of SVMs, as the kernel function type and the configuration of its parameters significantly determine the form of the higher-dimensional feature space, and consequently the quality of the generated solution. Since the Gaussian RBF kernel is defined by a single parameter γ, it is currently the most popular nonlinear kernel function for predicting PV power with SVR [22].

The output of an SVM is calculated as follows:

$$y = \sum_{i=1}^{n} \left(\theta_i k\left(x_i, x'\right) + \theta_0 \right) \tag{4.20}$$

where x' is the testing input vector and x_i is sample i of the training dataset input vectors. SVMs use the structural risk minimization principle to solve (4.20). In addition to the regression hyperplane, SVMs generate upper and lower bounds in the higher-dimensional feature space that form a tube around the hyperplane. The goal in classification tasks with SVMs is to maximize the margin between the classes, while simultaneously minimizing the classification error. The margin created by the upper and lower bounds is only affected by the input vectors that lie on the tube surface. These vectors are the support vectors. In SVR, the input vectors that lie inside the tube are not considered in the error minimization process. The support vectors in SVR are all the input vectors that lie outside or on the surface of the tube. Figure 4.10 illustrates the hyperplane and the tube generated by SVR for the same data samples shown in Figure 4.4. The blue line represents the hyperplane, the dashed lines correspond to the upper and lower bounds, and the red dots represent the support vectors.

In SVR, (4.20) is solved with LR by minimizing the regularized risk function:

$$R(C) = C\frac{1}{n} \sum_{i=1}^{n} L_\varepsilon(y_i, \widehat{y}_i) + \frac{1}{2} \|\theta\|^2 \tag{4.21}$$

Table 4.2 Common kernel functions used in SVR

Kernel function	Formula	Parameters
Linear	$k(x, x') = x^T x'$	–
Sigmoid	$k(x, x') = \tanh(\gamma x^T x' + r_t)$	γ: scaling parameter, $\gamma > 0$ r_t: controls the threshold mapping
Polynomial	$k(x, x') = (\gamma x^T x' + r)^\circ$	deg: degree of polynomial
Gaussian RBF	$k(x, x') = e^{-\gamma \|x - x'\|^2}$	γ: determines the spread of the RBF

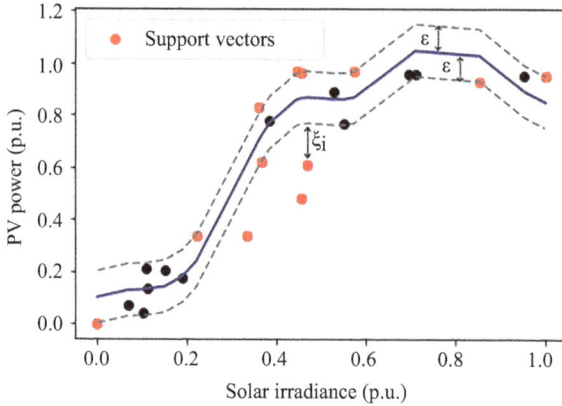

Figure 4.10   *SVR results for 24 random samples of a PV dataset. Measurements are taken from a 1.2 MW PV power plant located in Western Macedonia, Greece.*

where C determines the penalty for the prediction errors, ε is the distance between the hyperplane and the upper and lower bounds, and L_ε is the ε-insensitive loss function defined as follows:

$$L_\varepsilon(y_i, \widehat{y}_i) = \begin{cases} |y_i - \widehat{y}_i| - \varepsilon = \xi_\iota & \text{if } |y_i - \widehat{y}_i| \geq \varepsilon \\ 0 & \text{if } |y_i - \widehat{y}_i| < \varepsilon \end{cases} \tag{4.22}$$

where ξ_ι is the distance of the real output value y_i from the tube. The results presented in Figure 4.10 are generated for $C = 100$ and $\varepsilon = 0.1$.

SVMs have been extensively applied to PV power forecasting [1]. Examples of successful SVR implementation for PV power or solar irradiance forecasting can be found in [32–34]. In general, SVMs achieve a good tradeoff between accuracy and generalization ability while maintaining low levels of computational complexity. However, their overall performance significantly depends on the configuration of their parameters [22]. Parameters C and ε, as well as the parameters of the kernel function, are difficult to fine-tune. The sensitivity of SVMs toward their parameters is probably the reason behind contradicting conclusions regarding the effectiveness of SVR in forecasting PV power [1]. Therefore, parameter fine-tuning is a necessary step to secure the optimal regression performance by SVMs, which will be discussed in Chapter 7 [35]. Another drawback of SVMs is that they are based on similarity comparisons between all training instances. Even though the similarities are directly computed in the higher-dimensional feature space, without necessitating explicit instance transformations, the overall training duration is significantly affected by the size of the training dataset. The necessary comparisons for each instance increase with the increase in the total samples in the training dataset. This can quickly lead to a computational explosion, especially as PV-related fine-grained datasets are becoming increasingly available.

4.1.4   k-Nearest neighbors

The k-nearest neighbors (k-NN) algorithm was first developed in [36] and expanded in [37]. Although the k-NN algorithm has mainly been applied to classification tasks, it has also been used for regression and PV power forecasting. The k-NN algorithm first calculates the distance between the explanatory variable vector of the target instance and the explanatory variable vectors of all the other instances. Several distance functions can be used, such as the Euclidean distance, the Manhattan distance, and the Minkowski distance, given by (4.23), (4.24), and (4.25), respectively:

$$d_{\text{Eucl}}\left(\widehat{\boldsymbol{x}}, \boldsymbol{x}_j\right) = \sqrt{\sum_{i=1}^{K} \left(\widehat{x}_i - x_i\right)^2} \tag{4.23}$$

$$d_{\text{Manh}}\left(\widehat{\boldsymbol{x}}, \boldsymbol{x}_j\right) = \sum_{i=1}^{K} \left|\left(\widehat{x}_i - x_i\right)\right| \tag{4.24}$$

$$d_{\text{Mink}}\left(\widehat{\boldsymbol{x}}, \boldsymbol{x}_j\right) = \sqrt[p]{\sum_{i=1}^{K} \left|\left(\widehat{x}_i - x_i\right)\right|^p} \tag{4.25}$$

where $\widehat{\boldsymbol{x}}$ is the explanatory variable vector of the target instance, $\boldsymbol{x}_j$ is the explanatory variable vector of instance j, and p is a constant integer controlling the order of the Minkowski distance.

The output is generated considering only the k nearest explanatory variable vectors. This procedure is schematically illustrated in Figure 4.11. In regression tasks, the output is usually calculated as the weighted average of the dependent variable values corresponding to the k nearest explanatory variable vectors:

$$\widehat{\boldsymbol{y}} = \frac{1}{k} \sum_{i=1}^{k} w_i \boldsymbol{y}_i \tag{4.26}$$

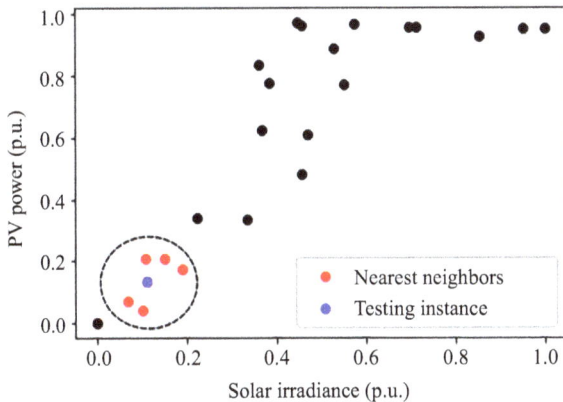

Figure 4.11   *Implementation example of the k-NN algorithm, with k = 5, on 24 random instances of a PV dataset. Measurements are taken from a 1.2 MW PV power plant located in Western Macedonia, Greece.*

where w_i and $\mathbf{y}_i$ are the weight and the output vector, respectively, corresponding to the ith nearest neighbor of the target instance. The weights are inversely proportional to the distances between the target instance and the k nearest instances.

In the context of PV power forecasting, the k-NN algorithm explores the instances of a historical dataset and extracts the instances with the closest weather-related or PV-related conditions (explanatory variable vectors) to the target prevailing conditions (explanatory variable vector of the target instance). Then, the target PV power output is calculated based on the extracted historical instances using (4.26). The k-NN algorithm is thus simpler than SVMs, as it performs comparisons in the original feature space and only requires the determination of the number of neighbors to consider. Furthermore, the performance of the k-NN algorithm does not rely on data preprocessing [19]. For its simplicity, the k-NN algorithm has been proven to perform particularly well for steady-state conditions; thus, it is commonly employed as a benchmarking model [38]. However, given that the k-NN algorithm relies only on local information to extract dependencies between variables, it tends to underperform for weather conditions of increased variability [38]. Furthermore, like SVMs, the k-NN algorithm is sensitive to the size of the training dataset. Therefore, in the context of PV power forecasting, the k-NN algorithm has mainly been used for classification tasks, such as weather classification.

4.1.5   Decision trees

Decision trees (DTs) are hierarchical top-down decision-based tree structures that are mainly used for classification purposes. DTs consist of nodes and branches. The first node of a DT is called the root node or the top node. Every node except for the root node is the successor of strictly one node, called the parent node. Nodes that have no successors are called leaf nodes or terminal nodes. Each nonleaf node contains a splitting criterion that generates exactly two successor nodes. Each split, starting from the root node, creates a new level of nodes. The total number of levels represents the depth of a DT.

The objective of DTs is to achieve a tradeoff between accuracy and computational efficiency. Building a DT involves two main subtasks: tree growing and pruning [39]. During tree growing, appropriate splitting criteria are chosen to divide the explanatory variable sample set into nodes. Starting from the root node, the initial sample set is divided into subsets based on the splitting criterion. If the criterion is met, the tree progresses toward the left successor node; otherwise, the tree progresses to the right successor node. A different splitting criterion is applied to the sample subset of each successor node to generate new successors. This recursive procedure continues until no additional splits are necessary. The number of leaf nodes in a DT is equal to the number of decision rules the DT can extract. Pruning aims at predetermining or readjusting the depth of the DTs. Shallow DTs, i.e., DTs with a limited number of levels, often fail to effectively capture the underlying relationships between data, while too deep DTs tend to overfit. Commonly employed algorithms for tree growing and pruning are ID3, C4.5, CART, and CHAID.

An essential part of tree growing is defining an optimal splitting criterion. An optimal splitting criterion consists of two parts: identifying the optimal splitting point for each explanatory variable and selecting the explanatory variable whose optimal splitting point leads to the highest classification purity. The metrics mainly used to measure the classification purity of each splitting criterion in DTs are the entropy and the Gini index.

With entropy, the aim is to define splitting criteria that maximize information gain. For each node, the entropy is calculated for each possible split of each explanatory variable as follows:

$$E = -(p_1 \log_2 p_1 + p_2 \log_2 p_2) \tag{4.27}$$

where p_1 denotes the proportion of samples that meet the splitting criterion and p_2 denotes the proportion of samples that do not meet the splitting criterion. The Gini index is calculated as follows:

$$G = 1 - (p_1{}^2 + p_2{}^2) \tag{4.28}$$

Information gain is then calculated as the difference between the classification purity of the parent node (entropy or Gini index) and the weighted average of the classification purities of the successor nodes.

DTs are called RTs when they are used to solve regression problems. In RTs, the leaf nodes correspond to continuous variables instead of class labels. The value of a leaf node is usually the average of all the dependent variables of the training set corresponding to that leaf node. The optimal splitting criterion is inferred by regression metrics, such as the MSE minimization or variance reduction.

An example of an RT grown from the data samples in Figure 4.4 is shown in Figure 4.12. The temperature of the PV modules is added as an explanatory variable in addition to solar irradiance. Each leaf node contains the optimal splitting criterion and the total number of samples belonging to it. The root node contains all 24 samples. The leaf nodes contain the number of samples belonging to them, as well as the dependent variable's average value. The optimal splitting criterion is derived by

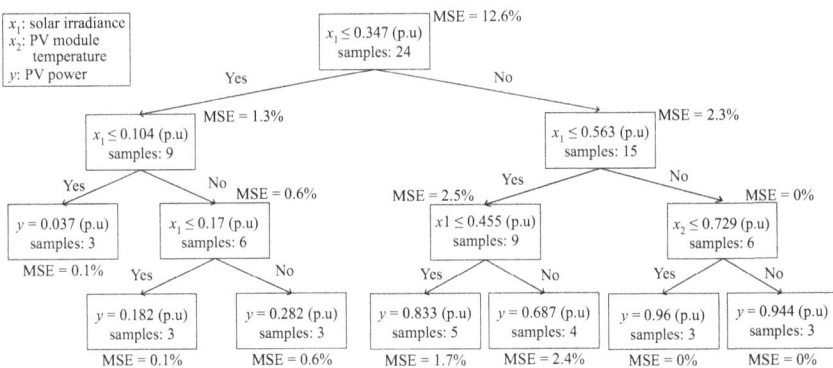

Figure 4.12   RT example grown for 24 random samples of a PV dataset. Measurements are taken from a 1.2 MW PV power plant located in Western Macedonia, Greece.

minimizing the MSE. The minimum MSE calculated in each step is also illustrated. It is evident that the MSE reduces on average as the RT grows. Furthermore, two constraints were applied during the tree growing process: the tree depth should not exceed four layers, and the minimum number of samples necessary for a node to be considered a leaf node was set to three.

DTs offer several advantages that make them an attractive option, mainly for classification tasks [17]. DTs are highly interpretable and offer great intuition regarding the extracted decision rules, the correlation between variables, and the importance of each explanatory variable. For example, in Figure 4.12, x_1 is chosen for the optimal splitting criterion in all nonleaf nodes except for one. This indicates that solar irradiance has a much higher correlation with PV power compared to the temperature of the PV modules, providing better splits during the RT growing process. Furthermore, DTs are highly efficient, with a computational complexity of order $O(n \cdot \log_2 n)$ [39].

While DTs can be very effective for certain applications, especially for classification tasks, they still have some drawbacks. Most notably, DTs are highly sensitive to the training dataset, which they tend to overfit. Small variations in the training dataset often lead to big variations in the performance of the DTs. Furthermore, RTs partition the continuous space of the dependent variable, often resulting in constant predictions under varying conditions. Therefore, RTs are rarely used to forecast PV power.

To overcome the limitations of DTs and RTs, various ensemble approaches have been proposed. Most notably, the random forest (RF) method, first proposed in [40] and expanded to its current form in [41], has been extensively used for PV power forecasting. The RF method combines multiple different DTs or RTs, generated for different subsets of the same initial training dataset, to reduce the overall variation of the results and increase the generalization capability. While RF is basically a DT ensemble method, its specific structure and its extensive use for PV power forecasting [42] deem it a stand-alone ML model; thus, the RF method is discussed in this chapter.

The basic structure of the RF method is presented in Figure 4.13, where B subsets are sampled from the initial training set using the bootstrap method. The bootstrap method consists of creating subsets of the same size by randomly sampling from an initial set. Bootstrap sampling includes replacement, i.e., each sample can be chosen more than once for the same subset. A DT is then created for each subset. Each DT differs not only because of the different training subset but also because different combinations of explanatory variables are considered during tree growing. The outcomes of all DTs are then combined to form the ensemble result. In the case of RTs, the ensemble result is calculated as the average of all the outcomes generated by the individual RTs.

The RF method maintains all advantages of DTs while simultaneously improving the generalization capability. The diverse DTs generated for different subsets of the training dataset introduce a stochastic component during the function inference process. Having multiple DTs in the RF ensures stability in the generated outcome, owing to the averaging effect across all individual results. Furthermore, the computational efficiency of the RF method remains at high levels, as each DT is independent

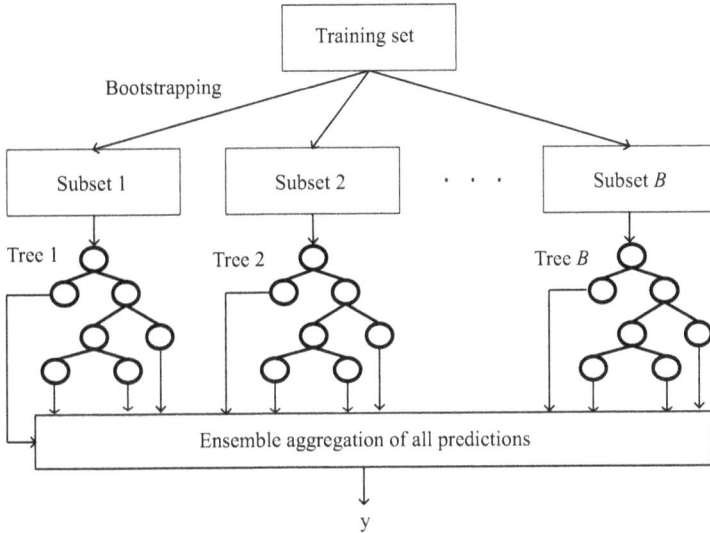

Figure 4.13 Schematic illustration of the RF method

of the others, enabling parallel computations. The high inference accuracy, performance stability, and computational efficiency of the RF method have constituted it an attractive option for predicting PV power generation. Numerous studies have suggested the use of the RF method, either as a stand-alone model or as part of a hybrid structure. For example, in [43], an RF was used to predict the 15-min-ahead output of a polycrystalline PV module.

4.1.6 Summary, comparison, and limitations of supervised learning methods

Various supervised ML methods, such as SVR, k-NN, LR, ANN, and RTs, have been proposed for PV power forecasting over the past few decades [1]. However, systematically ranking these supervised ML methods in terms of forecasting performance remains a challenging, if not impossible, task. PV power forecasting can be conducted with varying forecasting horizons, temporal resolutions, and spatial scales. Each supervised ML method offers unique characteristics, advantages, and limitations that make it more suitable for specific PV power forecasting applications. For example, LR models and the k-NN algorithm are often favored for low-resolution, day-ahead forecasts in power system operations, due to their simplicity, interpretability, and ease of implementation. In contrast, in ultra-short-term forecasting applications, such as microgrid control and battery scheduling, shallow MLPs constitute a practical option balancing between forecasting accuracy and inference time. Furthermore, the performance of supervised ML methods heavily depends on hyperparameter finetuning and varies with the type, quantity, and quality of the dataset used.

Several attempts have been made recently to systematically compare supervised ML methods for PV power forecasting [30,44–47]. However, the results and conclusions are contradictory due to differences in datasets, hyperparameter tuning strategies, and evaluation metrics. In general, nonlinear supervised ML methods tend to outperform linear methods, such as LR [45]. Despite this, LR remains valuable for certain forecasting applications due to its low computational complexity [16]. Some comparative studies suggest the use of MLPs for PV power forecasting [45,46]. However, MLPs typically require long training times and often suffer from local optima entrapment during optimization. Nevertheless, MLPs and ANNs, in general, remain the most popular supervised ML methods for PV power forecasting, owing to their adaptability and strong inference capabilities. RBFNNs and ELMs are generally preferred when lower training times are required, whereas the NARX model is particularly effective in short-term PV energy yield forecasting. RFs have also proven superior compared to other supervised ML models, such as SVMs and MLPs [30,44] and stand as an attractive choice for both classification and regression PV-related tasks, combining accuracy and runtime efficiency. In contrast, SVR performance varies with the PV power forecasting task and the hyperparameter tuning. Therefore, the use of SVMs in PV power forecasting may increasingly be limited to classification tasks in the near future. In [47], different ML methods are found to be better suited for PV power forecasting, depending on the evaluation metric, application, and the forecasting circumstances. It is therefore safe to assume that there is no supervised ML method that is superior in all the evaluation criteria.

Supervised ML methods suffer from some general disadvantages that can make them less suitable for certain PV power forecasting tasks. For example, supervised ML methods rely entirely on historical data to learn inherent relationships between the dependent and explanatory variables. Although historical datasets are increasingly abundant, they often do not capture the full range of possible weather conditions, particularly extreme or previously unseen conditions. As the impacts of the climate crisis intensify, there is a growing need for approaches that can generalize better to rare or previously unobserved weather patterns, rather than being strictly constrained by the historical data. Additionally, supervised ML methods struggle to handle data with extremely high temporal and spatial resolutions, i.e., when data is sampled at very short intervals, where the averaging effect is absent and they tend to smooth out the fine-grained, highly fluctuating PV power curves. However, accurately forecasting PV power using such high-resolution data (sampled in second-scale resolutions) is becoming increasingly essential in modern power systems, particularly for tasks that require minutes-ahead PV power forecasting. Consequently, in such scenarios, supervised ML methods should be combined with other methods, such as detailed physics-based PV models, hill-climbing optimization algorithms, and ground-based sky image processing techniques, to form hybrid models that can effectively capture intense PV power variations.

4.2   Unsupervised learning

Unlike supervised ML methods, unsupervised ML methods do not require labeled datasets. Hidden patterns and relationships are discovered between variables using only the variables themselves, without the need for labels or other dependent variables. It is the recent radical evolution of computing systems and ML technology that has enabled the use of such techniques on large datasets for regression or classification problems. Therefore, unsupervised ML generally constitutes a less explored field compared to supervised ML, in the context of PV power forecasting. In fact, unsupervised ML methods are merely used for data analysis and preprocessing tasks rather than direct forecasting of PV power. The primary objectives of unsupervised ML for PV power forecasting applications mainly include data clustering and dimensionality reduction tasks.

4.2.1 Clustering

PV power generation patterns can significantly differ depending on weather conditions and the forecasting horizon. Therefore, it is common to classify the sample sets of explanatory variables based on their similarity or correlation and separately train a PV power forecasting model for each class. However, classification requires ground truth labels, based on which the explanatory variables will be grouped. Reliable labels are usually extracted by combining several explanatory variables, rather than relying on a single explanatory variable. As reliable labels are rarely available in PV-related datasets, classification depends on manual labeling, which is sensitive to human error and often time-consuming to perform. Thus, in recent years, unsupervised clustering techniques have been increasingly used to automatically classify explanatory variable sample sets without the need for ground truth labels.

4.2.1.1 K-Means clustering

The most common clustering technique applied for PV power forecasting is k-means clustering [48]. K-means clustering has been applied to various PV-related classification tasks, such as solar irradiance time-series clustering [49], sky image clustering [50], and image feature clustering [51]. The k-means clustering method is an effective, efficient, and flexible clustering algorithm that aims to maximize the distance between clusters while minimizing the intra-cluster variation. It is an iterative algorithm that assigns each sample set of explanatory variables to one of k clusters, to minimize the objective function of (4.29):

$$J = \sum_{i=1}^{n} \sum_{k=1}^{K} u_{ik} \left\| \boldsymbol{x}_i - \boldsymbol{\mu}_k \right\|^2 \tag{4.29}$$

In (4.29), u_{ik} equals 1 if the vector $\boldsymbol{x}_i$ of sample set i is closest to the centroid $\boldsymbol{\mu}_k$ of cluster k compared to the centroids of the other clusters, otherwise u_{ik} equals 0. The k-means clustering method consists of two steps. First, it calculates the distances between the vectors of the sample sets and the centroids of the clusters and assigns each sample set to the cluster with the closest centroid. The distance calculation is usually based on the Euclidean norm. Then, it updates the centroids of the clusters by calculating the mean vector of all sample sets of each cluster. The cluster centroids are usually initialized randomly.

The k-means algorithm often leads to variable clustering results, depending on the initialization of the cluster centroids. Furthermore, the k-means clustering method is sensitive to outliers in the sample sets, as it relies on the calculation of the mean vector of the sample sets belonging to each cluster. Finally, determining the optimal

number of clusters is a challenging task with no obvious optimal solution. Therefore, the k-means clustering method often requires multiple runs with different *k*
configurations to ensure optimal clustering results.

Alternatives to the k-means clustering method, commonly employed in PV
power forecasting tasks to mitigate the sensitivity to outliers, include the k-medians
clustering method and the k-medoids clustering method [52]. Both clustering methods differ from k-means clustering in terms of calculating the cluster centroids. The
k-medians clustering method calculates the median vector of the sample sets included
in each cluster instead of the mean vector. In contrast, k-medoids clustering relies on
the estimation of each cluster's medoid, which is the sample set that has the minimum
average distance with the rest of the sample sets inside the cluster. While k-medians
clustering and k-medoids clustering reduce the impact of the outliers on the calculation of the cluster centroids, they are relatively less computationally efficient
compared to k-means clustering. Other alternatives to k-means clustering include
density-based clustering and the k-means++ algorithm [53,54].

Figure 4.14 illustrates the clustering results generated by different clustering
methods for the data samples shown in Figure 4.4. Besides solar irradiance, the

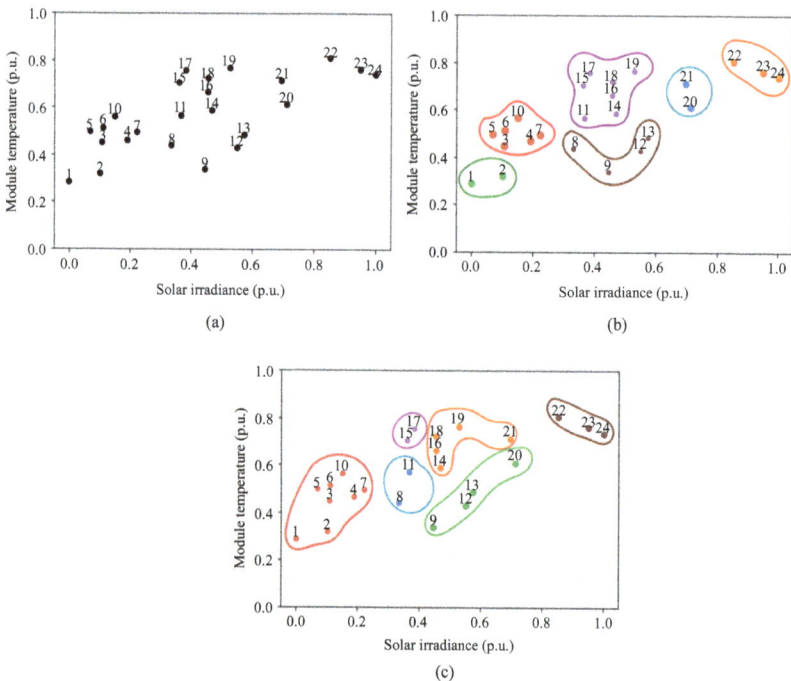

*Figure 4.14   Results of different clustering methods. (a) Scatter plot of 24 random
samples of a PV dataset. Measurements are taken from a 1.2 MW PV
power plant located in Western Macedonia, Greece. (b) k-Means
clustering results for six clusters. (c) k-Medoids clustering results for
six clusters.*

temperature of the PV modules is used as an additional variable to increase the reliability of clustering. The scatter plot of the 24 random sample sets is illustrated in Figure 4.14(a). The clustering results of the k-means and k-medoids methods are presented in Figure 4.14(b) and(c), respectively. The sample sets inside each cluster share the same color. Despite the small number of sample sets and their dense representation, the clustering results are not identical. For example, while sample sets 1 and 2 form an independent cluster with k-means clustering, the k-medoids clustering method includes them with sample sets {3,4,5,6,7,10}. This observation aligns with the reduced sensitivity to outliers of k-medoids clustering compared to k-means clustering.

4.2.1.2   Hierarchical clustering

The most popular hierarchical clustering algorithm employed for PV power forecasting is agglomerative clustering, also known as agglomerative nesting, which recursively merges pairs of clusters in the dataset. Agglomerative clustering is a bottom-up approach, in which each sample set is initially treated as a self-contained cluster. The most similar clusters are then successively merged until a single cluster containing all sample sets is formed. Similar to k-means clustering, the distance calculation is usually based on the Euclidean norm. This process is schematically represented by a dendrogram, as presented in Figure 4.15. There are several different linkage criteria to create a dendrogram. The most common linkage criteria are complete linkage, single linkage, average linkage, and Ward's method. The complete and single linkage criteria are based on the maximum and minimum distances between two clusters, respectively. The average linkage defines the distance between two clusters as the average distance between the sample sets of each cluster, while Ward's method aims to minimize the intra-cluster variance. The dendrogram of Figure 4.15 is constructed based on Ward's method.

Compared to k-means clustering, hierarchical clustering is more interpretable, as it provides a visual representation of the clustering process and the hierarchical

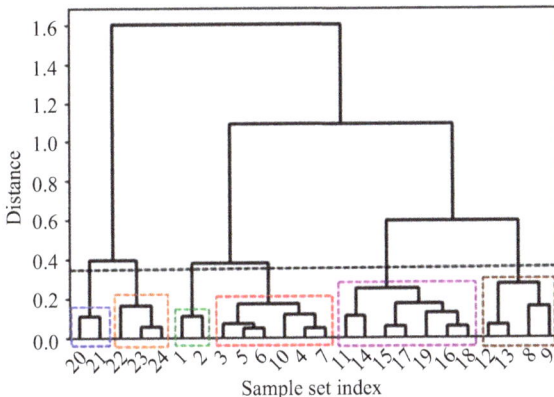

Figure 4.15   Hierarchical agglomerative clustering method. Dendrogram of hierarchical agglomerative clustering for the data shown in Figure 4.14(a). The dendrogram is cut at a specific height to create six clusters.

structure between the sample sets. Furthermore, instead of predefining the number of clusters, hierarchical clustering describes all possible clusters formed with the bottom-up approach. To obtain clusters, one only needs to cut the dendrogram at a specific hierarchical level (distance) and treat each branch below that level as an independent cluster. In Figure 4.15, the dendrogram is cut at a distance just below 0.4 (dashed line) to obtain six clusters. Note that for the sample sets, as shown in Figure 4.14(a), the clusters generated at this hierarchical level are identical to the clusters generated by k-means clustering.

Hierarchical agglomerative clustering has been proven particularly effective for PV power forecasting tasks. In some cases, it has been reported to outperform k-means clustering, capturing more complex structures [55,56]. However, hierarchical clustering is computationally intensive, with a typical complexity of $O(n^3)$. The scalability of hierarchical clustering is thus limited, deeming it ineffective for handling large volumes of sample sets.

4.2.2   Dimensionality reduction

The growing availability of large PV-related datasets, containing numerical information across multiple temporal and spatial resolutions, along with non-numerical data such as ground-based sky images and satellite images, often leads to high-dimensional input spaces with diverse features and potential explanatory variables. This increased dimensionality can introduce several challenges in ML, commonly referred to as the curse of dimensionality. These challenges include reduced generalization ability due to noise from irrelevant features, increased data sparsity, and higher computational complexity. Simple feature selection methods based on linear correlation metrics often fail to capture the complex nonlinear relationships between the explanatory and dependent variables, as well as the various interactions and multivariate interdependencies among the explanatory variables themselves.

Dimensionality reduction methods based on unsupervised ML effectively address the challenges posed by the curse of dimensionality. As unsupervised methods, they do not require predefined ground-truth labels or prior knowledge of the data structure. By simultaneously capturing nonlinear correlations, feature interactions, and multivariate interdependencies, they transform high-dimensional data into a lower-dimensional representation that is easier to process, cluster, visualize, and interpret. Thus, the reduced set of explanatory variables retains dense information that enhances the accuracy of PV power forecasting.

4.2.2.1   Principal component analysis

Principal component analysis (PCA) can be considered an unsupervised ML method that performs linear dimensionality reduction. PCA aims to transform higher-dimensional interdependent data into a lower-dimensional space while maintaining as much of the data variation as possible [57]. The features of the lower-dimensional space, which are called principal components (PCs), are ordered so that the first PCs contain most of the initial dataset variation.

Given a set of K variables $x_1, x_2, \ldots, x_K$, PCA transforms them into a new set $z_1, z_2, \ldots, z_{m_{pc}}$ comprising m_{pc} PCs, with $m_{pc} \leq K$, where z_i is defined as follows:

$$z_i = \mathbf{a}'_i\mathbf{x} = a_{i1}x_1 + a_{i2}x_2 + \cdots + a_{iK}x_K \tag{4.30}$$

In (4.30), $\mathbf{a}'_i$ is a transposed vector of constants that maximizes the variance at the ith PCA level. To calculate $\mathbf{a}'_i$ at each PCA level, the following objective function needs to be optimized [57]:

$$\text{Maximize} \quad \mathbf{a}'_i \Sigma \alpha_i$$
$$\text{s.t.} \quad \mathbf{a}'_i \alpha_i = 1 \tag{4.31}$$

where Σ is the covariance matrix of x. Using the technique of Lagrange multipliers λ, (4.31) can be represented as follows:

$$\text{Maximize} \quad a'_i \Sigma \alpha_i - \lambda(a'_i \alpha_i - 1) \tag{4.32}$$

Differentiating (4.32) with respect to α_i and equaling to 0 lead to the following expression:

$$(\Sigma - \lambda I)\,\alpha_i = 0 \tag{4.33}$$

Thus, λ is an eigenvalue of Σ, with α_i being the corresponding eigenvector. Each PC is calculated such that the variation at that level is maximized:

$$\max \, var\,[a'_i x] = \max a'_i \Sigma \alpha_i = \max a'_i \lambda \alpha_i = \max \lambda \tag{4.34}$$

Therefore, the ith PC of x requires only the calculation of the eigenvector of Σ that corresponds to the ith maximum eigenvalue.

To demonstrate the effect PCA can have on PV-related datasets and tasks such as clustering, the data samples shown in Figure 4.14(a) are enriched with three additional variables: inclined solar irradiance, relative humidity, and ambient temperature. Figure 4.16(a) illustrates the clustering results by directly applying the k-means method to the 24 random samples of all five variables. The sample sets are plotted with respect to their solar irradiance and module temperature values to facilitate comparisons with the plots in Figure 4.14(b) and (c). As expected, including the additional variables available in the PV dataset slightly alters the clustering results. Clustering

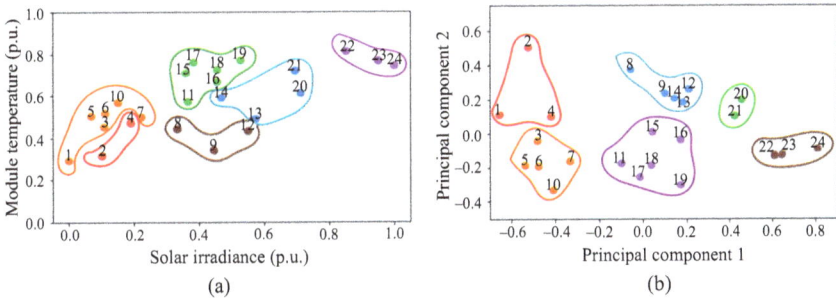

Figure 4.16   *Clustering results on 24 random samples of a PV dataset. Measurements are taken from a 1.2 MW PV power plant located in Western Macedonia, Greece. (a) Direct k-means clustering on the full dataset (five explanatory variables). (b) k-Means clustering results after applying PCA on the full dataset. Only the first two PCs are considered.*

is then reapplied to the sample sets after transforming them to a lower-dimensional space with PCA. Specifically, the sample sets are transformed into a two-dimensional space. In Figure 4.16(b), the sample sets are plotted with respect to the values of their first two PCs. The two-dimensional representation of the sample sets based on their first two PCs significantly differs compared to the representation generated by plotting with respect to any pair of explanatory variables. Furthermore, the clustering results after applying PCA differ compared to the results generated by direct clustering on any dimension of the original sample sets.

Even though the number of explanatory variables used in the example of Figure 4.16 is small, the effect PCA can have on PV-related data is visible. PCA can transform data into lower-dimensional spaces, where the representation of the sample sets can be more accurate, as structural interdependencies are considered while simultaneously reducing the effect of irrelevant or constant explanatory variables. Furthermore, PCA can significantly reduce the overall computational complexity without significantly compromising the information of the initial sample sets. For example, for the sample sets of Figure 4.16, the first two PCs maintain 91.3% of the initial explanatory variables' total variation. PCA has been widely employed in the context of PV power forecasting, either as a stand-alone data dimensionality reduction technique or paired with other methods for tasks such as classification. For example, PCA was used in [58] to identify crucial frequency features for predicting day-ahead solar irradiance in a spatiotemporal framework. In [59], PCA was combined with k-means clustering to effectively cluster similar input instances.

4.2.2.2 Shallow auto-encoders

While PCA has been proven particularly effective for several different applications, including dimensionality reduction of PV-related datasets, it still has some limitations as a linear transformation technique. In recent years, focus has been driven toward nonlinear dimensionality reduction techniques, the most prominent of which are based on auto-encoders (AEs). AEs are based on ANNs, and owing to their structure, they do not require labels and are therefore considered part of the unsupervised ML family. The most popular type of ANN used to construct AEs in the context of PV power forecasting is the convolutional neural network (CNN), due to its effectiveness in compressing PV-related images, such as sky images or satellite images. FCNNs, such as the MLP, can also be used as the basis of AEs. AEs based on FCNNs are mainly used to compress one-dimensional arrays of explanatory variables, such as PV-related measurements and image features. This chapter focuses only on shallow AEs based on FCNNs.

A fully connected AE with a single hidden layer is presented in Figure 4.17. An AE consists of two parts, the encoding part and the decoding part. The middle layer, called the bottleneck layer, contains the latent space representation of the explanatory variables. The output layer has the same number of neurons as the input layer and corresponds to the reconstructed input variables. Thus, the ground truth labels used at the output are the input variables themselves. AEs aim to encode and decode the variables with the minimum possible reconstruction error.

Fully connected AEs have been employed in PV power forecasting frameworks mainly for data preprocessing tasks, such as feature extraction and denoising. For

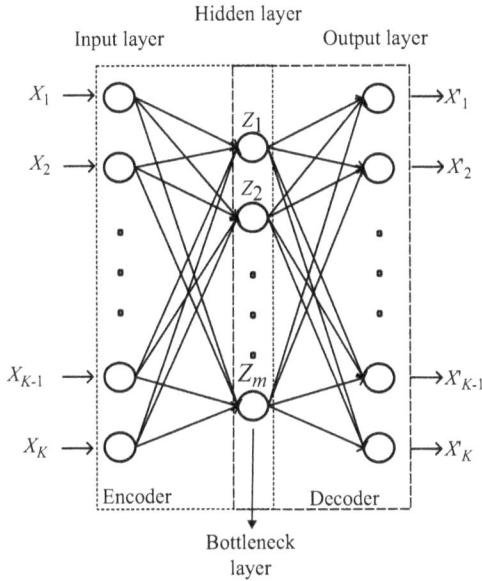

Figure 4.17   Fully connected AE with one hidden layer

example, a fully connected AE was used in [60] as part of the preprocessing stage of a short-term solar irradiance forecasting framework, for outlier detection and denoising. However, similar to MLPs, AEs often get stuck in the local optima during training due to gradient descent optimization. To minimize the risk of local optima entrapment, deeper fully connected AE architectures are usually preferred in the context of PV power forecasting, due to their capability in capturing complex nonlinear relationships and patterns between variables. For example, a deep fully connected AE was employed in [61] to forecast the energy production of large PV power plants. In contrast, it has been proven that shallow, fully connected AEs can achieve optimal encoding performance with minimum reconstruction errors when combined with linear dimensionality reduction techniques, such as PCA [62].

4.2.3   Self-organizing maps

A self-organizing map (SOM) is an unsupervised ANN-based structure that is used to visualize high-dimensional data by projecting the nonlinear interdependencies of the data to a simpler, lower-dimensional space [63]. The basic structure of an SOM is depicted in Figure 4.18. Each neuron of the input layer corresponds to an input variable. The neurons of the input layer are connected to a lattice-type matrix of neurons, which can be displayed in any two-dimensional shape. The neurons of the input layer and the lattice-type matrix are usually fully connected. The neurons of the lattice-type matrix do not perform computations; rather, they represent vectors whose elements are equal to the weight values of the synapses connected to them.

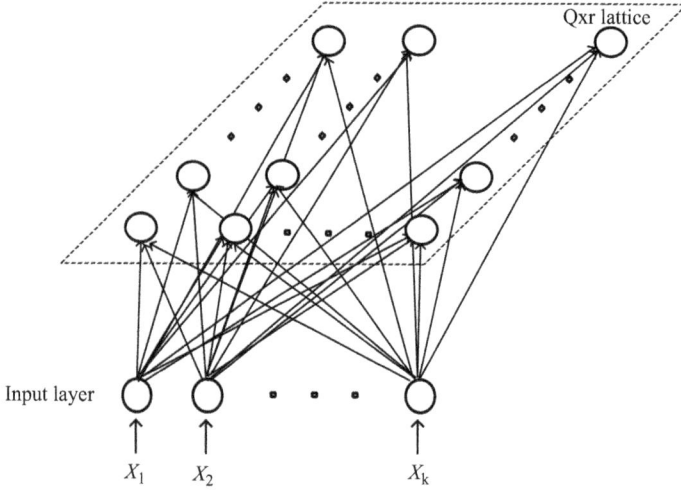

Figure 4.18   Schematic representation of an SOM

During training, the weights of an SOM are iteratively adjusted. Each sample set is assigned to the best-matching neuron of the lattice-type matrix, i.e., the neuron that lies closest to the sample set. The most common distance metric employed for training SOMs is the Euclidean distance. The weight values of the synapses connecting the neurons of the input layer to the neurons inside a neighborhood N_k around the best matching neuron, including the best matching neuron itself, are then updated based on (4.35):

$$w_{jk}(i+1) = w_{jk}(i) + h_{jk}(i)(x_j(i) - w_{jk}(i)) \tag{4.35}$$

where w_{jk} is the weight of the synapse connecting neuron j of the input layer to neuron k of the lattice type matrix and h_{jk} is the neighborhood function. The most common choice for the neighborhood function is described by (4.36) [63]:

$$h_{jk}(i) = \begin{cases} a(i), & k \in N_k \\ 0, & k \notin N_k \end{cases} \tag{4.36}$$

where α is the learning rate, taking values between 0 and 1. Another common choice for the neighborhood function is the Gaussian function [63]:

$$h_{jk}(i) = \alpha(i)e^{\left(-\frac{\|x_j(i) - w_{jk}(i)\|^2}{2\sigma^2(i)}\right)} \tag{4.37}$$

The choice of the optimal learning rate and neighborhood size depends on the problem requirements and the data type. In most cases, α and N_k start with relatively large initial values, which gradually decrease during the optimization process to allow convergence.

SOMs have been particularly useful for PV power forecasting applications, owing to their ability to cluster explanatory variables by projecting them into a lower-dimensional interpretable space, while preserving the topology of the data after transformation. SOMs were successfully employed in [64] as part of the classification stage of an hourly day-ahead PV power forecasting framework. The SOMs were used to extract features from daily PV power time series, which were then classified using learning vector quantization. SOMs suffer from a few drawbacks as well. For example, the performance of an SOM highly depends on the configuration of α and N_k, as well as the initial weight values. Besides parameter sensitivity, SOMs can be computationally intensive, as their computational requirements increase with the lattice matrix dimensions and the dataset size.

4.3   Conclusion

This chapter focuses on the most common shallow ML approaches used for PV power forecasting. Deep learning approaches will be discussed in the next chapters. ML approaches are divided into supervised and unsupervised learning algorithms. Over the past two decades, supervised learning algorithms have been extensively employed to forecast PV power generation. From simple regression models and DTs to diverse ANN-based architectures and RFs, supervised ML offers a wide variety of options to model PV power dependencies. The choice of a suitable supervised ML approach depends on several factors, such as the forecasting horizon, the spatiotemporal resolution, and the weather conditions. However, unsupervised learning algorithms have primarily been used for data analysis and preprocessing tasks, such as data representation, dimensionality reduction, and weather-related clustering.

The increasing availability of fine-grained PV-related data makes ML an inevitable direction for the future of PV power forecasting. Supervised ML algorithms such as SVMs, k-NN, and DTs have been proven particularly useful for classification tasks. Provided high-level historical data, supervised ML algorithms are also effective in regression tasks such as PV power forecasting. Unsupervised ML algorithms unlock a wide range of new possibilities, given their ability to discover complex nonlinear hidden patterns between variables without the need for labels. This is particularly important as it becomes increasingly challenging to label large datasets consisting of multiple variables with different spatial and temporal sampling resolutions.

Despite the numerous advantages offered by ML approaches, they still face several challenges, including data dependency, generalization and fine-tuning difficulties, as well as difficulties in modeling PV power in fine-grained spatiotemporal resolutions and over extended forecasting horizons. Acknowledging these limitations is essential for designing PV power forecasting models that balance forecasting performance and computational efficiency. Some of these challenges can be addressed with deep learning, which will be discussed in the following chapters. Additionally, combining ML with physics-based models, metaheuristic algorithms, advanced data preprocessing techniques, and human expertise enables the development of robust hybrid PV power forecasting models that help mitigate current limitations of ML.

References

[1]   A. Alcañiz, D. Grzebyk, H. Ziar, and O. Isabella, "Trends and gaps in photovoltaic power forecasting with machine learning," *Energy Rep*, vol. 9, pp. 447–471, 2023.

[2]   P. Gupta and R. Singh, "PV power forecasting based on data-driven models: A review," *Int J Sustain Eng*, vol. 14, no. 6, pp. 1733–1755, 2021.

[3]   L. Fahrmeir, T. Kneib, S. Lang, and B. D. Marx, *Regression*. Berlin: Springer, 2021.

[4]   S. H. Oudjana, A. Hellal, and I. H. Mahammed, "Power forecasting of photovoltaic generation," *Int J Electr Comput Eng*, vol. 7, no. 6, pp. 627–631, 2013.

[5]   T. Zhang, "Solving large-scale linear prediction problems using stochastic gradient descent algorithms," *Proceedings, Twenty-First International Conference on Machine Learning, ICML 2004*, pp. 919–926, 2004.

[6]   W. Lizhen, Z. Yifan, W. Gang, and H. Xiaohong, "A novel short-term load forecasting method based on mini-batch stochastic gradient descent regression model," *Electr Power Syst Res*, vol. 211, p. 108226, 2022.

[7]   V. K. Saini and A. Maity, "Random sample consensus in decentralized Kalman filter," *Eur J Control*, vol. 65, p. 100617, 2022.

[8]   M. A. Fischler and R. C. Bolles, "Random sample consensus: A paradigm for model fitting with applications to image analysis and automated cartography," *Commun ACM*, vol. 24, no. 6, pp. 381–395, 1981.

[9]   A. E. Hoerl and R. W. Kennard, "Ridge regression: Biased estimation for nonorthogonal problems," *Technometrics*, vol. 42, no. 1, p. 80, 2000.

[10]   R. Tibshirani, "Regression shrinkage and selection via the lasso," *J R Stat Soc Ser B: Stat Methodol*, vol. 58, no. 1, pp. 267–288, 1996.

[11]   G. Wang, Y. Su, and L. Shu, "One-day-ahead daily power forecasting of photovoltaic systems based on partial functional linear regression models," *Renew Energy*, vol. 96, pp. 469–478, 2016.

[12]   H. Mubarak, A. Hammoudeh, S. Ahmad, *et al.*, "A hybrid machine learning method with explicit time encoding for improved Malaysian photovoltaic power prediction," *J Clean Prod*, vol. 382, p. 134979, 2023.

[13]   J. H. Friedman, "Multivariate adaptive regression splines," *Ann Stat*, vol. 19, no. 1, 1991.

[14]   Y. Li, Y. He, Y. Su, and L. Shu, "Forecasting the daily power output of a grid-connected photovoltaic system based on multivariate adaptive regression splines," *Appl Energy*, vol. 180, pp. 392–401, 2016.

[15]   L. Massidda and M. Marrocu, "Use of multilinear adaptive regression splines and numerical weather prediction to forecast the power output of a PV plant in Borkum, Germany," *Sol Energy*, vol. 146, pp. 141–149, 2017.

[16]   E. Shirazi, I. Gordon, A. Reinders, and F. Catthoor, "Sky images for short-term solar irradiance forecast: A comparative study of linear machine learning models," *IEEE J Photovolt*, vol. 14, no. 4, pp. 691–698, 2024.

[17]   P. S. Georgilakis, *Spotlight on Modern Transformer Design*. London: Springer, 2009.

[18]   R. Lippmann, "An introduction to computing with neural nets," *IEEE ASSP Mag*, vol. 4, no. 2, pp. 4–22, 1987.

[19]   C. Paoli, C. Voyant, M. Muselli, and M.-L. Nivet, "Forecasting of preprocessed daily solar radiation time series using neural networks," *Sol Energy*, vol. 84, no. 12, pp. 2146–2160, 2010.

[20] A. Mellit and A. M. Pavan, "A 24-h forecast of solar irradiance using artificial neural network: Application for performance prediction of a grid-connected PV plant at Trieste, Italy," *Sol Energy*, vol. 84, no. 5, pp. 807–821, 2010.

[21] M. Ding, L. Wang, and R. Bi, "An ANN-based approach for forecasting the power output of photovoltaic system," *Procedia Environ Sci*, vol. 11, pp. 1308–1315, 2011.

[22] U. K. Das, K. S. Tey, M. Y. I. Idris, *et al.*, "Forecasting of photovoltaic power generation and model optimization: A review," *Renew Sustain Energy Rev*, vol. 81, pp. 912–928, 2018.

[23] J. Park and I. W. Sandberg, "Universal approximation using radial-basis-function networks," *Neural Comput*, vol. 3, no. 2, pp. 246–257, 1991.

[24] C. Chen, S. Duan, T. Cai, and B. Liu, "Online 24-h solar power forecasting based on weather type classification using artificial neural network," *Sol Energy*, vol. 85, no. 11, pp. 2856–2870, 2011.

[25] G.-B. Huang, Q.-Y. Zhu, and C.-K. Siew, "Extreme learning machine: Theory and applications," *Neurocomputing*, vol. 70, no. 1–3, pp. 489–501, 2006.

[26] M. K. Behera, I. Majumder, and N. Nayak, "Solar photovoltaic power forecasting using optimized modified extreme learning machine technique," *Eng Sci Technol: Int J*, vol. 21, no. 3, pp. 428–438, 2018.

[27] P. Tang, D. Chen, and Y. Hou, "Entropy method combined with extreme learning machine method for the short-term photovoltaic power generation forecasting," *Chaos Solitons Fractals*, vol. 89, pp. 243–248, 2016.

[28] D. Anagnostos, T. Schmidt, S. Cavadias, D. Soudris, J. Poortmans, and F. Catthoor, "A method for detailed, short-term energy yield forecasting of photovoltaic installations," *Renew Energy*, vol. 130, pp. 122–129, 2019.

[29] C. Cortes and V. Vapnik, "Support-vector networks," *Mach Learn*, vol. 20, no. 3, pp. 273–297, 1995.

[30] J. Gaboitaolelwe, A. M. Zungeru, A. Yahya, C. K. Lebekwe, D. N. Vinod, and A. O. Salau, "Machine learning based solar photovoltaic power forecasting: A review and comparison," *IEEE Access*, vol. 11, pp. 40820–40845, 2023.

[31] I. K. Bazionis, M. A. Kousounadis-Knousen, P. S. Georgilakis, E. Shirazi, D. Soudris, and F. Catthoor, "A taxonomy of short-term solar power forecasting: Classifications focused on climatic conditions and input data," *IET Renew Power Gener*, vol. 17, pp. 2411–2432, 2023.

[32] M. Pan, C. Li, R. Gao, *et al.*, "Photovoltaic power forecasting based on a support vector machine with improved ant colony optimization," *J Clean Prod*, vol. 277, p. 123948, 2020.

[33] K. Y. Bae, H. S. Jang, and D. K. Sung, "Hourly solar irradiance prediction based on support vector machine and its error analysis," *IEEE Trans Power Syst*, vol. 32, pp. 935–945, 2017.

[34] J. Zeng and W. Qiao, "Short-term solar power prediction using a support vector machine," *Renew Energy*, vol. 52, pp. 118–127, 2013.

[35] E. Shirazi, "Model tuning and performance evaluation in machine learning models for PV power forecasting: Case study of a BIPV system in Switzerland," *IEEE 52nd Photovoltaic Specialists Conference (PVSC)*, Seattle, Washington, USA, 2024.

[36] E. Fix and J. L. Hodges, "Discriminatory analysis. nonparametric discrimination: Consistency properties," *Int Stat Rev*, vol. 57, no. 3, p. 238, 1989.

[37] T. Cover and P. Hart, "Nearest neighbor pattern classification," *IEEE Trans Inf Theory*, vol. 13, no. 1, pp. 21–27, 1967.

[38] H. T. C. Pedro and C. F. M. Coimbra, "Assessment of forecasting techniques for solar power production with no exogenous inputs," *Sol Energy*, vol. 86, no. 7, pp. 2017–2028, 2012.

[39] L. A. Wehenkel, *Automatic Learning Techniques in Power Systems*. Boston, MA: Springer, 1998.

[40] Tin Kam Ho, "Random decision forests," *Proceedings of 3rd International Conference on Document Analysis and Recognition*, IEEE Computer Society Press, 1995, pp. 278–282.

[41] L. Breiman, "Random Forests," *Mach Learn*, vol. 45, no. 1, pp. 5–32, 2001.

[42] J. Antonanzas, N. Osorio, R. Escobar, R. Urraca, F. J. Martinez-de-Pison, and F. Antonanzas-Torres, "Review of photovoltaic power forecasting," *Sol Energy*, vol. 136, pp. 78–111, 2016.

[43] I. A. Ibrahim, M. J. Hossain, and B. C. Duck, "An optimized offline random forests-based model for ultra-short-term prediction of PV characteristics," *IEEE Trans Industr Inform*, vol. 16, no. 1, pp. 202–214, 2020.

[44] M. W. Ahmad, M. Mourshed, and Y. Rezgui, "Tree-based ensemble methods for predicting PV power generation and their comparison with support vector regression," *Energy*, vol. 164, pp. 465–474, 2018.

[45] D. Markovics and M. J. Mayer, "Comparison of machine learning methods for photovoltaic power forecasting based on numerical weather prediction," *Renew Sustain Energy Rev*, vol. 161, p. 112364, 2022.

[46] S. Theocharides, G. Makrides, G. E. Georghiou, and A. Kyprianou, "Machine learning algorithms for photovoltaic system power output prediction," *2018 IEEE International Energy Conference (ENERGYCON)*. Piscataway, NJ: IEEE, 2018, pp. 1–6.

[47] K. Mahmud, S. Azam, A. Karim, S. Zobaed, B. Shanmugam, and D. Mathur, "Machine Learning Based PV Power Generation Forecasting in Alice Springs," *IEEE Access*, vol. 9, pp. 46117–46128, 2021.

[48] J. MacQueen, "Some methods for classification and analysis of multivariate observations," *Proceedings of the 5th Berkeley Symposium on Mathematical Statistics and Probability, Volume 1: Statistics*, vol. 5. Berkeley, CA: University of California Press, 1967, pp. 281–298.

[49] S. Theocharides, G. Makrides, A. Livera, M. Theristis, P. Kaimakis, and G. E. Georghiou, "Day-ahead photovoltaic power production forecasting methodology based on machine learning and statistical post-processing," *Appl Energy*, vol. 268, p. 115023, 2020.

[50] Z. Zhen, J. Liu, Z. Zhang, *et al.*, "Deep learning based surface irradiance mapping model for solar PV power forecasting using sky image," *IEEE Trans Ind Appl*, vol. 56, no. 4, pp. 3385–3396, 2020.

[51] Z. Zhen, S. Pang, F. Wan, *et al.*, "Pattern classification and PSO optimal weights based sky images cloud motion speed calculation method for solar PV power forecasting," *IEEE Trans Ind Appl*, vol. 55, no. 4, pp. 3331–3342, 2019.

[52]   L. Kaufman and P. J. Rousseeuw, *Finding Groups in Data*. New York: Wiley, 1990.

[53]   M. Ester, H.-P. Kriegel, J. Sander, and X. Xu, "A density-based algorithm for discovering clusters in large spatial databases with noise," *Proceedings of the Second International Conference on Knowledge Discovery and Data Mining*, in KDD'96. Washington, DC: AAAI Press, 1996, pp. 226–231.

[54]   D. Arthur and S. Vassilvitskii, "k-Means++: The advantages of careful seeding," *Proceedings of the Eighteenth Annual ACM-SIAM Symposium on Discrete Algorithms*, in SODA'07. Philadelphia, PA: Society for Industrial and Applied Mathematics, 2007, pp. 1027–1035.

[55]   C. Pan and J. Tan, "Day-ahead hourly forecasting of solar generation based on cluster analysis and ensemble model," *IEEE Access*, vol. 7, pp. 112921–112930, 2019.

[56]   C. Feng, M. Cui, B.-M. Hodge, S. Lu, H. F. Hamann, and J. Zhang, "Unsupervised Clustering-Based Short-Term Solar Forecasting," *IEEE Trans Sustain Energy*, vol. 10, no. 4, pp. 2174–2185, 2019.

[57]   I. T. Jolliffe, *Principal Component Analysis*. New York: Springer, 2002.

[58]   H. Lan, C. Zhang, Y.-Y. Hong, Y. He, and S. Wen, "Day-ahead spatiotemporal solar irradiation forecasting using frequency-based hybrid principal component analysis and neural network," *Appl Energy*, vol. 247, pp. 389–402, 2019.

[59]   D. Liu and K. Sun, "Random forest solar power forecast based on classification optimization," *Energy*, vol. 187, p. 115940, 2019.

[60]   M. Neshat, M. M. Nezhad, S. Mirjalili, D. A. Garcia, E. Dahlquist, and A. H. Gandomi, "Short-term solar radiation forecasting using hybrid deep residual learning and gated LSTM recurrent network with differential covariance matrix adaptation evolution strategy," *Energy*, vol. 278, p. 127701, 2023.

[61]   G. Ramesh, J. Logeshwaran, T. Kiruthiga, and J. Lloret, "Prediction of energy production level in large PV plants through AUTO-encoder based neural-network (AUTO-NN) with restricted Boltzmann feature extraction," *Future Internet*, vol. 15, no. 2, p. 46, 2023.

[62]   T. J. Kärkkäinen and J. Hänninen, "Additive autoencoder for dimension estimation," *Neurocomputing*, vol. 551, p. 126520, 2023.

[63]   T. Kohonen, *Self-Organizing Maps*, vol. 30. Berlin: Springer, 2001.

[64]   H.-T. Yang, C.-M. Huang, Y.-C. Huang, and Y.-S. Pai, "A weather-based hybrid method for 1-day ahead hourly forecasting of PV power output," *IEEE Trans Sustain Energy*, vol. 5, no. 3, pp. 917–926, 2014.

Chapter 5
Deep learning approaches for PV forecasting

*Emanuele Ogliari[1], Maciej Sakwa[1], Silvana Matrone[1]
and Sonia Leva[1]*

Deep learning (DL) is a part of the more general artificial intelligence (AI) field that focuses on building and optimizing complex "black-box" mathematical models. These models can learn and later recognize the descriptive features of complex phenomena happening in the real world, at the cost of a large amount of required data and a very high computational burden. However, due to their inherent flexibility and capabilities, DL models have found an increased interest among researchers.

This chapter provides a clear outlook on how DL models are used in the field of PV forecasting, starting with general definitions of the used models and followed by case studies that demonstrate an example of how they can be utilized successfully. However, beforehand, it is important to identify the forecasting task and its variations, its role in the entire power system, and the typologies of used and available data. The definition will be followed by a short introduction to the basics of DL, with an overview of the training procedure and common challenges in the field such as overfitting.

After the introduction, the chapter continues with three sections that share a similar general structure. In each section, a subset of DL models is introduced, including the necessary but concise theoretical explanations of the elementary building blocks, an analysis of the state-of-the-art applications recently proposed by researchers from around the globe, and a case study showcasing a detailed look at intricacies of the model application in a real-life scenario.

The first section focuses on artificial neural networks (ANNs), also called fully connected neural networks (FCNNs), a typology that is by far the simplest of the bunch yet the most useful due to its flexibility. The next section addresses image-based forecasting, which in the PV field usually revolves around sky images and satellite images. Commonly, the convolution operation is used to extract the descriptive features from the images, hence the name of the model: convolutional neural network (CNN). The final section deals with temporal data and sequence analysis, focusing on recurrent models (RNNs) and their close relatives, such as long short-term memory (LSTM) and gated recurrent units (GRUs). Their structure allows for the recognition of both short- and long-term patterns that are often embedded in real-life temporal data.

[1]Department of Energy, Politecnico di Milano, Italy

5.1   Introduction

5.1.1   The role of PV forecasting

For generations, the development of industrial and digital society has been interconnected with an ever-increasing energy demand. In recent years, policymakers have
introduced new laws and regulations that have led to a larger spread of power generation in which renewable energy sources (RESs) play a crucial role. Green, renewable
technologies are being pushed forward by ambitious climate targets announced at
COP26, along with the passing global energy crisis caused by geopolitical instabil-
ities. Both reasons contribute to the acceleration of work toward a greener future
through the transition to green and clean energy. This trend is reflected in the worldwide annual RES capacity additions, which increased by 6% year-on-year in 2021,
reaching almost 295 gigawatts of installed power [1]. This growth was expected to
rise further despite the continuation of pandemic-driven supply chain challenges and
record-level prices for raw materials, as shown in Figure 5.1.

Solar photovoltaic (PV) is expected to account for over 60% of the total growth
[2], due to its versatility in multiple applications, such as residential, commercial, or
industrial usage, either via grid-connection or stand-alone networks. Hence, solar
energy is becoming one of the most promising RESs, which soon will occupy a
major share in energy production. However, its integration into existing or future
energy supply structures is a major challenge due to its nature, which is dynamic and
dependent on geographic location rather than on sudden climate changes [3].

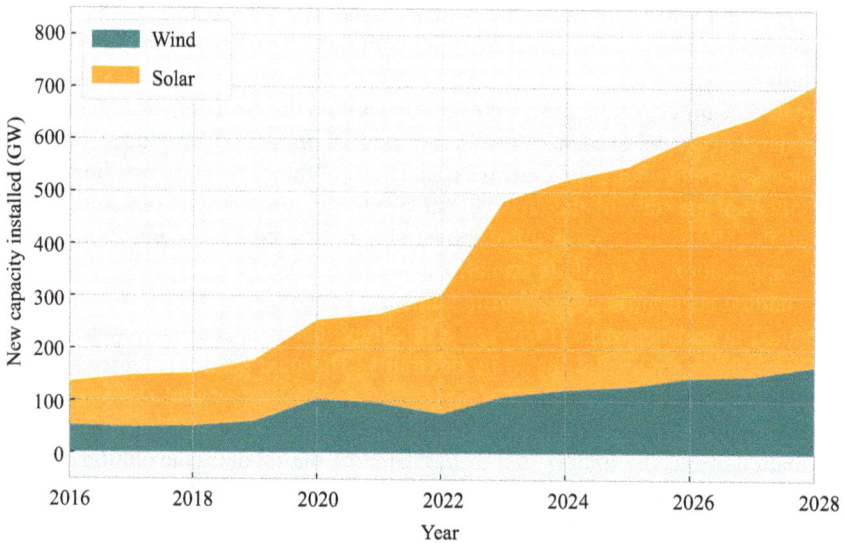

Figure 5.1   *Global net installed capacity of wind and solar energy systems in
recent years, with a forecast to 2028 [2]*

The weather-dependent, intermittent, and unpredictable nature of solar energy could generate relevant problems in balancing power generation and load demand. Sudden drops in production can induce a power loss of up to 80% within a few minutes, causing numerous problems in the grid, such as voltage fluctuations, unreliability or instability of the system, reactive power compensation, or delayed frequency response [4]. The main approaches proposed to mitigate these problems include increased storage capabilities, improved accuracy of resource and load forecasting, and demand response [5,6]. These solutions, singularly or combined within the same system, could contribute to integrating higher levels of volatile RESs. Among the listed solutions, power generation forecasting is becoming an indispensable tool for the energy market and the grid. Solar forecasting offers advantages beyond grid operators. Accurate forecasts enable adaptability to changing weather conditions, minimizing disruptions and operational costs through optimized hardware usage and control strategies. Moreover, precise solar forecasting is crucial for integrating larger shares of variable renewable energy sources, mitigating solar intermittency, and ensuring grid stability.

5.1.2 The forecasting task definition

According to Hyndman [7], the forecasting task "is about predicting the future as accurately as possible, given all of the information available, including historical data and knowledge of any future events that might impact the forecasts." Mathematically, the task can be described as follows:

$$y_{t+n} = f(X_t, X_{t-1}, X_{t-2}, \ldots, X_{t-m}, \text{error})$$ (5.1)

where the vectors of historical values $X_t, X_{t-1}, X_{t-2}, \ldots, X_{t-m}$, with a lookback window of m, are chosen as the best descriptors of the forecasted value y_{t+n}. The *error* term allows for random variations that are not modeled by the proposed approach. In the case of PV forecasting, the input feature vector X is usually related to the recorded or forecasted weather parameters, such as irradiation, tilt irradiation, ambient temperature, or wind speed. Moreover, the task itself is commonly simplified to the prediction of related meteorological parameters, with the closest and simultaneously most important in the case of PV being the GHI, due to its high correlation with the power produced [3,8].

In contrast, the n parameter of the searched value is also called the *forecast horizon*. According to the definition, the forecast horizon is "the future period for output forecasting or the time duration between the actual and effective time of prediction" [9]. Its temporal resolution can vary from a few minutes to hundreds of hours, depending on the set goal, and the selection of the forecast horizon should be strongly motivated by the objective to be achieved with the forecast. Nowadays, researchers generally refer to four horizon categories [10]: very short-term, short-term, medium-term, and long-term forecasting, as listed in Table 5.1, each serving a different application in PV load forecasting.

Table 5.1   Main forecast horizon categories

	Forecast horizon	Applications
Very short-term	1–30 min	Facilitates balancing of electricity marketing through power smoothing, monitoring of real-time electricity dispatch, and PV storage control
Short-term	30 min–6 h	Manages load dispatch and plant operations
Medium-term	6 h–1 day	Maintenance scheduling and high inertia load scheduling
Long-term	1 day–4 weeks	Long-term power balancing, transmission, and distribution through understanding seasonality

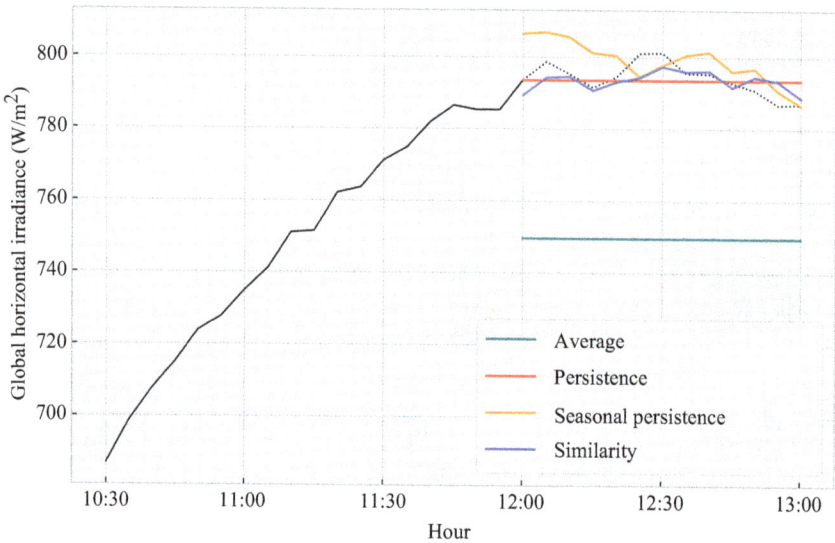

Figure 5.2   Example of implementation of naïve forecasters in GHI prediction

5.1.3   Benchmarks and metrics

When describing the results or benefits of a given forecasting approach, it is common to talk in terms of *relative values* of accuracy instead of *raw values*. The reason behind this is that many simple methods yield surprisingly good results for tasks considered highly complicated and are sufficiently considered accurate for many applications, especially when strong patterns are present in the temporal structures of the data. Particularly, Naïve methods are often used as benchmark methods due to their simplicity, which leads to both low computational and time costs. Among them, average, persistence, seasonality, and similarity methods are the most popular. The selection of a given method should be motivated by the task at hand. An example of all these methods can be seen in Figure 5.2.

- *Average* method – based on the concept of predicting all future values equal to the average of the historical data [7]. Considering the notation for historical data as $(y_1 + \cdots + y_t)$, the prediction is defined as follows:

$$\hat{y}_{t+h|t} = \bar{y} = (y_1 + \cdots + y_t)/t \tag{5.2}$$

where t is the instant of time, and h is the forecast horizon.
- *Persistence* method – used to benchmark more complex forecasting techniques. The persistence forecast is based on the concept that today equals tomorrow, as it assumes preservation of the temporal variable and predicts the last observed quantity as the model output [7]. Formally, it can be stated as follows:

$$\hat{y}_{t+h|t} = y_t \tag{5.3}$$

Prediction based on the last recorded observation yields fair accuracy in short-term forecasting, particularly given the uncertainty of weather forecasts [11]. However, it cannot account for weather variability, such as clouds' motion and rainfall within the forecast window.
- *Seasonal naïve* method – useful for highly seasonal data [7]. Each forecast is set to be equal to the last observed value from the same period of the interval considered. Mathematically, the form is presented as follows:

$$\hat{y}_{t+h|t} = y_{t+h-m(k+1)} \tag{5.4}$$

where m is the seasonal period and k is the integer part of the ratio $(h-1)/m$, which represents the number of complete time intervals (day, week, month, year, etc.) that have passed in the forecast window prior the instant $t + h$.
- *Similarity* method – based on investigations into the past searching for patterns in historical data that match the current conditions. It is done under the assumption that they can be more informative than the entire past behavior of the variables of interest [12]. In particular, analogy among several boundary conditions in the historical data can be analyzed to assess the target through *similarity-weighted averaging* [13,14].

The performance of the selected forecaster can be evaluated using typical metrics used for regression analysis to assess the error committed by the tested model. Some of the most common metrics include the mean absolute error (MAE), root mean squared error (RMSE), or mean absolute percentage error (MAPE), defined as follows:

$$\text{MAE} = \frac{1}{n}\sum_{j=1}^{n}|y_i - \hat{y}_i| \tag{5.5}$$

$$\text{RMSE} = \sqrt{\frac{1}{n}\sum_{i=1}^{n}(y_i - \hat{y}_i)^2} \tag{5.6}$$

$$\text{MAPE} = \frac{1}{n}\sum_{j=1}^{n}\left|\frac{y_i - \hat{y}_i}{y_i}\right| \tag{5.7}$$

These error metrics are often converted into relative values to assess the improvement in performance over the selected benchmark method. Usually, the *forecast skill* (FS) is used, which is defined as follows:

$$\text{Forecast Skill} = 1 - \frac{\text{error}_{\text{forecast}}}{\text{error}_{\text{benchmark}}} \quad (-) \tag{5.8}$$

where $\text{error}_{\text{forecast}}$ and $\text{error}_{\text{benchmark}}$ are the values of the selected error metric of the model and the benchmark, respectively.

5.2   Deep learning – state-of-the-art in PV forecasting

In practice, the forecasting task is usually performed using one of the two available techniques: the first centers around the usage of physics-based equations to perform numerical weather predictions (NWPs), while the second adopts the "black-box" design of the modern computational techniques using machine learning (ML) and, even more often, deep learning (DL) models [15]. For the very short-term forecasts, classical NWPs are often considered lacking in terms of both spatial and temporal resolution of the predictors [16,17]. Regarding the second approach, forecasting using ML techniques such as artificial neural networks (ANN) or support vector machine (SVM) [18–20] has been widely studied with confirmed results.

Learning methods are based on the ability of artificial intelligence to learn from historical data and further improve predictive abilities via training runs. An ANN mimics the information-processing mechanism of the human brain and approximates nonlinear functions with high fidelity and accuracy, thanks to properties such as self-learning, adaptivity, and fault tolerance [21].

In general, their purposes are far and wide, and their modern implementations find applications in fields ranging from mathematics and economics to neuroscience and psychology [22]. With this in mind, it is not surprising that ML techniques have been chosen as a solution for PV forecasting tasks, and their usability is often justified in accuracy, given their complexity and computational times [23].

Supervised learning plays the leading part in modern ML and DL development, as the majority of recent progress can be attributed solely to this category of models. In *supervised learning*, tasks by definition are based on two typologies of data: the inputs – various numerical or categorical descriptors of the process that help understand the governing principles (often written as X), and the labels – the output parameters that have to be estimated (often written as y). The role of the ML or DL algorithm is to find a function that approximates the labels $\hat{y}$ using the inputs X and optimize a set of internal weights W so that the approximation error is minimal $- \min(y - \hat{y})$.

$$\hat{y} = f(X, W) \quad \text{so that} \quad \min(y - \hat{y}) \tag{5.9}$$

Deep learning provides a complex and robust mathematical and statistical framework to solve these types of problems [24]. The emergence of DL had a profound impact on forecasting in the field of renewable energy, with research almost completely pivoting toward these new and emerging techniques [25,26].

In the field of DL, a few different approaches have been tested by researchers. In [27], a LSTM model is used to perform a time series forecast based on the temporal weather data. In [28,29], on the other hand, the authors change the approach, proposing hybrid models with additional convolution layers responsible for feature extraction. As for image-based predictions, typically sky-dome images from an All-Sky Imager (ASI) camera are employed for irradiation forecasting in combination with, e.g., ANNs [30] or more commonly CNNs [31–33]. Satellite images have been used for predictions of cloud coverage [34] and GHI [35]; in this case, authors used the GHI satellite maps as the model input. In [36], authors propose an Omnivision model that combines the two approaches with ASI images and satellite images as the input and successfully performs a short-term GHI forecast (the methods will be described in detail later in this chapter).

5.3 Artificial neural networks

5.3.1 Perceptron

The origins of ANNs can be traced to the concept of an *artificial neuron*, which was first proposed by McCulloch and Pitts to reproduce a biological neuron. It was based on binary signals, in which the neuron takes one or more inputs and provides a single output. In particular, the proposed artificial neuron is activated only when determined logical conditions (i.e., not, and, or) on the inputs are met. Since its introduction, it has served as a pillar for ANNs, as even with such a simple structure, it is possible to create any desired logical proposition.

The idea was pushed further by Frank Rosenhalt in 1957 when the *perceptron* was introduced. It is based on a different approach to an artificial neuron, where the input and the output are numbers instead of binary 0–1 values. The *perception* calculates a weighted sum of the inputs ($\sum_{i=1}^{n} x_i \cdot w_i + x_b \cdot w_b$) and passes them through a specified *activation function* to obtain the output (as seen in Figure 5.3).

Even a single *perceptron* can be used for simple cases of regression analysis and even classification. It computes a linear combination of the inputs, which can serve as an output for regression analysis or be compared against a threshold to decide on class membership once the threshold is exceeded.

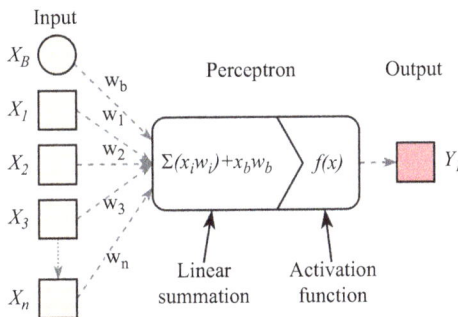

Figure 5.3   The perceptron

To solve more advanced problems involving nonlinearity, the perceptions can be easily connected to form *layers*. Moreover, these layers can be stacked on top of each other by connecting all the inputs of the new layer with all the outputs of the previous one. Hence, *a fully connected layer* is created (also called *a dense layer*) since it links all neurons of the previous layer to the neurons of the next layer. Once a few of these layers are connected, the multilayer perceptron is created.

5.3.2   Multilayer perceptron

Multilayer perceptron (MLP) also often called ANNs or deep feedforward neural networks (DFFNN) are the staple of deep learning models. In theory, they operate by learning a function F that approximates the output y using the provided inputs X. The learning procedure is performed by optimizing the weights associated with the connections between layers W to end up with the best possible approximation of the output $\hat{y} = F(X, W)$. The optimization of weights follows a selected cost function that judges how closely the approximated output matches the original one $L = y - \hat{y}$.

The networks are called *feedforward* as the information flows in a single direction through evaluation of the function values on X to eventually approximate the output with $\hat{y}$. ANNs are normally composed of multiple linear layers of neurons that sequentially pass the inputs X to the output applying the selected transformations, according to the equation:

$$y_i = f_i(\mathbf{W}_i^T x_i + b_i) \tag{5.10}$$

Where y_i is the output of the ith layer, $f_i(z)$ is the activation function, and W_i and b_i are weights and bias matrices, respectively. This chain of layers includes the input layer, the middle layers (also called hidden, as normally we do not have access to their inputs and outputs), and an output layer (as seen in Figure 5.4) that transforms the input into a desired output using a chain of functions. Hence, the F can be approximated as $F = f_3(f_2(f_1(X)))$, where f_1 is the first layer activation function, f_2 the second, and so on. This chain allows the ANN to model complex nonlinear dependencies between the input and the output. ANNs are generally used in a variety of tasks due to their inherent simplicity and flexibility. The main parameters that have to be set are the number and size (in terms of neurons) of hidden layers and the said activation functions $f_1, f_2, \ldots, f_n$ that decide the output of neurons. The sizes of the input and output layers are usually imposed by the structure of the available data and the desired model output.

5.3.3   Activation functions

Before moving to the case study, it is important to look at the typical activation functions ($f_i(z)$ from (5.10)) used in various typologies of networks. There is no clear choice of which works best every time, as the selection procedure is usually case-dependent and each case study might benefit from a different approach. However, as a rule of thumb, it is common to start with ReLU and switch to others if it proves inaccurate.

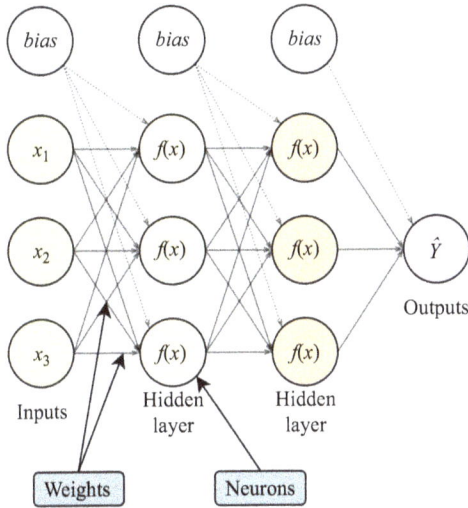

Figure 5.4   A simple visualization of a multilayer perceptron, with a single input layer, two hidden layers, and a single output layer. The arrows represent the passing of information between singular neurons with weights attached to them.

The most commonly used are as follows:

- Single step function – a step from 0 to 1 at a given value of x (usually 0):

$$f(x) = \{0 \quad \text{if } x < 0 \quad 1 \quad \text{if } x \geq 0 \tag{5.11}$$

- Sigmoid (also called logistic) – an S-shaped soft step function from 0 to 1:

$$f(x) = \frac{1}{1 + e^{-x}} \tag{5.12}$$

- Hyperbolic tangent – similar to sigmoid, an S-shaped soft step function from -1 to 1:

$$f(x) = \tanh(x) \tag{5.13}$$

- Rectified linear unit (ReLU) – the most commonly used function due to its simplicity and quickness, a linear function that flatlines all values below 0:

$$f(x) = max(0, x) \tag{5.14}$$

All the mentioned activation functions along with their derivatives are presented in Figures 5.5 and 5.6.

These functions and their derivatives have a crucial part in the part of the training procedure called the *backpropagation algorithm*. This algorithm calculates the gradient of the network's error in two steps: forward and backward. First, it makes a prediction (forward pass) and measures its error, then it goes through each layer in reverse to find out how each connection weight and each bias term affected the error.

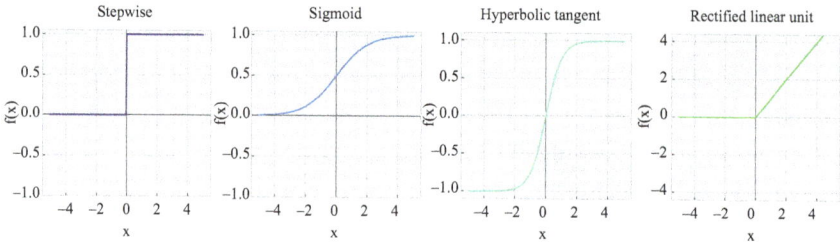

Figure 5.5   *The most commonly used activation function in the development of custom DL algorithms*

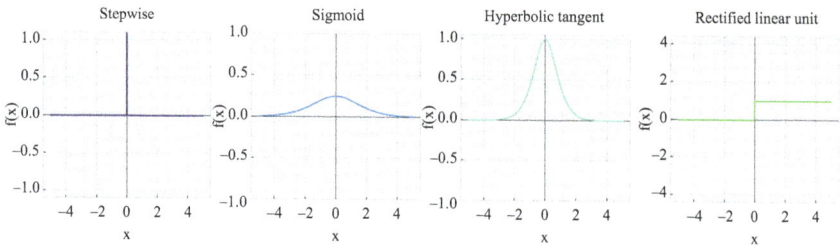

Figure 5.6   *The derivatives of the most commonly used activation function in the development of custom DL algorithms.*

Once it has these gradients, it performs a regular gradient descent step, and the whole process is repeated until the network converges to the solution.

Calculating the gradient of each node is a crucial step in training the network; hence, the activation function has to be easily derivable. The initially used step function is indeed constructed of two flat segments, so there is no gradient to work with. The solution was to replace the step function with other continuous functions characterized by a nonzero derivative (especially for positive values), allowing gradient descent to make some progress at every step of the training.

5.3.4   Examples in the literature

Due to their simplicity and versatility, the ANNs have been tested in various tasks centered on PV forecasting. In [37], Ogliari *et al.* proposed a method for establishing the best fit of an ANN for a day-ahead hourly PV forecasting. The implemented method used various historical weather parameters recorded at a 15-min resolution at the location of interest. The study demonstrates greatly the potential of implementing ANNs in PV forecasting, establishing the sizes of the networks necessary for accurate forecasts. Furthermore, Dolara *et al.* in [38] experimented with different feature sets that can be provided to the ANN to improve the accuracy of the forecast.

In [39], the authors evaluated the usability of ANNs in PV forecasting by assessing the performance of a simple ANN model against a physical-deterministic method. The results show an increasing potential of ML techniques with the *black-box* model slightly outperforming the *white-box* mathematical model. However, the requirement for local production data in ANNs renders them less useful in newly installed PV power plants.

In [40], Leva *et al.* proposed a *hybrid* ANN in which the PV forecast is supported by a mathematical model of theoretical sky radiation. The model is later used for both day-ahead PV forecasts that are further refined by a secondary short-term predicting ANN. For the test year, the improvement in performance was staggering compared to a normal ANN, as the normalized MAE dropped by 36.6%. Moreover, the performance was best in clear sky conditions, with improvements in performance reaching 58%. This also coincides with findings presented in [41]. In [42], the authors tested the proposed hybrid model using a real-life operating microgrid. The obtained results showed a great increase in forecast performance. The error committed, evaluated in terms of normalized MAE, was reduced from 5.28% to 3.81% on a monthly basis.

In [43], Moncada *et al.* tested the possibility of using an ANN to predict the GHI in the very short term using image-based input. They experimented with various methods of image downscaling and cropping, with the MLP slightly outperforming the traditional ML-based methodologies such as the Random Forest or Gradient Boost Trees. Their results demonstrate the potential applicability of sky dome images for GHI nowcasting. A similar, image-based study has been performed by [44], in which the authors compared the ANN's performance in very short-term PV output power estimation. The authors compared satellite image input with highly correlated weather data, proving that image-based techniques can be highly effective for very short-term forecasts.

In [45], Matteri *et al.* tested the possibility of dividing the training dataset into specific groups based on either sky clearness indices or unsupervised similarity-based methods (such as clustering). The results did not show any improvement in performance for grouped datasets, as dividing the data into smaller more specialized groups decreased the ability of the model to generalize.

5.3.5   Case study

This case study focuses on the forecasting of GHI using a time-series-based approach. GHI is a critical parameter, closely correlated with PV output, making it a primary input for forecasting PV power generation. The objective of this study is to predict GHI up to 6 h in advance by leveraging local weather station data. Accurate 6-h forecasts enable engineers to proactively manage load balancing and secure alternative energy sources when PV-based generation is compromised by adverse weather conditions.

As discussed in Section 5.1.3, simple models such as persistence are often employed as benchmarks for very short-term forecasting. However, in the context of GHI prediction, seasonal persistence is a more suitable benchmark due to the highly seasonal nature of GHI, characterized by distinct day–night cycles. Predicting deviations from the clear-sky GHI profile allows for estimating potential power losses in the installed PV system. The primary goal of this case study is to evaluate the performance of the proposed model in comparison to the seasonal persistence benchmark. Forecast skill, as defined by (5.8), will serve as a key metric for assessing model performance.

5.3.5.1   Dataset

The dataset employed for model training and testing was derived from real-time weather station data collected at the Department of Energy, Politecnico di Milano. The recorded meteorological parameters included humidity levels, ambient temperature, rainfall levels, cloud cover, GHI, tilted irradiance, and diffuse irradiance. While the raw data was initially captured at a 10-s interval, it was subsequently resampled to a 15-min frequency for model development. This downsampling step was deemed necessary to mitigate the effects of noise inherent in high-frequency data, which could potentially hinder model generalization. The 15-min resolution was determined to strike a balance between model robustness and overall performance, providing a suitable compromise for the intended application. The dataset spans nearly a year of continuous recording and was checked for missing data and artifacts. After the check, the final shape of the used dataset was over 35 thousand samples for each of the 10 recorded weather features.

To facilitate the prediction process, the dataset was partitioned into moving windows. Each window consisted of a lookback period, in this case set to 24 h (or 96 data samples), and a forecast horizon of 6 h (or 24 data samples). This windowing approach provides the model with relevant historical weather data, enabling it to accurately capture GHI trends. Figure 5.7 illustrates the windowing methodology in detail.

5.3.5.2   Proposed model

To solve the task, a simple deep ANN model can be used. The model consists of a stack of layers including an input layer, three hidden layers, and a single output layer. The hidden layers are all identical, with 20 neurons each and a *ReLU* activation function (see Figure 5.5). The final output layer has to be shaped according to the model output. In this case, it is a sequence of GHI values of 24 samples (or 6 h), therefore,

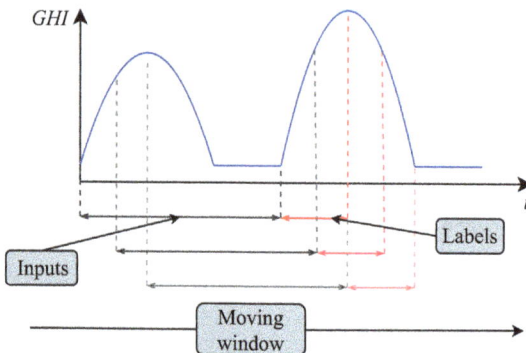

Figure 5.7   The windowing operation – a sliding window goes along the time series, extracting the inputs (the last 24 h) and the outputs (the next 24 h). The windowing is continuous, or in other words, the same input weather parameters can be a part of many windows that pass through the same region (more specifically, they are a part of 96 windows).

the output layer must be built of 24 neurons. Moreover, the suggested activation function at the output layer for regression tasks is a simple linear function, so no further nonlinear modifications are introduced in the output. This allows the final output to be shaped in the full range of values, compared to $(-1, 1)$ for the hyperbolic tangent as an example. The final model has 20,564 trainable parameters.

With such a robust approach and only a single year of data available to perform training and testing, the trained model becomes prone to overfitting. Overfitting is a complex phenomenon that is a result of *overtraining* the model so that it learns too many details about the training dataset. In other words, the proposed model learns the *noise* present in the training dataset, undermining the ability of the model to generalize on new datasets. Overfitting is usually caused by one of the three reasons:

- The proposed model is too big for the studied dataset and easily learns to predict the general trend. Therefore, an obvious solution is to decrease the size or scope of the model, for example, the number of hidden layers or the number of neurons per layer. Another approach is to change the entire model and focus on a simpler approach, for example, maybe the problem can be described using a simple regression model, or a random forest would better suit it.
- The used dataset is too small and does not entirely describe the ongoing process. Sometimes the datasets used for training and testing the models are simply too small for the model to learn the trends governing the studied process. It is important to ensure that the data fed to the model is sufficient and of the highest quality.
- The model is trained for too long. Even if the model is a good fit for the dataset, the model can be overtrained to the point that it learns the intrinsic random patterns in the training dataset. An early stopping mechanism can be adapted.

In practice, the most common ways to counteract the overfitting problem are model regularization and early stopping. Regularization includes actions such as putting restrictions on model weights or limiting the gradient range at each training loop iteration. Early stopping is by far the most common way to avoid overfitting, yet it comes at the cost of slightly reducing the size of the training dataset. It requires a *third* partition in the dataset to be extracted – the validation dataset, which is constantly being used to evaluate the model output at each iteration of the training loop. It should never be used for the training itself; it is only used to check how good the generalization skills of the model are at each epoch. Normally, both the training error and the validation error should decrease gradually. In contrast, when entering the overfitting territory, the validation error starts to rise while the training error continues to drop. This phenomenon can be seen in Figure 5.8. The early stopping mechanism notices when the validation error starts to grow and terminates the training procedure accordingly.

To avoid overfitting, the model is trained with early stopping enabled.

5.3.5.3 Results

The analysis of the model's performance is based on the mean absolute error metric for predictions for selected samples from the test dataset that has been separated and unseen in training. In other words, the model's performance is evaluated with

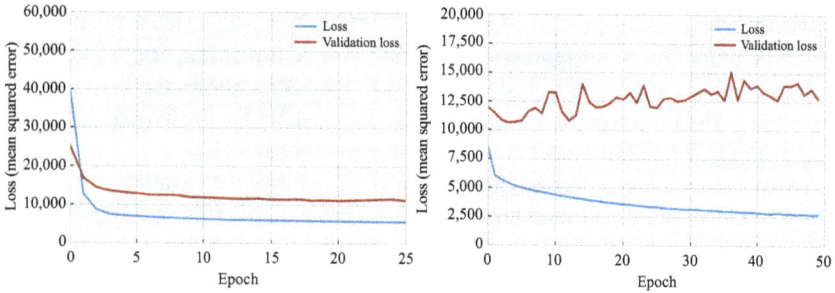

Figure 5.8   *Example of the training curves for a correctly conducted training procedure (actual curve from the case study) and an example of an overfitted model (a bigger model is used and early stopping is disabled).*

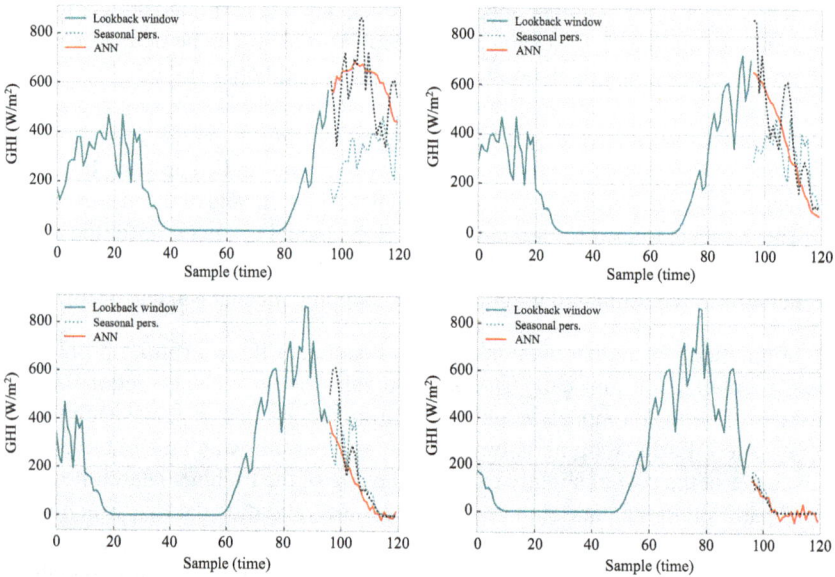

Figure 5.9   *Result examples of the ANN for GHI forecasting at different hours of the same day. As can be seen, the ANN follows the general trend quite accurately, resulting in a decent prediction of GHI 6 h ahead. The poor performance of seasonal persistence is a result of a high day-to-day deviation in cloud coverage.*

regard to the ability to predict the time series of GHI in the near future (6 h) based on the multiparameter weather inputs. In detail, the result examples can be seen in Figure 5.9, where different time periods of the same day are tested. In this example, the seasonal persistence is severely underperforming due to the high difference in cloud coverage between the days.

Table 5.2 Detailed results of the ANN for PV forecasting

Period	MAE$_{pred}$ W/m^2	MAE$_{pers}$ W/m^2	FS %
1	116.49	275.20	58
2	105.12	158.43	34
3	68.50	68.90	0.6
4	16.97	7.48	−130

In more detail, the exact numerical results are presented in Table 5.2. As can be seen, for most of the periods except for the last one, the proposed model outperforms the selected benchmark metric. However, it is interesting to notice that the error generally decreases with the progress of the day, and the MAE becomes almost negligible close to the sunset. This indicates that stochastically predicting the actual peak of the GHI (and thus the PV generation), which has to be done in the morning is significantly harder than the prediction of afternoon slopes.

Finally, the overall performance of the model on the entire test dataset can be evaluated. In this case, the overall MAE is equal to 85.93 W/m^2, the seasonal persistence MAE is equal to 232.05 W/m^2, and the forecast skill to 62.9%. The designed ANN is a significant improvement over the simple benchmark method.

5.4 Convolutional neural networks

CNNs, first introduced in 1998 by Yann LeCun [46], are specialized networks designed to handle grid-like structures. The most common examples are images that are a 2D grid of pixels, but time series data sampled at regular intervals can easily be modeled as a grid of 1D pixels. As the name implies, CNNs employ the *convolution* operation in their layers instead of traditional matrix multiplications to extract and process data along with the *pooling* operation that reduces the dimensions of increasingly big feature spaces.

5.4.1 The convolution operation

To describe mathematically the convolution operation, let us assume a 1D **input** vector $x(t)$ (e.g., a time-series recording of a signal). Next, a weighted filter $w(l)$ (or *kernel*) of a length l is passed along the initial signal. The filter is multiplied along the way with the corresponding $x(t)$ values, and the results are summed together to obtain the output signal $y(t)$, also called a *feature map* [24]:

$$y(t) = \int x(l)w(t-l)dl = (x*w)(t) \tag{5.15}$$

Commonly, the operation is denoted with an asterisk. A visual representation of the operation can be seen in Figure 5.10

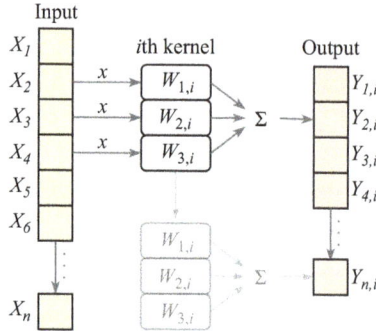

Figure 5.10   *The convolution operation for a 1D signal. The kernel W passes over the input X to obtain the feature map Y.*

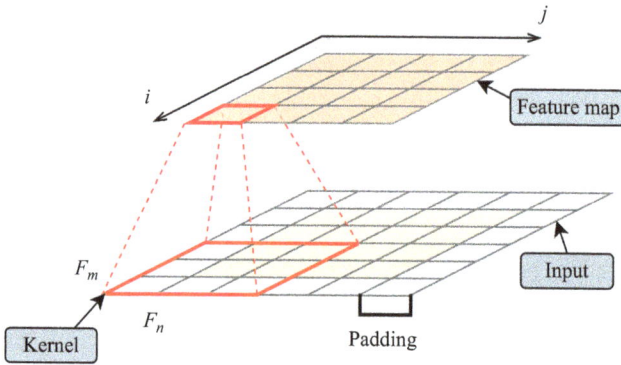

Figure 5.11   *The convolution operation for a 2D. The kernel of a size $F_m \times F_n$ passes over the input X to obtain the feature map Y.*

However, the most common use for the convolution operation in the domain of ML is to extract features from multidimensional arrays (also called tensors) such as images. In fact, images are usually represented as three-dimensional tensors of (W, H, C) shape, where W is the image width, H is the height, and C is the number of channels (usually representing different colors). In the case of higher dimensionality, the convolution operation works similarly with the kernel sliding along the input across the additional dimensions. A visual representation of a 2D 3 by 3 kernel sliding along a 2D input image can be seen in Figure 5.11. If the output feature map should be of the same shape as the original input, additional *padding* can be added at the edges of the input. Usually, these are additional pixels with 0-values in them. Moreover, the filter can *stride* while moving by committing a number of pixels at each step of the pass over the input image.

Mathematically, the initial equation (5.15) becomes

$$S(i,j) = (I * K)(i,j) = \sum_m \sum_n I(m,n)K(i-m,j-n) \qquad (5.16)$$

where I is the input image of size i by j, and K is the passed kernel of size m by n.

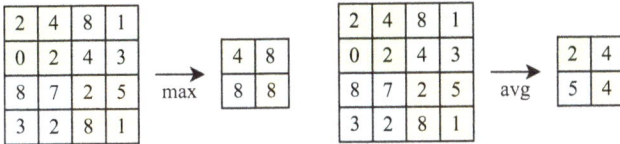

Figure 5.12   Examples of pooling operation performed with a 2 × 2 filter. On the left, there is the most commonly used max; on the right, mean pooling

A single convolution layer usually employs a high number (i.e., between 32 and 128, but even higher values are sometimes employed) of convolution filters to extract feature maps from inputs. Each filter is characterized by an individual set of trainable weights that through training become specialized in the detection of specific features embedded in the presented input. For example, a single filter can be trained to detect vertical lines in an image. Another will be specialized in horizontal lines, and yet another in curves. In other words, the information on what is present in the input array is extracted through a series of dot-product operations between the input image pixels and the weights of the filter. Moreover, each filter creates a single corresponding feature map; however, as the number of filters grows, the number of extracted maps also quickly becomes significant.

The majority of modern CNNs are created by stacking convolution layers on top of each other. As result, the number of feature maps extracted from the input array can skyrocket quickly. Moreover, each layer and each filter represents a set of trainable parameters that perform significantly more operations in a single forward pass compared to a simple *fully connected layer* (that represents a single matrix multiplication). This leads to a significant increase in the computational load (in particular RAM or VRAM if the process is run on a GPU) needed to train the model and run the inference later. Therefore, an information reduction method has to be used – *pooling*.

5.4.2   Pooling operation

Pooling layers [47] are designed to gradually reduce the dimensionality of the extracted input representation. In doing, they reduce the number of parameters, the memory usage, and the computational burden. The operation of a pooling layer is somewhat similar to the one described for convolution as its size, stride, and padding can be easily tweaked in the same way.

In contrast, a pooling filter has no trainable weights and is tasked with unifying the input through a selected aggregation function, such as a maximum or a mean. Figure 5.12 shows the result of the pooling operation for the max and mean. The example shows a typical 2 × 2 pooling operation that reduces the size of an extracted feature map by fourfold, highlighting the efficiency by dropping the excessive parameters.

5.4.3   Examples in the literature

CNN-based models for PV forecasting usually center around image analysis. The two main sources of images that are commonly used are satellite imaging (for tasks such as cloud tracking or global irradiation predictions) and All-Sky Imagers (ASI) sky dome photography (for local predictions). Works based on satellite images guarantee good accuracy in prediction since they focus on large spatial and temporal resolutions, which smooth sudden solar radiation changes. However, they have some disadvantages as they usually rely on measurements of the Cloud Based Height (CBH) – a fundamental parameter to define cloud types and their thickness. The time to receive and process these data from the satellite is high, which decreases the usability of short-term forecasting. As a result, other studies utilize and investigate the uses of ASI.

Kazantzidis *et al.* [48] used a digital camera with a fish-eye lens and a hemispheric dome for automatic detection of the clouds. The percentage cloud coverage was determined by applying thresholds to each pixel of the images based on RGB colors. Cloudy pixels were then classified into seven categories by a k-Nearest-Neighbor method, which achieved an average accuracy of 87.9%. Schmidt *et al.* [49] applied a cloud detection method based on color thresholding to classify seven different categories of sky conditions to develop a multistep model to forecast the GHI up to 25 min ahead. Historical data of solar irradiance were provided by pyranometers and CBH by a ceilometer. The Sun position, along with clear sky irradiance, was computed by a theoretical model for ray tracing and subsequent cloud shadow mapping. Cloud movement was determined by applying an optical flow algorithm. Despite the numerous modeling efforts taken, the forecast was not able to outperform the persistence for all cloud conditions. These methods were limited in their usability and accuracy, hence the CNN was introduced to this task a specific type of deep learning model particularly suited to deal with image-based predictions.

Shi *et al.* [50] exploited CNN models for ground-based cloud classification. Two different configurations for the CNN were defined to compare a fast and simple model with a more complex one. Both outperform conventional handcrafted feature spaces. The faster model got a best accuracy of 98.20% and 89.90%, respectively, for the first and second datasets. The more complex one had the best accuracy percentage, respectively, of 98.20% and 90.32%. A similar study was made by Zhang *et al.* [51], who built a dataset containing 11 categories of clouds, including contrails for the first time. The CNN took inputs from a series of fixed-size red–green–blue cloud images and provided a sequence of label predictions as output. The network was optimized by subtracting the mean RGB value of each pixel to augment the contrast and improve training speed and accuracy. The results show good accuracy for such specific classification, ranging from 86% to 90% for all the categories. In [52], the authors proposed an approach to CBH estimation with CNNs, with different approaches to the classification strategy. However, despite the inclusion of other parameters for CBH estimation (such as the Pressure of Lifted Condensation Level), the results achieved did not exceed 63% accuracy due to the low availability of high-quality data. This demonstrates the biggest disadvantage of DL-based techniques – an excessive amount of data is required to properly train and set up the models. As can be seen, in solar forecasting, CNNs are becoming a promising technology for cloud classification. However, their ability to extract features from images makes them also relevant for directly predicting solar irradiance. Sun *et al.* [53] used a CNN to correlate the power output of two PV panels with simultaneous all-sky images. The authors made a sensitivity analysis to get the most suitable CNN architecture for their specific study, and the results demonstrated that images contain significant information for solar prediction. The resulting normalized RMSE is equal to 28%. In [54], Pothineni *et al.* presented an image-based approach to anticipate the irradiance state both 5 and 10 min ahead of time through a CNN. The future irradiance state ("occluded" and "clear") is defined considering a threshold of 80% on the clear-sky irradiance computed using the Perez–Ineichen clear-sky model [55]. Two large datasets of images taken in Italy and Switzerland were used to validate the model, achieving accuracies of 92.93% and 91.34%, respectively. These values demonstrate that the network is not influenced by local site-specific properties.

In recent years, Infrared (IR) ASI imaging has become increasingly popular in the forecasting context due to its capabilities to provide cloud properties and atmospheric data [56,57] and to show better visual contrast between clouds and clear sky. In fact, visible cameras display a significant Sun flare, whereas the Sun disk affects very low infrared images [58]. An application of the thermal infrared ASI in the forecasting field is presented in the work of Liandrat *et al.* [59]. The features of the images are extracted via two steps: pixel thresholding to define cloud/clear sky and an optical flow algorithm to detect cloud motion. Four months of data derived from images and irradiance measurements are used by a random forest ANN for forecasting. The authors demonstrated the benefits of solar prediction with respect to the "no-forecast" scenario. In fact, the load balance error (difference between load and production) decreases for an equal number of blackouts in the two cases, resulting in lower fuel consumption up to 17%. Another application is presented in [31], where the authors proposed a CNN-based approach for very short-term irradiance forecasting. It implements a stack of historical ASI images to predict the future state of the GHI over different time horizons. Moreover, the proposed model benefits from embedding actual measured values of GHI into the stacked images to boost the accuracy of the prediction. In a case study, the *enhanced* model with embeddings outperforms a raw CNN and a benchmark persistence by a clear margin, achieving an FS of 19%. In [60], the authors implemented a CNN-based forecasting scheme to evaluate the variability of the GHI in very short-term time horizons by assigning a GHI-variation class to the output. The IR images were preprocessed by isolating the sun position region to enhance the quality of the prediction. After the elimination of rainy days from the initial dataset, the overall classification accuracy achieved surpassed 65% for a 5-min forecast horizon, with the results dropping as the horizon increased.

5.4.4   Case study

The case study presented in this section describes an implementation of an approach to PV forecasting using image-based processing and DL. The approach uses a huge dataset of IR sky dome images collected at the Department of Energy at Politecnico di Milano. In more detail, an advanced CNN-based ML model is used for forecasting GHI values in a very short-time horizon, i.e., between 5 and 15 min ahead. This forecasting technique (also called *nowcasting*) helps in short-term load balancing, as it allows for the detection of huge spikes and drops in PV generation resulting from the instability of weather and the volatility of the local cloud cover.

However, the prediction of an exact value of GHI, especially without the entire knowledge of the historical trend of value and using only contemporaneously available data, might be difficult for the model, as the values for the GHI normally range between 0 W/m^2 at night and 1000 W/m^2 on sunny days. Therefore, the values are normalized using the theoretical values of the GHI, the Clear Sky GHI (GHI CS). GHI CS for a given location can be easily acquired using any open-source PV analysis package. In this case, it is computed through the Ineichen and Perez clear sky model, which can be simulated in the PVlib Python package. The resulting parameter is called the Clear Sky Index (CSI), a dimensionless parameter that directly informs on the weather conditions. In most cases, the value of this parameter ranges between 0 and 1.

$$CSI = \frac{GHI}{GHI\ CS} \tag{5.17}$$

As previously described in Section 5.1.3, the most common benchmarks for very short-term forecasting are the simple models such as persistence. For such a small-term horizon, these models are characterized by almost unbeatable accuracy, as reality is full of inertia and usually the assumption of permeance of reality serves as a good enough predictor. Therefore, the main goal of the case study is to find out whether the proposed model is capable of outperforming the persistence benchmark. Performance can be measured using the metrics introduced earlier, such as the forecast skill, introduced earlier in (5.8).

5.4.5 Dataset

The dataset used in this case study comes in two parts. The first is a sequence of infrared all-sky images recorded at a rooftop lab in the Department of Energy at Politecnico di Milano. The second is a sequence of corresponding measured local weather parameters, including temperature, humidity, and also GHI acquired by a pyranometer localized in close proximity to the ASI camera. The acquisition campaign spans almost a year, starting from late June 2023 to early June 2024. During this period, the camera took pictures at a frequency of 1 per minute, resulting in a sequence of almost half a million sky dome images (due to occasional shutdowns and images, some periods are missing data). However, this number is significantly reduced after eliminating incorrect images and images taken during the night. In fact, when considering only the images taken in hours of perceived sunshine (so with a sun elevation angle higher than 20°), the final number of images that can be used for model training and testing is reduced to 145 thousand. Before feeding them into the DL model, the images are preprocessed accordingly to enhance the relative differences between regions with clear skies and regions with cloud coverage. This can be achieved using various filtering options, such as scaling or log filtering. Next, the images are cropped to get a better focus on the area of interest and downscaled from 480 × 480 to 128 × 128. The initial and the final image can be seen in Figure 5.13.

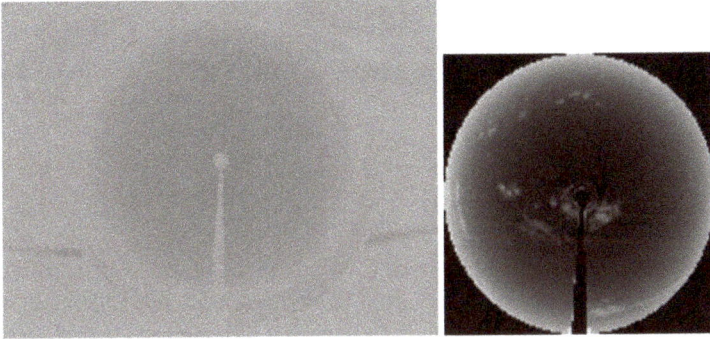

Figure 5.13   Initial and processed IR images for use in the CNN-based forecaster

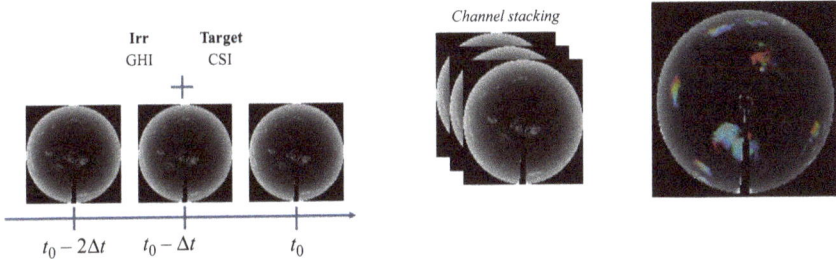

Figure 5.14   Image transformation pipeline for the CNN input. The flat grayscale images are stacked on top of each other creating an RGB image, with various temporal layers encoded in different channels.

The biggest disadvantage of using a raw CNN is that it can process a single image at a time. In other words, it does not understand the sense of temporality of acquired sequential data. Its output depends solely on analysis of the passed image, and neither short-term nor long-term patterns in sequences of images are observed or applied to the final result. In reality, the images could also be shuffled in input and still, the result would stay the same. To counteract this disadvantage, some temporal information can be smuggled into the model. In the second step of preprocessing, the acquired images are flattened to one channel. Once flattened, the input image is stacked on top of two historical images that were taken at $t_0 - \Delta t$ and $t_0 - 2\Delta t$ to form a new three-channel RGB image. For the purpose of this case study, the selected Δt is equal to 2 min, as this small delta improves short-term traceability of cloud movement and thus prediction accuracy. This stacked "temporal" image will serve as the input in training and testing of the selected CNN model. The complete preprocessing pipeline can be seen in Figure 5.14. As for the division between train and test data, the majority of the dataset is used for training, and solely, the final days recorded in early June 2024 are used as a test. However, due to the high variability of weather in that period in Milan, they serve as a good enough approximation of real-life conditions.

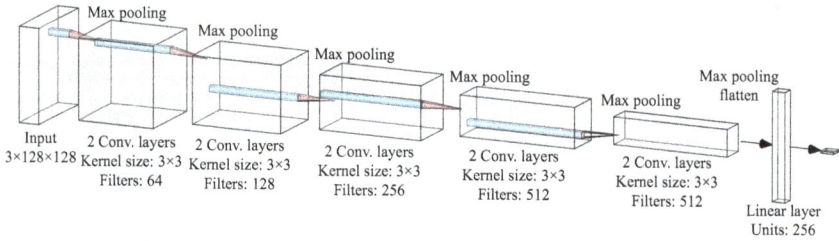

Figure 5.15   A schema of the proposed CNN model – a modified VGG-16 network with five double convolution blocks, this relatively simple network already has over 15 million trainable parameters.

5.4.6   Proposed model

To accomplish the task, a modified VGG-16 CNN architecture has been used. It is composed of a series of double-stacked convolution layers divided by max-pooling layers to decrease the size of the model and accelerate convergence in training. The model has five said blocks, with an increasing number of filters (from 64 in the first block to 512 in the last), but equal among the layers inside the block. All layers use the same size of the kernels (3×3) and the same activation function. Once the feature maps are extracted and downscaled with max pooling, they are flattened and transferred to fully connected layers for prediction. The output is a single number describing the predicted CSI in the near future. The choice of exact values of hyperparameters (i.e., number of layers or number of neurons per layer) is a matter of optimization of the model output. While there are many methods to achieve an optimal result, most of them are very time- and resource-consuming. Here, hyperparameters have been fine-tuned using a simple random search.

5.4.7   Results

The results analysis is based on the model accuracy for selected days of the test dataset. In detail, the forecasting model is used to predict the 15-min ahead value of the GHI for the first five days of June 2024. Detailed results in terms of the achieved forecast along with the general description of the weather conditions perceived can be seen in Table 5.3. The $FS(avg)$ value represents the accuracy of a smoothed trend of the rolling average of the forecast, as minute data leads to noisy prediction. The overall calculated FS for the entire test dataset is equal to 18.8%. Detailed results for Day 1 and Day 3 are shown in Figure 5.16.

5.5   Recurrent neural networks

Recurrent neural networks (RNNs) are a family of neural networks designed for processing sequential data. Similar to CNNs, RNNs process grid-like structures of data; however, their cyclical design allows for easier extraction of important trends or pieces of information regardless of their position in the sequence, e.g., a position

Table 5.3   Forecast skill for five selected test
days in June 2024

Day	Weather	FS	FS (avg)
01/06/24	Sunny	−0.005	−0.069
02/06/24	P. cloudy	0.144	0.296
03/06/24	P. cloudy	0.222	0.273
04/06/24	Cloudy	−0.007	−0.015
05/06/24	Shady	0.233	0.276

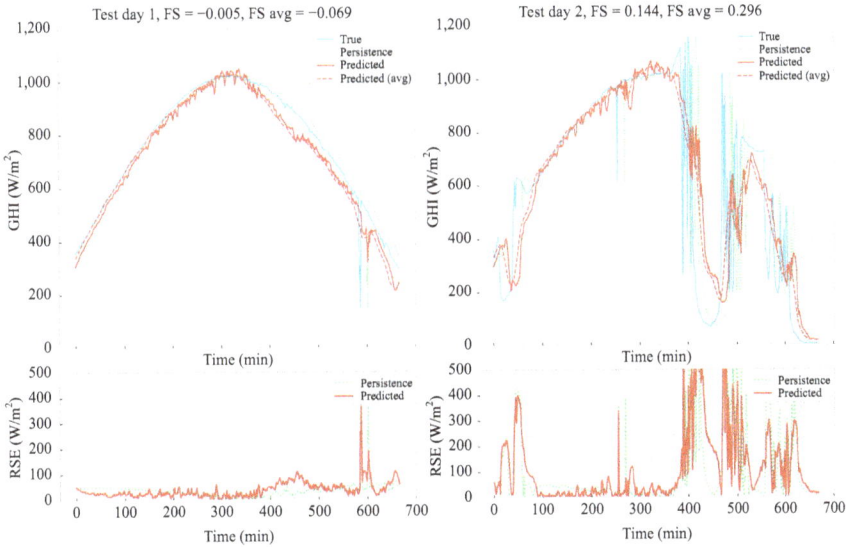

Figure 5.16   Detailed results for CNN test day 1 and test day 2. Test day 1
represents a typical cloudless sunny summer day, with measured GHI
quite indistinguishable from theoretical curves obtained from the CS
model. Under these stable conditions, the persistence benchmark is
exceedingly hard to beat, which is reflected in slightly worse results
compared to benchmark and negative FS values. In contrast, on test
day 2, when the weather conditions start to worsen in the second half
of the day, the CNN provides a smoothing effect to the GHI, obtaining
significantly better prediction accuracy compared to the simple
model.

of a specific information can be at the beginning or the end of a sentence. This is
done by incorporating a hidden state variable **h** that is used as an additional variable
to perform predictions.

$$h^t = f(h^{t-1}, x^t, \theta) \tag{5.18}$$

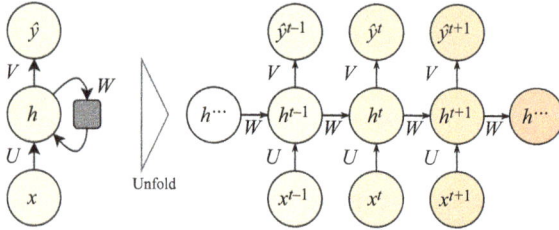

Figure 5.17   The folded and unfolded RNN computation graph

When performing predictions, the RNN is trained to use the hidden state variable as a lossy summary of the past states of the network, remembering important task-related aspects of the processed sequence. Depending on the task, this lossy summary might preserve some aspects of the sequence with higher precision than others.

RNNs are often represented as either folded or unfolded computational graphs, as shown in Figure 5.17. The presented RNN maps the vector of inputs x values to a corresponding sequence of output values $\hat{y}$. Training such a network would be centered on optimizing the weights matrices: input-to-hidden matrix U, hidden-to-hidden matrix W, and hidden-to-output matrix V.

The forward pass equations in such a network take the following form:

$$h^t = \phi(b + Wh^{(t-1)} + Ux^t)$$
$$\hat{y}^t = \theta(c + Vh^t) \tag{5.19}$$

where b and c are bias parameters, and U, V, and W are the weight matrices mentioned before. The $\phi()$ and $\theta()$ represent respective activation functions that are selected accordingly for the task to be performed.

The circular behavior pattern of RNNs helps solve numerous sequential problems with DL techniques. However, it also introduces new issues to the design of RNNs itself. Compared to previous networks (both ANNs and CNNs), a single layer of RNN applies the activation function many times, not just one. When passing the entire sequence through the RNN's neuron, the output is dependent not only on the input but also on the hidden state calculated from the previous input. The hidden state also depends on the previous hidden state:

$$h^t = W^\top h^{t-1} = (W^\top)^t h^0 \tag{5.20}$$

Therefore, a small tweak of the weights in gradient-based optimization carries resonating consequences in the state of the RNN. Small values (below 1) become exponentially accelerated toward 0, and larger values grow at the same pace. The problem is typical for RNNs and is called the *exploding and vanishing gradients* problem. The biggest consequence of this problem is uneven learning between short-term and long-term dependencies in sequences, as the order of magnitude of the weights representing these dependencies varies significantly. The most common solution to this problem usually lies in network regulation, for example, the activation function can be replaced with a saturating one (such as hyperbolic tangent or sigmoid), or the values of the weights can be rescaled, through normalization among others. However, in the space of applied ML, many researchers turn to a ready-made solution, i.e., the LSTM units or GRUs.

5.5.1   *Long short-term memory unit*

The LSTM architecture and the general concept of *memory cells* and *gate units* were originally proposed by Hochreiter and Schmidhuber in 1997 [61] and, with minor modifications, is nowadays widely used in several fields of application concerning the analysis of sequences, such as natural language and translations [62]. LSTM introduces a new processing unit, the memory cell. To control the cell, a number of gates are needed with internal mechanisms that can regulate the flow of information coming in and out of the cell. The common architecture of the LSTM cell can be observed in Figure 5.18.

The core concept of the model is the cell state $\mathbf{c}_t$, represented by the horizontal line running through the top of the diagram. It acts as a conveying belt that stores and transfers condensed information down the sequence chain. The content of the cell state changes through time, thanks to the interaction of the previously computed output h_{t-1} and the current external input x_t. The rectangular units in Figure 5.18 represent either sigmoid or hyperbolic tangent activation functions and transformations. The operations represented in the squares are the point-wise multiplications ($\times$) and point-wise addition (+).

In (5.21), the governing principles are given.

$$
\begin{aligned}
\mathbf{i}_t &= \sigma\left(\mathbf{W}_{xi}^{\top}\mathbf{x}_t + \mathbf{W}_{hi}^{\top}\mathbf{h}_{t-1} + \mathbf{b}_i\right)\\
\mathbf{f}_t &= \sigma\left(\mathbf{W}_{xf}^{\top}\mathbf{x}_t + \mathbf{W}_{hf}^{\top}\mathbf{h}_{t-1} + \mathbf{b}_f\right)\\
\mathbf{o}_t &= \sigma\left(\mathbf{W}_{xo}^{\top}\mathbf{x}_t + \mathbf{W}_{ho}^{\top}\mathbf{h}_{t-1} + \mathbf{b}_o\right)\\
\mathbf{g}_t &= \tanh\left(\mathbf{W}_{xg}^{\top}\mathbf{x}_t + \mathbf{W}_{hg}^{\top}\mathbf{h}_{t-1} + \mathbf{b}_g\right)\\
\mathbf{c}_t &= \mathbf{f}_t \otimes \mathbf{c}_{t-1} + \mathbf{i}_t \otimes \mathbf{g}_t\\
\mathbf{y}_t &= \mathbf{h}_t = \mathbf{o}_t \otimes \tanh\left(\mathbf{c}_t\right)
\end{aligned}
\tag{5.21}
$$

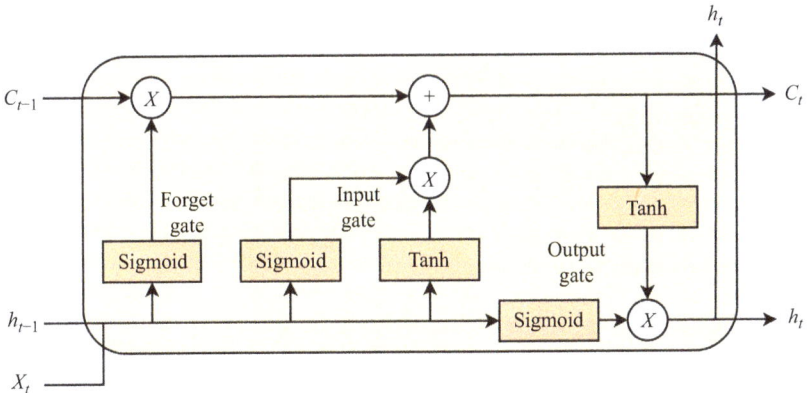

Figure 5.18   The LSTM unit

Here, $\mathbf{i}_t$, $\mathbf{f}_t$, $\mathbf{o}_t$, and $\mathbf{g}_t$ are the input, the forget, the output, and the update gate, respectively. The weight matrices for the connection to the input $\mathbf{x}_t$ are $\mathbf{W}_{xi}$, $\mathbf{W}_{xf}$, $\mathbf{W}_{xo}$, and $\mathbf{W}_{xg}$, while $\mathbf{W}_{hi}$, $\mathbf{W}_{hf}$, $\mathbf{W}_{ho}$, and $\mathbf{W}_{hg}$ are the weight matrices of each layers to the previous step output $\mathbf{h}_{t-1}$. Finally, $\mathbf{b}_i$, $\mathbf{b}_f$, $\mathbf{b}_o$, and $\mathbf{b}_g$ are the bias terms for each of the four layers. An LSTM memory cell allows the encoder to retain the dependencies across multiple data points, reducing the dimensionality of the input space.

5.5.2   Gated recurrent unit

The GRU architecture is a simplified version of the LSTM unit. In most tasks, it performs with similar accuracy, making the choice between using the LSTM and the GRU relatively undecisive, as in some tasks, one outperforms the other, and in some, the opposite is true. However, GRU is often preferred due to simplifications in design. The design of the unit can be observed in Figure 5.19.

As can be seen, the main simplifications include the merging of state vectors into a single vector h_t, the merging of the input and forget gates into a single gate controller z_t, and the removal of the output gate; hence, the whole state vector is output at all times. However, there is one new element added – the gate controller r_t that controls, which part of the previous state vector $h(t-1)$ is passed to the update gate g_t. Mathematically, the unit is governed by the equations below:

$$z_t = \sigma \left(W_{xz}^\top x_t + W_{hz}^\top h_{t-1} + b_z \right)$$
$$r_t = \sigma \left(W_{xr}^\top x_r + W_{hr}^\top h_{t-1} + b_r \right)$$
$$g_t = \tanh \left(W_{xg}^\top x_t + W_{hg}^\top (r_t \otimes h_{t-1}) + b_g \right)$$
$$c_t = z_t \otimes h_{t-1} + (1 - z_t) \otimes g_t$$

(5.22)

5.5.3   Examples in the literature

LSTMs and GRUs are widely adopted by researchers in PV forecasting due to their robust usability in the modeling of time-series data. In [63], Sala *et al.* argued that LSTMs can be a particularly useful tool for the prediction of PV power output, especially when high quantities of high-quality data are available. In fact, in [64], the importance of having high-quality data has been studied and proven. In their study, the authors compared RNN-based prediction with classic ML-based approaches, such

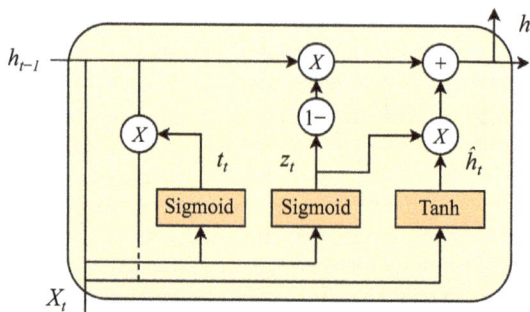

Figure 5.19   The GRU unit

as linear regression, random forest, and a basic MLP. The LSTM proves to be particularly accurate for longer time horizon predictions, as it models better the trends governing the sequence.

In [65], Jayalakshimi *et al.* proposed a novel multitask learning algorithm, implemented with the LSTM neural network, which demonstrated highly consistent performance for all time-scale forecasting with improved metric results. The model is trained using an hourly solar irradiance dataset, and forecasting is performed on hourly, daily, and weekly scales with resource-sharing capabilities. The hyperparameter estimation of the proposed LSTM model, using a hybrid chicken swarm optimizer and grey wolf optimization algorithm, resulted in improved metric results for single-time-scale forecasting and multi-time-scale forecasting.

In [66], the authors proposed a novel CNN-boosted LSTM for improved feature extraction from the input data for short-term GHI prediction. They tested the proposed model against a variety of benchmarks and achieved a significant improvement in terms of accuracy for univariate time-series forecasting. The proposed hybrid LSTM–CNN model is a reliable alternative for short-term GHI prediction due to its high predictive accuracy under diverse climatic, seasonal, and sky conditions.

In [29], the authors proposed a similar model with an LSTM–CNN hybrid for improved extraction of the spatiotemporal features from the dataset. The model is trained with meteorological data from 23 locations in California, USA. Performance is evaluated for a year, four seasons, and under three sky conditions, showing high predictive accuracy for diverse weather and climate conditions. The proposed model shows forecast skill scores in the range of approximately 37%–45%.

In [28], the authors also proposed a hybrid WPD–CNN–LSTM–MLP multi-branch model for hybrid DL-based hourly solar irradiance forecasting. Unlike typical approaches, the model additionally uses wavelet decomposition to extract frequency domain features from the GHI time series. the proposed WPD–CNN–LSTM–MLP model outperforms others in prediction performance, the minimum RMSE is 32.1 W/m^2, the minimum nRMSE is 15.5%, the maximum *s* is 0.44, and the maximum FS is 0.66, benchmarked against the general persistence model. The conducted study also demonstrates the potential of utilizing branching and multi-input networks to improve the accuracy of the prediction, especially in the very short term.

5.5.4   Case study

The case study for the RNN revisits the case study presented in the first section with the ANN. It again focuses on forecasting GHI using a time-series-based approach. Here, an additional RNN-based model will be tested and compared with the same benchmarks to give the reader an idea of the advantages and disadvantages of RNN implementations. Specifically, the case study will be centered on an LSTM-based model that is trained to predict the GHI values 6 h ahead using the 24-h history of the recorded weather parameters.

5.5.4.1   Dataset

As the idea behind this case study is to replicate the same procedure presented before, the dataset used is exactly the same as described in Section 5.3.5. The transformations applied, data cleaning procedures, and outlier elimination are also performed in the same way. This is done to ensure an identical testing environment and justly compare the performance of two different DL models.

5.5.4.2   Proposed model

Here, the actual deviations from the previous case study start to show. The implemented model is based on LSTM cells, discussed in detail in Section 5.5.1. The proposed model consists of an input layer, two hidden layers, and a single output layer. The hidden layers are composed entirely of LSTM units, 96 each, and have the same activation function – hyperbolic tangent to eliminate the possibility of exploding and vanishing gradients in training. The output layer is a standard fully connected layer with 24 neurons that correspond to the length of the output sequence. These exact parameters are a result of a hyperparameter optimization procedure to achieve the best possible results and avoid excessive use of resources (as RNNs take significantly more computational burden to train). Moreover, similarly to the ANN, and the CNN, early stopping is implemented to avoid overfitting. In total, the proposed LSTM has 117,528 parameters.

5.5.4.3   Results

The model's performance is assessed using MAE on unseen test data. Its ability to predict near-future (6 h) GHI time series, based on multiparameter weather inputs, is evaluated. Figure 5.20 illustrates example predictions for different time

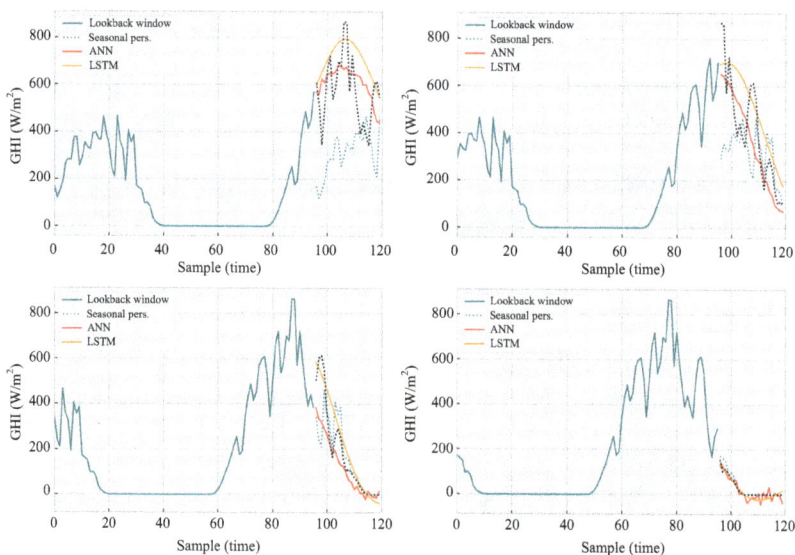

Figure 5.20   Result examples of the LSTM for GHI forecast with additional comparison to the previously tested model. As can be seen, the LSTM follows the general trend quite accurately, resulting in an accurate prediction of GHI 6 h ahead; moreover, it exhibits a smoothing effect, resulting in a trend similar to theoretical GHI values.

Table 5.4   Detailed results of the RNN for PV forecasting

Period	MAE$_{pred}$ ANN W/m^2	MAE$_{pred}$ RNN W/m^2	MAE$_{pers}$ %	FS ANN	FS rNN
1	116.49	146.07	275.20	58	FS: 47
2	105.12	135.23	158.43	34	15
3	68.50	58.26	68.90	0.6	15
4	16.97	10.82	7.48	−130	−45

periods within the same day. In these examples, seasonal persistence underperforms significantly due to varying cloud cover conditions.

In more detail, the exact numerical results are presented in Table 5.4. The proposed model outperforms the benchmark metric for most periods, except for the final one. Interestingly, the error generally decreases throughout the day, becoming negligible near sunset. Predicting the GHI peak (and thus PV generation) in the morning is stochastically challenging compared to predicting afternoon slopes. The LSTM's smoothing effect improves the performance in the late hours of the afternoon, when the GHI values approach 0.

Finally, the overall performance of the model on the entire test dataset can be evaluated. In this case, the overall MAE is equal to 92.48 W/m^2, the seasonal persistence MAE is equal to 232.05 W/m^2, and the forecast skill is 60.1%. The designed RNN represents a significant improvement over the simple benchmark method; however, it does not outperform the simple ANN. Not always a bigger model.

5.6   Conclusions

DL is a subset of AI that focuses on constructing and optimizing complex mathematical models. These models can learn and recognize patterns in complex real-world phenomena, often requiring substantial data and computational resources. Despite these demands, DL's flexibility has made it a popular choice among researchers due to its broad range of applications.

This chapter explored the application of DL models to PV forecasting. It begins by defining the forecasting task, its role within the power system, and the available data types. Subsequently, it provides a gentle introduction to DL fundamentals, model typologies, and common issues such as overfitting.

The chapter then delves into three primary DL model categories, each with its own theoretical explanations, state-of-the-art applications, and a real-world case study:

- Artificial neural networks (ANNs): The simplest but highly versatile DL models, ANNs or FCNNs, are explored in the first section.

- Image-based forecasting: This section focuses on CNNs, which are used to extract features from the sky or satellite images for PV forecasting. Convolutional operations are integral to this approach.
- Temporal data and sequence analysis: The final section highlights recurrent models, such as RNNs, LSTMs, and GRUs, which are adept at recognizing both short-term and long-term patterns in time-series data.

By understanding these DL models and their applications, researchers can advance PV forecasting capabilities and contribute to a more sustainable and efficient energy grid.

References

[1]   IEA. Global EV Outlook 2022 – Securing supplies for an electric future. *Global EV Outlook 2022*, p. 221, 2022.

[2]   IEA. Renewables 2023 – Analysis and forecast to 2028. *Renewables 2023*, p. 221, 2024.

[3]   Ahmed R., Sreeram V., Mishra Y., and Arif M. D. A review and evaluation of the state-of-the-art in PV solar power forecasting: Techniques and optimization. *Renewable and Sustainable Energy Reviews*, 124:109792, 2020.

[4]   Sobri S., Koohi-Kamali S., and Rahim N. A. Solar photovoltaic generation forecasting methods: A review. *Energy Conversion and Management*, 156:459–497, 2018.

[5]   Kaur A., Nonnenmacher L., Pedro H. T. C., and Coimbra C. F. M. Benefits of solar forecasting for energy imbalance markets. *Renewable Energy*, 86:819–830, 2016.

[6]   Sinsel S. R., Riemke R. L., and Hoffmann V. H. Challenges and solution technologies for the integration of variable renewable energy sources—a review. *Renewable Energy*, 145:2271–2285, 1 2020.

[7]   Hyndman R. J. *Forecasting: Principles and Practice*. OTexts, 2018.

[8]   Rajagukguk R. A., Ramadhan R. A. A., and Lee H.-J. A review on deep learning models for forecasting time series data of solar irradiance and photovoltaic power. *Energies*, 13(24):6623, 2020.

[9]   Das U. K., Tey K. S., Seyedmahmoudian M., *et al.* Forecasting of photovoltaic power generation and model optimization: A review. *Renewable and Sustainable Energy Reviews*, 81:912–928, 2018.

[10]   Raza M. Q., Nadarajah M., and Ekanayake C. On recent advances in PV output power forecast. *Solar Energy*, 136:125–144, 2016.

[11]   Dutta S., Li Y., Venkataraman A., *et al.* Load and renewable energy forecasting for a microgrid using persistence technique. vol 143. Amsterdam: Elsevier, pp 617–622, 2017.

[12]   Dendramis Y., Kapetanios G., and Marcellino M. A similarity-based approach for macroeconomic forecasting. Journal of the Royal Statistical Society Series A: Statistics in Society, 183(3):801–827, 2020.

[13]   Gilboa I., Lieberman O., and Schmeidler D. A similarity-based approach to prediction. *Journal of Econometrics*, 162(1):124–131, 2011.

[14]   Mu Q., Wu Y., Pan X., Huang L., and Li X. Short-term load forecasting using improved similar days method. *2010 Asia-Pacific Power and Energy Engineering Conference*, pages 1–4. IEEE, 2010.

[15]   Antonanzas J., Osorio N., Escobar R., *et al.* Review of photovoltaic power forecasting. *Solar Energy*, 136:78–111, 2016.

[16]   Dolara A., Leva S., and Manzolini G. Comparison of different physical models for PV power output prediction. *Solar Energy*, 119:83–99, 2015.

[17]   Brabec M., Pelikán E., Krc P., Eben K., Maly M., and Juruš P. A coupled model for energy production forecasting from photovoltaic farms. *ES1002: Workshop March 22nd–23rd*, 2011.

[18]   Yona A., Senjyu T., Saber A. Y., *et al.* Application of neural network to 24-hour-ahead generating power forecasting for PV system. *2008 IEEE Power and Energy Society General Meeting-Conversion and Delivery of Electrical Energy in the 21st Century*, pp 1–6. IEEE, 2008.

[19]   Palacios L. E. O., Guerrero V. B., and Ordoñez H. Machine learning for solar resource assessment using satellite images. *Energies*, 15(11), 2022.

[20]   Shi J., Lee W.-J., Liu Y., Yang Y., and Wang P. Forecasting power output of photovoltaic systems based on weather classification and support vector machines. *IEEE Transactions on Industry Applications*, 48(3):1064–1069, 2012.

[21]   Abiodun O. I., Jantan A., Omolara A. E., *et al.* State-of-the-art in artificial neural network applications: A survey. *Heliyon*, 4:938, 2018.

[22]   Stuart Russell and Peter Norvig. *Artificial Intelligence: A Modern Approach.* Prentice Hall Press, USA, 3rd edition, 2009.

[23]   Pedro H. T. C. and Coimbra C. F. M. Assessment of forecasting techniques for solar power production with no exogenous inputs. *Solar Energy*, 86:2017–2028, 2012.

[24]   Goodfellow I., Bengio Y., and Courville A. *Deep Learning.* Cambridge: MIT Press, 2016.

[25]   Kumari P. and Toshniwal D. Deep learning models for solar irradiance forecasting: A comprehensive review. *Journal of Cleaner Production*, 318:128566, 2021.

[26]   Wang H., Lei Z., Zhang X., Zhou B., and Peng J. A review of deep learning for renewable energy forecasting. *Energy Conversion and Management*, 198:111799, 2019.

[27]   Jayalakshmi N. Y., Shankar R., Subramaniam U., *et al.* Novel multi-time scale deep learning algorithm for solar irradiance forecasting. *Energies*, 14(9), 2021.

[28]   Huang X., Li Q., Tai Y., *et al.* Hybrid deep neural model for hourly solar irradiance forecasting. *Renewable Energy*, 171:1041–1060, 2021.

[29]   Kumari P. and Toshniwal D. Long short term memory–convolutional neural network based deep hybrid approach for solar irradiance forecasting. *Applied Energy*, 295:117061, 2021.

[30]   Moncada A., Richardson W., and Vega-Avila R. Deep learning to forecast solar irradiance using a six-month UTSA skyimager dataset. *Energies*, 11(8), 2018.

[31]   Ogliari E., Sakwa M., and Cusa P. Enhanced convolutional neural network for solar radiation nowcasting: All-sky camera infrared images embedded with exogeneous parameters. *Renewable Energy*, 221:119735, 11 2023.

[32]   Zhen Z., Liu J., Zhang Z., *et al.* Deep learning based surface irradiance mapping model for solar PV power forecasting using sky image. *IEEE Transactions on Industry Applications*, 56(4):3385–3396, 2020.

[33]   Rajagukguk R. A., Kamil R., and Lee H.-J. A deep learning model to forecast solar irradiance using a sky camera. *Applied Sciences*, 11(11):5049, 2021.

[34]   Son Y., Yoon Y., Cho J., and Choi S. Cloud cover forecast based on correlation analysis on satellite images for short-term photovoltaic power forecasting. *Sustainability*, 14(8), 2022.

[35]   Pérez E., Pérez J., Segarra-Tamarit J., and Beltran H. A deep learning model for intra-day forecasting of solar irradiance using satellite-based estimations in the vicinity of a PV power plant. *Solar Energy*, 218:652–660, 2021.

[36]   Paletta Q., Arbod G., and Lasenby J. Omnivision forecasting: Combining satellite and sky images for improved deterministic and probabilistic intra-hour solar energy predictions. *Applied Energy*, 336:120818, 2023.

[37]   Ogliari E., Gandelli A., Grimaccia F., Leva S., and Mussetta M. Neural forecasting of the day-ahead hourly power curve of a photovoltaic plant. *2016 International Joint Conference on Neural Networks (IJCNN)*, pp 654–659. IEEE, 2016.

[38]   Dolara A., Grimaccia F., Leva S., Mussetta M., and Ogliari E. Comparison of training approaches for photovoltaic forecasts by means of machine learning. *Applied Sciences*, 8:228, 2 2018.

[39]   Ogliari E., Dolara A., Manzolini G., and Leva S. Physical and hybrid methods comparison for the day ahead PV output power forecast. *Renewable Energy*, 113:11–21, 2017.

[40]   Leva S., Nespoli A., Pretto S., Mussetta M., and Ogliari E. G. C. PV plant power nowcasting: A real case comparative study with an open access dataset. *IEEE Access*, 8:194428–194440, 2020.

[41]   Leva S., Dolara A., Grimaccia F., Mussetta M., and Ogliari E. Analysis and validation of 24 hours ahead neural network forecasting of photovoltaic output power. *Mathematics and Computers in Simulation*, 131:88–100, 2017.

[42]   Nespoli A., Ogliari E., Leva S., *et al.* Day-ahead photovoltaic forecasting: A comparison of the most effective techniques. *Energies*, 12:1621, 2019.

[43]   Moncada A., Richardson W., and Vega-Avila R. Deep learning to forecast solar irradiance using a six-month UTSA skyimager dataset. *Energies*, 11(8), 2018.

[44]   Son Y., Yoon Y., Cho J., and Choi S. Cloud cover forecast based on correlation analysis on satellite images for short-term photovoltaic power forecasting. *Sustainability*, 14(8), 2022.

[45]   Matteri A., Ogliari E., and Nespoli A. Enhanced day-ahead PV power forecast: Dataset clustering for an effective artificial neural network training. *The 7th International conference on Time Series and Forecasting*, p. 16. MDPI, 2021.

[46] Y. Lecun, L. Bottou, Y. Bengio, and P. Haffner. Gradient-based learning applied to document recognition. *Proceedings of the IEEE*, 86(11):2278–2324, 1998.

[47] O'Shea K. *An introduction to convolutional neural networks*. arXiv:1511.08458, 2015.

[48] A. Kazantzidis, P. Tzoumanikas, A. F. Bais, S. Fotopoulos, and G. Economou. Cloud detection and classification with the use of whole-sky ground-based images. *Atmospheric Research*, 113:80–88, 9 2012.

[49] Schmidt T., Kalisch J., Lorenz E., and Heinemann D. Evaluating the spatiooral performance of sky-imager-based solar irradiance analysis and forecasts. *Atmospheric Chemistry and Physics*, 16:3399–3412, 2016.

[50] Shi C., Wang C., Wang Y., and Xiao B. Deep convolutional activations-based features for ground-based cloud classification. *IEEE Geoscience and Remote Sensing Letters*, 14:816–820, 2017.

[51] Zhang J., Liu P., Zhang F., and Song Q. Cloudnet: Ground-based cloud classification with deep convolutional neural network. *Geophysical Research Letters*, 45:8665–8672, 2018.

[52] Ogliari E., Nespoli A., Collino E., and Ronzio D. Cloud-base height estimation based on cnn and all sky images. *ITISE 2022*, p. 5. MDPI, 2022.

[53] Sun Y., Szucs G., and Brandt A. R. Solar PV output prediction from video streams using convolutional neural networks. *Energy and Environmental Science*, 11:1811–1818, 2018.

[54] Pothineni D., Oswald M. R., Poland J., and Pollefeys M. KloudNet: Deep learning for sky image analysis and irradiance forecasting. In: Brox T, Bruhn A, and Fritz M (eds.), *Pattern Recognition. GCPR 2018*. Lecture Notes in Computer Science, Springer, Cham, pp. 535–551, 2019.

[55] Perez R., Ineichen P., Moore K., *et al.* A new operational model for satellite-derived irradiances: Description and validation. *Solar Energy*, 73(5):307–317, 2002.

[56] Wang Y., Liu D., Xie W., *et al.* Day and night clouds detection using a thermal-infrared all-sky-view camera. *Remote Sensing*, 13:1852, 2021.

[57] Thurairajah B. and Shaw J. A. Cloud statistics measured with the infrared cloud imager (ICI). *IEEE Transactions on Geoscience and Remote Sensing*, 43:2000–2007, 2005.

[58] Bertin C., Cros S., Saint-Antonin L., and Schmutz N. Prediction of optical communication link availability: real-time observation of cloud patterns using a ground-based thermal infrared camera. *Proceedings of SPIE*, 9641:96410A, 2015.

[59] Liandrat O., Braun A., Buessler E., *et al.* Sky-imager forecasting for improved management of a hybrid photovoltaic-diesel system, 2018. *Proceedings of 3rd International Hybrid Power Systems*, Tenerife, Spain, Available: https://reun iwatt.com/wp-content/uploads/sites/3/2018/09/hybridPV-Dieselsimulateur_ Reuniwatt.pdf

[60] Niccolai A., Orooji S., Matteri A., Ogliari E., and Leva S. Irradiance nowcasting by means of deep-learning analysis of infrared images. *Forecasting*, 4:338–348, 3 2022.

[61]   Hochreiter S. and Schmidhuber J. Long short-term memory. *Neural Computation*, 9(8):1735–1780, 1997.

[62]   Wang S. and Jiang J. Learning natural language inference with LSTM. *2016 Conference of the North American Chapter of the Association for Computational Linguistics: Human Language Technologies, NAACL HLT 2016 - Proceedings of the Conference*, pp. 1442–1451, 2016.

[63]   Sala S., Amendola A., Leva S., Mussetta M., Niccolai A., and Ogliari E. Comparison of data-driven techniques for nowcasting applied to an industrial-scale photovoltaic plant. *Energies*, 12:4520, 11 2019.

[64]   Nespoli A., Ogliari E., Pretto S., Gavazzeni M., Vigani S., and Paccanelli F. Data quality analysis in day-ahead load forecast by means of LSTM. In *2020 IEEE International Conference on Environment and Electrical Engineering and 2020 IEEE Industrial and Commercial Power Systems Europe (EEEIC/I&CPS Europe)*, IEEE. pp 1–5, 2020.

[65]   Jayalakshmi Y., Subramaniam U., Baranilingesan I., Karthick A., Rahim R., and Ghosh A. Novel multi-time scale deep learning algorithm for solar irradiance forecasting. *Energies*, 14:2404, 2021.

[66]   Jalali S. M. J., Ahmadian S., Kavousi-Fard A., Khosravi A., and Nahavandi S. Automated deep CNN-LSTM architecture design for solar irradiance forecasting. *IEEE Transactions on Systems, Man, and Cybernetics: Systems*, 52(1):54–65, 2021.

Chapter 6

Hybrid and ensemble models for solar energy forecast

Emanuele Ogliari[1], Silvana Matrone[1],
Binh Nam Nguyen[1] and Sonia Leva[1]

Hybrid forecasting models explore a cooperative approach to enhance forecasting accuracy by combining diverse modeling paradigms, including statistical, artificial intelligence, and physics-based models. This strategy leverages the strengths of each model type to overcome individual limitations, ultimately achieving superior predictive performance. The discussion encompasses the integration of time-tested statistical methods such as auto-regressive models with the adaptability and pattern recognition capabilities of AI models, creating a comprehensive framework. Furthermore, the incorporation of physics-based models adds domain-specific knowledge and constraints to the forecasting process. This chapter not only highlights the theoretical underpinnings of hybrid models but also provides practical insights into their implementation, emphasizing their relevance in improving accuracy across a range of forecasting applications. A case study is provided that demonstrates the application of a hybrid convolutional neural network (CNN) and long short-term memory (LSTM) model, utilizing infrared all-sky imager (ASI) data to improve short-term photovoltaic (PV) forecasting.

PV power forecasting is critical for integrating solar energy into power grids, enabling efficient energy resource management and reducing fossil fuel reliance. Given solar power's intermittent nature due to varying weather conditions, accurate forecasting is essential for grid stability, optimized power flows, and reliable energy supply. PV forecasting involves predicting solar energy production over short-, medium-, and long-term horizons based on irradiance, temperature, and historical generation data. Traditionally, this uses physical models relying on mathematical equations and solar radiation patterns or statistical models using historical data to infer trends [1]. Recently, hybrid models have gained traction in PV forecasting to improve accuracy and address stand-alone model limitations. Hybrid forecasting models integrate physical models with advanced machine learning (ML) techniques, leveraging both approaches' strengths [2]. Physical models, based on deterministic principles, excel at capturing predictable solar irradiance and atmospheric patterns but struggle with weather changes' nonlinear and stochastic nature. Conversely, ML

---

[1]Department of Energy, Politecnico di Milano, Italy

models adapt to new data and complex patterns but may overfit or underperform with inadequate training data or unseen conditions.

The hybrid approach seeks to combine these methodologies to create a more resilient and adaptive forecasting framework. By incorporating both deterministic physical components and data-driven ML techniques, hybrid models can achieve a balance between robustness and flexibility, providing more accurate and reliable forecasts across various time scales and weather scenarios. For instance, hybrid models often use artificial neural networks (ANNs), support vector machines (SVMs) [3], or ensemble learning techniques to adaptively learn from historical data, while physical models provide the foundational understanding of solar dynamics [4].

Classical hybrid modeling literature [5] typically begins by examining the structure and underlying assumptions of mechanistic models when integrating physics-based and data-driven approaches in different fields. In Figure 6.1, the traditional hybrid model architecture, representing a combination of the two main approaches, is depicted. However, hybridization can refer to any possible combination of two different models (possibly also belonging to the same class of approach) combined to achieve an enhanced result. It is noted that adjusting forecasts is quite different from a hybrid modeling approach, which employs two or more models at various steps to develop a single forecasting model [6].

This chapter will explore the general framework of PV forecasting, examining how hybrid models are constructed and the advantages they offer in terms of performance and adaptability. Various types of hybrid models, including those that combine physical solar radiation models with ML algorithms [7] and those that employ optimization techniques to fine-tune forecasting outputs will be discussed. Through case studies and comparative analyses, it will be demonstrated how hybrid models are transforming PV forecasting, making it a vital component of modern renewable energy management systems. Additionally, key challenges will be addressed, such as data integration, model interoperability, and the computational complexity involved in hybrid forecasting, offering insights into future trends and innovations in this evolving field [8]. Finally, the chapter also presents a case study focused on enhancing short-term PV forecasting by integrating CNN and LSTM models, using infrared ASI data. By incorporating LSTM into the hybrid approach, the model aims to more effectively capture temporal dependencies in solar irradiance, thereby improving accuracy and performance under rapidly changing weather conditions.

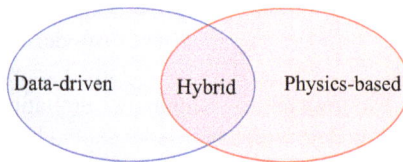

Figure 6.1   Classical hybrid approach viewed as a combination of data-driven and physics-based models

6.1 Hybrid models

A hybrid model refers to a system that integrates different models, methodologies, or technologies to create a more robust and effective solution. These models are widely applied across diverse domains, such as ML, finance, engineering, and scientific research, where combining complementary approaches can lead to more accurate and reliable outcomes. In the realm of PV energy generation forecasting, hybrid models play a crucial role by combining different techniques to improve prediction accuracy and reliability. The core principle behind these approaches is to harness the unique strengths of various forecasting methods while compensating for their individual weaknesses. For instance, numerical weather prediction (NWP) models provide critical atmospheric data for solar irradiance forecasts by analyzing broad meteorological patterns. While these models are essential for understanding general weather trends, they often lack the precision needed to account for local variations that significantly influence PV energy production. To address this, statistical models can be integrated into the hybrid system to refine these predictions by leveraging local solar irradiance patterns and historical data from specific PV installations. This dual approach allows for a more precise and location-specific forecast, as local environmental factors can vary widely from generalized predictions. By combining global atmospheric data with locally calibrated models, hybrid models deliver more nuanced and accurate predictions of PV output, optimizing energy management in dynamic conditions [9].

6.1.1 The principles at the base of hybrid models

Combining diverse forecast models has emerged as a crucial strategy for improving the accuracy and reliability of predictions, especially in renewable energy generation. By integrating different forecasting models, this approach capitalizes on the complementary strengths inherent in various methodologies, thereby addressing the biases and variances that individual models may introduce [10,11]. For example, while statistical models excel at capturing historical patterns, ML techniques can adeptly identify complex, nonlinear relationships within the data. This interplay allows forecasters to better capture the inherent variability in energy production, leading to a more comprehensive understanding of potential outcomes [12]. Ensemble learning techniques, such as bagging and boosting, further exemplify the effectiveness of this integration by combining the outputs of multiple models to enhance overall performance. Bagging reduces variance by training multiple models on different subsets of the data, while boosting focuses on sequentially improving performance by addressing the errors of previous models. The principles underlying the combination of forecasts highlight the importance of a multifaceted approach, enabling stakeholders to navigate the complexities and uncertainties associated with energy production and consumption more effectively. As the demand for renewable energy increases, such integrated forecasting strategies will be essential for optimizing grid management and enhancing decision-making processes in energy markets.

6.1.2   Building hybrid forecasting models

The creation of hybrid forecasting models typically begins with comprehensive data collection. This involves gathering extensive datasets, including historical solar irradiance measurements, and meteorological variables, such as temperature, humidity, wind speed, and actual PV output data. These datasets are sourced from diverse locations, including satellite imagery, ground-based weather stations, and historical performance logs from solar power plants. Such diverse data is necessary to capture the myriad factors influencing solar energy generation accurately.

Once the data is collected, the next step involves selecting appropriate models, often encompassing a combination of techniques. For instance, regression analysis can be used to identify correlations between different variables. At the same time, ML algorithms like ANNs and SVMs analyze complex nonlinear relationships within the data. By employing such a diverse array of methodologies, researchers can develop models that not only reflect historical patterns but also adapt to changing conditions.

Integration is a key aspect of hybrid models, achieved through techniques such as ensemble methods and data fusion. Ensemble methods combine the outputs of multiple models, enhancing overall accuracy by reducing the likelihood of error associated with any single model. Data fusion integrates data from multiple sources to improve the input quality for the models. This unified approach maximizes the models' forecasting capabilities, ultimately leading to more precise short- and medium-term forecasts for PV energy output. These forecasts are essential for grid management, as accurate predictions enable operators to balance supply and demand effectively.

The training and validation phases are critical to the hybrid model's development. The model is trained using historical data, optimizing its parameters to improve prediction accuracy. Validation techniques, such as cross-validation, are employed to ensure that the model performs robustly against unseen data. This iterative process not only enhances the model's reliability but also helps fine-tune its predictive capabilities over time.

Moreover, one of the defining characteristics of hybrid forecasting models is their adaptability. As new data becomes available, these models can continuously improve, refining their algorithms based on performance metrics and adjusting to the dynamic nature of renewable energy generation. This adaptability is crucial for addressing the challenges posed by the intermittent nature of solar energy and for ensuring that forecasts remain relevant and accurate in real-time applications.

Overall, hybrid forecasting models are not merely advantageous; they are essential for optimizing the management and integration of renewable energy resources into modern power systems. By improving forecasting accuracy, these models facilitate more effective energy trading, enhance grid reliability, and ultimately contribute to a more sustainable energy future. The growing reliance on renewable sources necessitates innovative forecasting techniques, making hybrid approaches indispensable for navigating the complexities of energy generation in a rapidly evolving landscape.

6.1.3   Hybrid forecast for PV production

Hybrid forecasting models have gained prominence in PV energy generation due to their ability to enhance prediction accuracy by integrating multiple forecasting techniques. These techniques combine different forecasting methods to improve the accuracy and reliability of solar energy predictions. Due to the inherent variability of solar power generation, which is influenced by weather conditions, temperature, and irradiance, hybrid models are particularly effective in addressing the limitations of individual approaches.

Hybrid models exploit the strengths of different approaches, leading to more accurate and reliable PV forecasts. This is crucial for addressing the nonlinear and complex relationship between weather variables and PV output. They can adapt to sudden weather changes or shifts in patterns, as ML models continuously learn from new data while physical models provide stability based on known principles. Accurate PV forecasting is essential for optimizing grid operations, particularly in regions with a high share of renewable energy. Hybrid models help grid operators manage supply and demand more effectively by providing reliable power predictions.

Hybrid models for PV forecasting combine multiple methodologies to enhance prediction accuracy, leveraging the strengths of statistical, physical, and ML approaches. Statistical and ML hybrid models integrate traditional techniques to capture historical patterns with ML algorithms that model complex, nonlinear relationships. Physical and data-driven hybrid models merge physical models, which use meteorological data and solar radiation physics, with data-driven models to adjust predictions based on real-world conditions. These models also often combine short-term and long-term forecasting techniques, integrating immediate predictions with longer-term weather forecasts. Ensemble methods, such as bagging and boosting, further improve forecast accuracy by combining the outputs of multiple models, reducing variance, and correcting prediction errors.

For instance, Brester *et al.* [13] explored the integration of NWP data and real-time weather observations as inputs to enhance forecast accuracy. This study demonstrated significant improvements in forecast accuracy through this hybrid approach, which harnesses the strengths of both data-driven neural networks and physics-based weather forecasting models. The integration allows for better handling of nonlinear relationships and variabilities in solar generation, leading to more reliable forecasts.

In their research, Alrashidi *et al.* [14] utilized a hybrid approach that combines support vector regression (SVR) with time series analysis to enhance the prediction of solar energy output. This model effectively captured both nonlinear relationships and seasonal variations, which are critical for accurate short-term forecasts. Their findings suggest that this hybrid approach can significantly improve the accuracy of PV output predictions.

Furthermore, Massucco *et al.* [15] showcased the effectiveness of ensemble learning techniques that incorporated clear-sky models, which estimate solar radiation under ideal weather conditions, an ensemble of ANNs, and a decision tree algorithm. Their approach further enhanced overall forecasting performance by aggregating the strengths of different models, thus minimizing prediction errors. This method of combining various learning algorithms allows for a more robust model that can adapt to diverse datasets and conditions.

Belmahdi *et al.* [16] highlighted the successful integration of autoregressive integrated moving average (ARIMA) models with ML techniques. By utilizing ARIMA's capabilities in capturing historical trends and cycles alongside ML's proficiency in handling complex, nonlinear patterns, their study illustrated a marked improvement in prediction capabilities for solar energy generation. This hybrid strategy effectively bridges the gap between traditional statistical approaches and modern ML methods.

A new approach termed physical hybrid artificial neural network (PHANN) is proposed in [7], which combines the strengths of ANNs with physical models such as clear sky curves to enhance forecasting accuracy. This hybrid model aims to address the intrinsic errors typically encountered in both purely physical and purely data-driven methods, ultimately improving the reliability of energy forecasts crucial for the integration of renewable energy sources into smart grids.

Overall, these examples illustrate how hybrid models can effectively combine various methodologies, significantly improving the reliability and accuracy of PV forecasting. The continued evolution and application of these hybrid approaches are crucial in meeting the growing demand for precise energy forecasts, which are essential for effective grid management and the integration of renewable energy sources into existing energy systems. As the renewable energy landscape becomes increasingly complex, the use of hybrid forecasting models will likely expand, contributing to a more sustainable energy future.

6.2   Ensemble models

Integration strategies for hybrid forecasting models are essential for enhancing the accuracy and reliability of predictions, particularly in PV energy generation. Integration strategies for hybrid forecasting models play a crucial role in enhancing the accuracy and reliability of predictions, particularly in the context of PV energy generation. Given the inherent variability and complexity of solar energy generation influenced by factors, such as weather conditions, geographic location, and system performance, it is essential to employ advanced integration strategies that leverage the strengths of multiple forecasting models. Ensemble methods refer to a class of techniques in ML that combine the predictions of multiple models to enhance overall performance. These methods often result in better accuracy, robustness, and generalization compared to relying on a single model. The key idea behind ensemble methods is that by aggregating the outputs of several models, the weaknesses of individual models can be compensated for by the strengths of others, allowing for more reliable predictions. According to [4], ensemble methods can be categorized as competitive or cooperative. In competitive ensemble forecasting, different predictors are trained individually using various datasets or parameters, and the final prediction is obtained by averaging their decisions. In cooperative ensemble forecasting, the prediction task is divided into subtasks, with specific predictors selected for each, and the final decision is the combined output of all base predictors. Different cooperative hybrid models combine traditional statistical techniques with neural networks to enhance forecasting performance. One common approach integrates ARIMA with LSTM models, while another combines SARIMA, an extension of ARIMA that accounts for seasonality, with LSTM. Other models use neural networks like back propagation neural networks (BPNN) or generalized regression neural networks (GRNN) alongside ARIMA. Additionally, some hybrid models incorporate wavelet transform decomposition to separate linear and nonlinear components of a time series before making predictions, improving the overall accuracy of forecasts [17]. Competitive ensemble methods can be broadly categorized into three main types: averaging, bagging, boosting, and stacking. This diversity of models ensures that errors made by one model can be mitigated by others, leading to a more balanced and stable prediction system. As a result, ensemble models are particularly valuable when working with complex datasets or when individual models have limitations in handling certain types of data. Ensemble learning is highly effective in addressing common challenges in ML, such as overfitting, where a model performs well on training data but poorly on unseen data, underfitting, where a model is too simple to capture underlying patterns in the data, or high variance in predictions, where model outputs are highly sensitive to fluctuations in the data [18]. Ensemble techniques leverage multiple models to smooth out these inconsistencies and offer a more accurate representation of the true underlying patterns.

6.2.1   Model averaging

Model averaging is an ensemble learning technique that enhances predictive performance by combining the outputs of multiple models to produce a more reliable and accurate prediction. The fundamental concept behind model averaging is to leverage the strengths of different models, mitigating the impact of individual model errors and leading to more stable predictions. This approach can be implemented through simple averaging, where the arithmetic mean of predictions from several models is calculated, or weighted averaging, where different weights are assigned to each model based on their performance. By doing so, model averaging increases accuracy, reduces variance, and creates a more robust framework that can better handle variability in data. The simplicity of model averaging makes it relatively easy to implement, requiring less computational effort than more complex ensemble techniques. Its applications span various domains, including finance, healthcare, and renewable energy, where it is particularly valuable in improving forecasts for stock prices, patient outcomes, and solar or wind energy production. Overall, model averaging serves as a powerful tool in ensemble learning, enhancing predictive capabilities and supporting effective decision-making in complex, data-driven environments. For instance, in their research [19], the authors explored the effectiveness of model averaging alongside various ML algorithms to predict solar irradiance. The findings demonstrated that averaging predictions significantly reduced overall forecasting errors compared to relying on individual models. This technique is particularly advantageous as it can capitalize on the strengths of diverse models while mitigating the impact of their weaknesses, thereby resulting in more reliable forecasts for solar energy generation.

In a hybrid modeling framework, various models are combined to enhance prediction performance, and assigning appropriate weights to each model is crucial for achieving optimal results. The weighting strategies chosen significantly influence the outcome by balancing the strengths and weaknesses of each model, thereby affecting key metrics, such as accuracy, bias, variance, and computational complexity [20]. Simple averaging assigns equal weights to all models, offering ease of use and minimal computational complexity. While this approach can perform reasonably well when individual models exhibit similar performance, it often leads to suboptimal accuracy because it fails to capitalize on the strengths of better-performing models. The bias introduced by equal weighting arises from poorly performing models still contributing equally to the final prediction, which can dilute the effectiveness of high-performing models. Performance-based weighting in contrast to simple averaging improves accuracy by assigning higher weights to models with stronger performance metrics, such as accuracy or precision. This approach reduces bias, as the best-performing models have a greater influence on the final forecast. However, it can increase variance if certain models that excel in specific scenarios are overly empha-sized. Additionally, the complexity of performance-based weighting increases when multiple evaluation metrics are considered, requiring careful tuning and adjustment to maintain robustness across diverse data subsets. Lee and Sun [21] presented a novel ensemble method that adjusts the contributions of different ML models based on their historical forecasting performance. The study compared various models, demon-strating that this performance-based weighting significantly enhanced forecasting reliability compared to traditional methods.

Bayesian model averaging (BMA) introduces a probabilistic framework that assigns weights based on the posterior probability of each model being the cor-rect one given the observed data. This method improves accuracy by reducing bias through the incorporation of uncertainty into the weighting process. However, BMA is computationally intensive, particularly when dealing with a large number of mod-els or complex likelihood functions. Additionally, its performance can be sensitive to prior assumptions about model correctness, necessitating careful consideration dur-ing implementation. For example, a study by [22] explored a synthetic BMA approach for solar energy forecasting. This research emphasizes how model averaging can effectively combine multiple predictions to enhance forecasting accuracy, mitigat-ing the impact of individual model errors. Specifically, the authors demonstrated that employing Bayesian model averaging with various machine-learning algo-rithms, such as LSTM networks, significantly improved the reliability of predictions compared to using any single model alone.

Weighted majority voting is an extension of the standard majority voting method. It is particularly effective in classification tasks, as it allows each model to "vote" for a class, with the final decision determined by a weighted majority. This approach can enhance accuracy, especially when models exhibit diverse strengths and weak-nesses. However, it may struggle in edge cases or regression problems, where the concept of voting may not be applicable, and careful weight tuning is required to avoid skewing results toward less reliable models. The article [23] presents a novel ensemble approach for improving power quality disturbance (PQD) classification in microgrids. The system incorporates both hydrogen-based microgrids and PV sys-tems, which generate complex PQDs due to their variable energy production. To enhance classification accuracy, multiple deep learning models (e.g., CNNs) are combined using a modified weighted majority voting algorithm, assigning higher weights to more reliable models for different disturbance types. This technique signif-icantly improves the precision and robustness of PQD detection in renewable energy setups. Error-based weighting focuses on reducing the influence of poorly performing models by assigning higher weights to those with lower errors on a validation dataset. This method typically leads to improved accuracy by effectively reducing bias. However, if certain models perform well only on specific data subsets, this weighting scheme might inadvertently increase variance as those models become overemphasized [24]. Optimization-based weighting involves solving an optimization problem to assign weights that minimize a specific cost function, such as mean squared error (MSE). This method allows for precise tuning of weights to achieve the best possible performance based on the available data. Although it generally results in high accuracy, particularly in complex forecasting scenarios, the risk of overfitting is significant, especially with small datasets or a large number of models. An interesting application of optimization weighting is presented in [25]. It features a novel method to enhance phishing website detection by utilizing particle swarm optimization (PSO) for optimizing feature weights. PSO is employed to assign optimal weights to extract features, improving the accuracy of ML classifiers in identifying phishing sites. Relevant features are identified for phishing detection, and various ML models are trained using weighted features to enhance differentiation between phishing and legitimate websites. The PSO-based approach significantly boosts accuracy and precision compared to traditional methods, demonstrating the importance of effective feature prioritization in phishing detection. In summary, each of these weighting strategies impacts forecasting performance differently. More sophisticated methods like stacked generalization, BMA, and optimization-based approaches tend to offer better accuracy and adaptability to various data conditions but come at the cost of increased complexity and computational demands. Therefore, the choice of weighting scheme should take into account the specific characteristics of the data, the diversity of the models being combined, and the desired balance between performance, robustness, and computational efficiency.

6.2.2   Bagging (Bootstrap Aggregating)

Bagging, short for Bootstrap Aggregating, is an effective ensemble technique widely used in ML to enhance the stability and accuracy of predictive models. The primary goal of bagging is to reduce variance, which is particularly beneficial in scenarios where a model may be overly sensitive to fluctuations in the training data. This technique operates by creating multiple versions of a base model, each trained on different subsets of the training dataset that are generated through a process known as bootstrapping. Bootstrapping involves random sampling with replacement, meaning that some instances from the original dataset may appear multiple times in a subset, while others may be omitted entirely. This random sampling strategy ensures that each model learns from a unique perspective of the data, effectively capturing diverse patterns and characteristics. Once the individual models are trained, each model independently generates its predictions for the task at hand. For regression tasks, the final output is typically calculated by averaging the predictions of all the models, resulting in a more stable estimate [26]. By effectively averaging out the noise and capturing a broader range of data characteristics, bagging significantly enhances the robustness of the final predictions. Additionally, bagging is particularly effective for high-variance models, such as decision trees, where small changes in the training data can lead to significant fluctuations in predictions. Overall, bagging is a powerful ensemble technique that not only improves the accuracy of machine-learning models but also enhances their reliability and generalization capabilities. A notable example of bagging is the random forest algorithm, which constructs multiple decision trees and combines their outputs for more robust predictions [27].

In [28], the authors developed a bagging ensemble model that integrates past PV output data to enhance prediction performance. A variety of base models are trained on different subsets of historical output data using a bagging technique, which helps reduce overfitting and increases robustness. The bagging ensemble method aggregates the results of these individual models to produce a more reliable forecast. This model demonstrated superior forecasting accuracy compared to individual ML models, particularly in capturing the nonlinearities and time-dependent patterns in PV power generation.

6.2.3   Boosting

Boosting is a powerful ensemble method that enhances model performance through a sequential training approach. Unlike bagging, which trains multiple models independently, boosting focuses on refining predictions by training a series of models where each new model aims to correct the errors made by its predecessor. This process involves adjusting the weights of the training data so that misclassified instances receive greater emphasis, allowing subsequent models to focus on challenging cases. The final prediction is calculated as a weighted sum of the outputs from all individual models, with more accurate models contributing more to the final decision. Techniques such as AdaBoost and Gradient Boosting are effective in transforming weak learners – models that perform slightly better than random guessing – into strong, high-performance predictors. Boosting is particularly beneficial for complex datasets, as it reduces both bias and variance, making it a versatile and widely used tool in various applications, including finance, healthcare, and natural language processing. Overall, boosting significantly improves predictive accuracy and robustness, making it an essential technique in ML [29].

The article [30] presents a hybrid approach to improve the accuracy of short-term PV power forecasting. The model integrates three techniques: similar-day selection, BP-AdaBoost, and the Grey–Markov model. BP-AdaBoost is a combination of backpropagation (BP) neural networks with the AdaBoost algorithm, enhancing the performance of the neural network by reducing prediction errors from previous iterations. Combining these methods allows the model to handle the stochastic nature of PV power generation more effectively, reducing forecasting errors and improving the reliability of predictions.

6.2.4   Stacking

Model stacking, also known as stacked generalization, is an advanced ensemble learning technique that significantly enhances predictive performance by combining multiple models into a hierarchical structure. The process begins with the training of several base models, which may include diverse algorithms such as linear regression, decision trees, support vector machines, or neural networks. Each base model is trained on the same dataset, allowing them to capture different patterns and relationships within the data independently. After training, these models generate predictions on a separate validation dataset – a portion of the original training data that was held out to prevent overfitting [31]. This step ensures that the meta-model learns from unbiased predictions. The predictions from the base models form a new dataset, with each model's output serving as a feature. This new dataset is then used to train a meta-model, which learns how to combine the predictions to produce a final output optimally. The meta-model identifies which base models are most reliable in various scenarios, effectively weighing their contributions based on performance. By leveraging the strengths of multiple algorithms, stacking improves accuracy and generalization, making it particularly effective in complex forecasting scenarios such as solar energy generation.

For example, Ardabili *et al.* [18] proposed a stacked ensemble model called Stack-ETR, which combines multiple ML techniques – random forest (RFR), extreme gradient boosting (XGBoost), and adaptive boosting (AdaBoost) – as base learners, with extra trees regressor (ETR) acting as the meta-learner. RFR helps with fitting noisy data, XGBoost captures data characteristics well, and AdaBoost reduces bias without overfitting.

By employing these integration strategies, hybrid forecasting models can achieve higher levels of accuracy and reliability, thereby providing stakeholders in the renewable energy sector with the critical insights needed for effective grid management, energy trading, and resource allocation. As the energy landscape continues to evolve, integrating diverse forecasting approaches will be paramount in optimizing the management and utilization of renewable energy resources.

6.3   Adaptive hybrid models

Adaptive hybrid models are an innovative approach in ML that integrates multiple algorithms or models while dynamically adjusting the contribution of each model based on changing conditions, such as variations in data distribution or the performance of individual models. This adaptability is crucial in applications where the underlying data or the environment may change over time, leading to variations in model performance. Adaptive hybrid models adjust the weights of individual models based on their recent performance. This practice is called dynamic weighting; for example, if a specific model excels at predicting under certain conditions, its influence on the final prediction can be increased, while underperforming models may have their weights reduced. This process helps in mitigating overfitting and improving generalization. This can be possible only when adaptive hybrid systems utilize mechanisms that allow them to learn continuously from new data. This real-time learning feature is crucial in environments where data patterns can change rapidly, such as in financial markets or online user behavior. Real-time learning ensures that the model remains relevant and effective as new data is introduced. These models often have contextual awareness [32], meaning they incorporate contextual information to inform model selection and weighting. For instance, contextual bandit algorithms can help choose which model to apply based on the current state of the environment, thus maximizing performance based on situational factors. Adaptive hybrid models can combine various modeling approaches, such as statistical methods, ML algorithms, and even deep learning techniques. This integration allows them to leverage the strengths of different models, leading to improved overall performance. These models are particularly effective in dynamic environments where conditions fluctuate [33], such as in time-series forecasting, fraud detection, and resource allocation in smart grids. Adaptive hybrid models often integrate ensemble learning techniques such as boosting, bagging, and stacking. In boosting, for instance, the algorithm places more emphasis on data points that were misclassified by previous models, adapting to the performance of each model iteratively. Stacked generalization (stacking) can also be adapted dynamically by changing the meta-model based on the performance of base learners under different conditions.

The article [34] introduces a two-stage hybrid forecasting model that integrates ARIMA, SVM, ANN, and ANFIS techniques, with a genetic algorithm (GA) used to optimally combine their outputs. The model is applied to 1-h ahead PV power predictions across three sites in Taiwan and Malaysia. Results show that the hybrid approach outperforms individual models, offering lower prediction errors and better handling of nonlinearities and variability in PV output caused by weather conditions. The study demonstrates the robustness and adaptability of GA-based hybridization in enhancing short-term solar forecasting performance. Another example can be found in [35], where a sophisticated hybrid approach is introduced. This model combines ML techniques, such as neural networks, with optimization algorithms like glowworm swarm optimization (GSO), effectively leveraging meteorological data to predict hourly energy output. The study applies the hybrid model to real-world PV and wind energy datasets, showing that this method can outperform traditional techniques, particularly when forecasting in diverse weather conditions. It adapts the weights of different forecasting models over time to reflect the performance of each method, which is particularly useful in complex, variable environments like wind and solar power generation. Its adaptive nature allows for real-time learning, dynamically adjusting model parameters and the weights of different forecasting methods based on their performance metrics and changing environmental conditions. This flexibility enables the model to continuously recalibrate predictions in response to fluctuating weather, optimize its forecasting strategy by integrating multiple models, and implement a feedback mechanism that assesses accuracy over time. Ahmed *et al.* [36] introduced a hybrid ensemble method for forecasting PV power output. The model uses LSTM networks and combines them with an adaptive weighting mechanism. This adaptive mechanism assigns varying importance to different forecasting models based on performance metrics over time. One unique aspect of the approach is the data segmentation technique, which splits input data into more homogeneous subsets. By processing these segmented data streams, the model improves its responsiveness to local variations in weather patterns and solar irradiance, which are crucial for accurate PV forecasting.

Adaptive approaches in various fields, particularly in ML and forecasting, are highly valued for their flexibility and responsiveness to changing conditions. They can adjust in real time based on incoming data, ensuring optimal performance even in dynamic environments. This adaptability is crucial in sectors like renewable energy forecasting, where factors such as weather conditions can significantly influence energy generation predictions. These models can learn continuously from new data, enabling them to recalibrate their parameters and improve accuracy. Additionally, adaptive models enhance generalization, making them effective across diverse datasets and situations. Overall, the ability to respond quickly to changes and maintain robust performance is a key advantage of adaptive methods.

6.4   Challenges for hybrid forecasting

Model compatibility is crucial for ensuring that different forecasting models can work together effectively without conflicts. Hybrid models often integrate diverse algorithms, including statistical methods, ML techniques, and neural networks, each governed by its own set of assumptions and requirements. This diversity can lead to compatibility issues, especially when attempting to combine outputs from models that operate under differing assumptions; for instance, a traditional time-series model might rely on linear relationships, while an ML model might capture more complex nonlinear patterns. Harmonizing predictions from such models can be challenging. Furthermore, models may operate on different scales or utilize distinct units, complicating the aggregation of results. Normalization techniques may be necessary to reconcile these differences, but they can introduce additional complexity and potential biases, making the ensemble less effective [37].

Interoperability is another critical aspect that deals with the ability of different systems and models to exchange and utilize information seamlessly. In hybrid forecasting scenarios, models often require data in various formats or structures, necessitating extensive preprocessing and transformation efforts. For example, one model might require continuous time-series data, while another may require it to be represented in discrete intervals. This inconsistency in data requirements can significantly hinder the implementation of hybrid systems. Additionally, hybrid forecasting often relies on various software tools and frameworks, which may not be designed to work together. The lack of integration can result in increased development time and costs, making it more challenging to implement hybrid models in practical scenarios [38].

Another significant challenge associated with hybrid forecasting is computational complexity. Combining multiple forecasting models can lead to increased computational demands, requiring more processing power and time. This complexity can make real-time forecasting difficult, especially in environments where timely predictions are crucial, such as in finance or demand forecasting for supply chains. The need for high computational resources can be a barrier to the practical application of hybrid forecasting models. This is particularly true when integrating models with high-dimensional data or those that involve complex architectures, such as deep learning models combined with statistical methods. The need for extensive hyperparameter tuning and handling of autocorrelations and cross-dependencies further complicates the process, making hybrid models computationally expensive [8].

Interpretation and transparency are also vital considerations. When multiple models are combined, understanding the final output can become complex. Stakeholders may find it difficult to discern how each model contributes to the overall forecast, which can lead to trust issues. Ensuring transparency in model selection and contribution is vital for effective communication among team members and stakeholders, promoting confidence in the forecasting process. Finally, hybrid forecasting models may inherit ensemble bias from individual models, particularly when certain models consistently perform better or worse than others. Managing this bias while ensuring that all models contribute appropriately to the final forecast can be a significant challenge. If certain models dominate the ensemble, it can lead to skewed predictions, undermining the benefits of using a hybrid approach [12].

Addressing these challenges in hybrid forecasting requires careful design considerations, including the selection of compatible models, effective data preprocessing, and robust integration strategies. Employing standardized data formats and using middleware solutions can facilitate interoperability, while clear documentation and practices that enhance model explainability can help improve stakeholder trust. By acknowledging and proactively managing these challenges, practitioners can harness the full potential of hybrid forecasting to deliver more accurate and reliable predictions across various applications, from economics to environmental forecasting.

6.5   Case study

The primary objective of this case study is to develop an accurate short-term forecasting technique utilizing a hybrid CNNs and LSTM model based on infrared all-sky imager (ASI) data. While CNNs and ASI have shown promise in this domain, they often face challenges in outperforming the simple yet effective persistence method, while hybridization of the method involves combining LSTM with the previous one.

The dataset was acquired from a thermal infrared ASI and a meteorological station situated at SolarTech^{LAB}, Politecnico di Milano (latitude: 45.50°N; longitude: 9.16°E). Infrared images of the sky and numerical meteorological measurements were collected between July 1, 2023 and June 20, 2024 (so they cover almost a year). With a sampling frequency of 1 min, the final number of images used is almost 500 thousand.

The Sky InSight^{TM} infrared imager captured 640×480 images every 60 s. Mounted on a mast, it observed a hemispherical mirror, providing a 180° field of view. Infrared radiation, particularly in the 8–13 μm range, is sensitive to water vapor and offers superior cloud detail compared to visible images [39]. Figure 6.2 demonstrates the clear differences between images captured in the (a) visible spectrum and (b) infrared spectrum.

A significant limitation of infrared imaging arises in overcast conditions. When clouds completely cover the sky, the infrared camera may perceive the sky dome as having a uniform temperature distribution. This homogeneous temperature field makes it challenging to differentiate between the sky and clouds based solely on infrared imagery, unlike in partly cloudy scenarios.

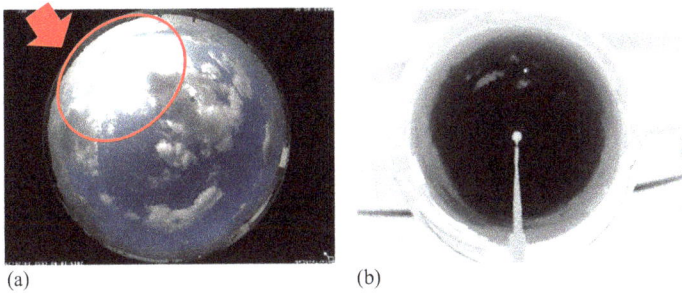

Figure 6.2 Examples of (a) visible spectrum and (b) IR spectrum images. Red is marked as the "sun glare" region that makes the operation on visual spectrum images significantly more difficult.

Apart from the images, environmental data were collected at a weather station within the Department of Energy at Politecnico di Milano. The weather station included solar irradiance sensors, temperature and humidity sensors, a wind speed and direction sensor, and a rain collector. Global irradiance was measured using two secondary standard pyranometers, one positioned horizontally and the other at a 30° tilt. Measurements were recorded every 10 s and later rescaled (mean) to minute data.

6.5.1 Proposed hybrid forecasting models

The proposed nowcasting technique utilizes a CNN that processes a sequence of three consecutive images as input features. This approach enables the CNN to detect cloud movement and predict its subsequent impact on irradiance. Moreover, an LSTM will be implemented on the sequence of the residuals corresponding to the results of the CNN, and additional weather parameters, to try to predict the CNN error based on the contemporary ambient conditions. The idea behind this case study is that typically, the CNN tends to decrease in performance in difficult weather conditions, as it has no historical knowledge of the weather processes. By hybridizing it with a CNN–LSTM ensemble prediction better accuracy should be reachable.

However, before training and testing the CNN, the image dataset undergoes preprocessing to enhance performance and execution time.

6.5.1.1 Image preprocessing

To enhance forecast accuracy, the image dataset was initially filtered to exclude instances with unclear weather conditions. Additionally, clear sky thermal radiance increases with air mass. At sunrise and sunset, the Sun's position limits the observable sky region, making cloud detection challenging in those areas. Therefore, images with a Sun elevation angle below 20° were also excluded. This filtering process reduced the total number of available images from almost 500 thousand to 145 thousand. No additional data augmentation techniques were employed.

After the general image removal, some image-transforming filters were applied to boost the visibility of borders between the clouds and the clear sky. For that purpose, two filters were used:

- Standard scaling:

$$\text{image}_{\text{scaled}} = \frac{\text{image} - \mu(\text{image})}{\sigma(\text{image})} \qquad (6.1)$$

- Log filter:

$$\text{image}_{\text{filtered}} = c \cdot (\log_{10}(\text{image} + 1)) \qquad (6.2)$$

where c is equal to:

$$c = \frac{255}{\log_{10}(1 + \max(\text{image}))} \qquad (6.3)$$

In the second step, a mask was applied to remove non-sky regions from the images. Subsequently, the images were cropped to a square area of 384×384 pixels. To further reduce image size, scaling was performed to reduce each dimension by a third (Figure 6.3). This process eliminates unnecessary data and yields a final image size of 128×128 pixels.

6.5.1.2   CNN architecture

Figure 6.4 illustrates the CNN architecture, inspired by the VGG-16 design [40]. This well-established architecture is known for its strong performance in image classification and low complexity. The CNN consists of five Feature Learning Blocks (FLBs). Each FLB comprises two or three convolutional layers with identical parameters and a max-pooling layer. The first FLB utilizes 64 filters, followed by 128, 256, and 512 filters in subsequent FLBs. All layers employ a 3×3 kernel size with the *Same* padding. After the convolutional layers, two linear fully connected layers are included. The first layer has 256 neurons, while the second has 1 neuron, responsible for predicting the final value. The ReLU activation function was used in convolutional layers, and the linear activation function was applied to fully connected layers. To mitigate overfitting, a 0.2 dropout was introduced before each linear layer. The model

Figure 6.3   Example of a postprocessed image. The visibility is boosted through filtering. Subsequently, the image is cropped, and unnecessary parts are removed with maps.

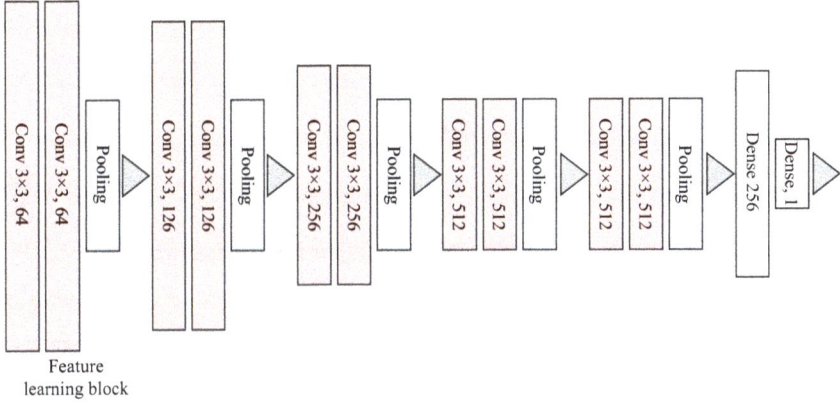

Figure 6.4   The modified VGG-16 architecture used in the case study, with five feature learning blocks

was trained with an initial learning rate of 0.0001, using the Adam optimizer and the Mean Absolute Error loss function, and early stopping enabled to avoid overfitting.

To preserve data structure and encourage the model to learn temporal patterns, the images were processed in sequences of three. Each image was assigned to a unique sequence by concatenating them along the channel dimension. For a given forecast horizon (Δt_{pred}), an appropriate time interval (Δt_{back}) was selected between the sequenced images. This interval has been set to 2 min for a 15 min forecast horizon. Equation (6.4) describes the input concatenation operation:

$$I_{input} = (I_{t-2\Delta t_{back}} \oplus I_{t-\Delta t_{back}} \oplus I_t) \tag{6.4}$$

In this equation, $\oplus$ represents image concatenation (I), t is the current time, and Δt_{back} is the selected time interval between images. The created sequences were then shuffled to enhance model generalization. The resulting sequence dataset was subsequently divided into training, validation, and test subsets, with proportions of 64%, 16%, and 20%, respectively.

With the stacked sequences, the CNN model identifies cloud movement patterns from the image sequences and correlates them with the Global Horizontal Irradiance (GHI) through the provided training labels.

Besides, GHI Clear Sky (GHI_{cs}) is calculated using the Ineichen and Perez clear sky model [41]. For a sequence of images captured at t, $t - \Delta t_{back}$, and $t - 2\Delta t_{back}$, the target GHI is measured at $t + \Delta t_{pred}$.

6.5.1.3   LSTM architecture

The LSTM model utilized for ensembling in this study adopts a relatively simple architecture. It consists of a single LSTM layer equipped with 128 hidden units, followed by two fully connected layers tasked with generating predictions. The first fully connected layer incorporates 128 hidden neurons, while the output layer, whose size is determined by the desired output dimensionality (in this case, 15 units), directly

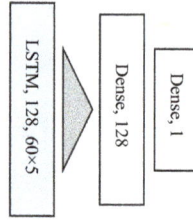

Figure 6.5   A simple LSTM-based model for the residual prediction

produces the forecasts. A ReLU activation function is employed between the internal layers, contributing to the overall simplicity of the LSTM model. The detailed view of the model used for residual prediction can be seen in Figure 6.5.

The primary goal of this LSTM model is to accurately predict the error introduced by the CNN, considering the fluctuating nature of weather conditions. To achieve this, the input data comprises a multivariate time series consisting of five distinct features: temperature, humidity, measured GHI, theoretical GHI calculated using a white-box mathematical model, and the CNN residual time series. This comprehensive input representation enables the LSTM model to effectively capture the complex interactions between weather variables and CNN error, leading to more precise and reliable predictions. The lookback window for the LSTM model is set to 60 min (or 60 data samples), providing the model with sufficient historical context. The prediction window, on the other hand, is set to the next 15 min (or 15 data samples), aligning with the desired forecast horizon. This configuration results in an input shape of 60×5 and an output size of 15×1. The model's predictions are subsequently postprocessed to extract the final output sample, which represents the residual of GHI 15 min into the future.

6.5.1.4   Result concatenation

The proposed method is an example of cooperative ensemble hybridization, as the results of two models are used to create the final prediction of the GHI value. The first model (CNN) is designed to predict the measured GHI 15 min in advance, and the second learns the typical residuals corresponding to various weather parameters and predicts the error the first model commits. Moreover, the LSTM model is also supported by a mathematical computation of the GHI_{cs}. Initially, the solution can be written as follows:

$$GHI_{pred} = GHI_{CNN} + \varepsilon \tag{6.5}$$

where the GHI_{pred} is the predicted value of the GHI, the GHI_{CNN} is the GHI predicted by the proposed CNN and the ε is the prediction error. In this case, the initial residual of the prediction is used to train a novel DL model based on temporal sequences that are used for the estimation of its own future values along with other weather parameters. The idea behind this hybridization method is that the error committed by CNN has a semblance of some consistency. For example, in typically clear sky conditions with high temperature and irradiation, the CNN tends to underestimate the measured GHI. However, in case of severe overcast, the estimation is overvalued.

Training a secondary model that is able to benefit from the weather measurement might be beneficial for the overall estimation of the GHI. Finally, the prediction can be mathematically described as follows:

$$\text{GHI}_{\text{pred}} = \text{GHI}_{\text{CNN}} + \Delta\text{GHI}_{\text{LSTM}} + \varepsilon_1 \tag{6.6}$$

where $\Delta\text{GHI}_{\text{LSTM}}$ is the residual predicted by the LSTM and ε_1 is the secondary error after the combined prediction.

6.5.2 Results

The results analysis focuses on the model's accuracy in predicting the 15-min ahead GHI value during the first five days of June 2024. Upon processing the results for the raw CNN, they are used to train the ensemble network and to predict the final value of GHI. Table 6.1 presents detailed results, including the achieved forecast and corresponding weather conditions. The result analysis is performed in terms of errors committed by the raw model and the upgraded ensemble hybrid model. For the five selected days, the Mean Absolute Error (MAE) has been calculated and compared against persistence to form the Forecast Skill (FS), also known as Skill Score, for each model defined as follows:

$$\text{FS} = \frac{\text{MAE}_{\text{ref}} - \text{MAE}_{\text{CNN}}}{\text{MAE}_{\text{ref}}} \tag{6.7}$$

where the subscript "ref" is the reference forecast (i.e., persistence).

As it can be observed, the simple ensemble is able to significantly outperform the standard CNN in a variety of weather conditions. The Forecast Skills achieved by the hybrid model are significantly higher than the raw CNN for the majority of the test days. The obvious outlier is the second test day – June 2. Here, the new method falls short with a decrease in performance and accuracy. In detail, day 2 represents particularly challenging and changing weather conditions with many clouds passing above the visor of the ASI camera and the supposed PV field. These fluctuating weather conditions can also have an impact on the performance of the trained LSTM, as the model cannot easily identify the trends governing the test dataset delivered to correctly infer the residual. Indeed, it can be observed that the smallest performance increase can be noted on partly clouded and shady days, which are characterized by dynamic weather patterns and fluctuating cloud behavior. In contrast, the biggest improvement is noted on sunny days, when the raw model's behavior can be easily estimated and improved upon. Detailed examples of the model results can be observed in Figure 6.6. As it can

Table 6.1 Error metrics for five selected test days in June 2024

Day	Weather	MAE$_{\text{CNN}}$ (W/m^2)	MAE$_{\text{Hyb}}$ (W/m^2)	FS$_{\text{CNN}}$ (%)	FS$_{\text{Hyb}}$ (%)
01/06/24	Sunny	38.89	21.29	−17.4	35.7
02/06/24	P. cloudy	105.46	114.18	2.5	−5.5
03/06/24	P. cloudy	150.94	138.69	14.2	21.2
04/06/24	Cloudy	186.43	159.31	−42.0	−21.3
05/06/24	Shady	99.81	88.49	9.2	19.5

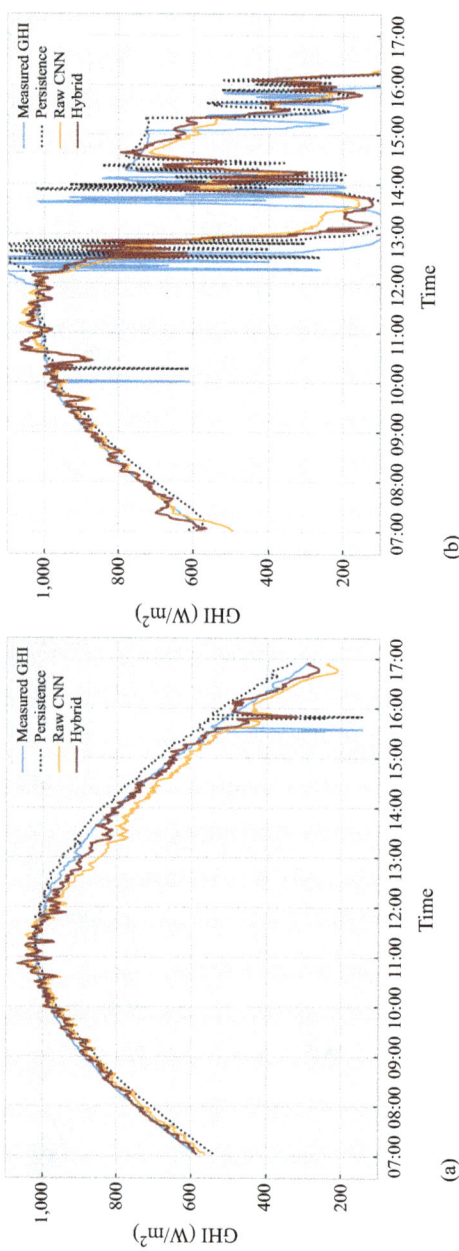

Figure 6.6   Examples of results of the raw CNN and hybrid model: (a) a sunny day, where the raw CNN model tends to underestimate the measured GHI values and (b) a day with dense clouds passing in front of the PV plant (and the sensor) in the afternoon.

be seen, on (a) June 1, which was sunny day, the raw CNN model tends to underestimate the measured GHI values; hence, the residual prediction significantly improves the performance. On the day after (b), which was turning to a cloudy day in the afternoon, the residual slightly improves the performance during the drop of GHI, where the minimum value of GHI is overestimated by the raw CNN. However, apart from that improvements are negligible.

6.6 Conclusions

Hybrid and ensemble forecasting models offer a synergistic approach to enhance forecasting accuracy by combining various modeling paradigms. These models leverage the strengths of statistical, machine learning, and physics-based methods to overcome individual limitations and achieve superior predictive performance.

This chapter explores the integration of time-tested statistical methods like auto-regressive models with the adaptability and pattern recognition capabilities of machine learning models. Additionally, the incorporation of physics-based models adds domain-specific knowledge and constraints to the forecasting process.

The chapter not only discusses the theoretical foundations of hybrid models but also provides practical implementation insights. It emphasizes their relevance in improving accuracy across various forecasting applications. The theoretical discussion is followed by a robust case study that demonstrates the application of a hybrid CNN and LSTM ensemble model. It utilizes infrared all-sky imager data to enhance short-term photovoltaic forecasting and incorporates additional multivariate time-series prediction for additional context-driven decisions and improved accuracy.

References

[1] Mellit A, Massi Pavan A, Ogliari E, *et al.* Advanced methods for photovoltaic output power forecasting: A review. *Applied Sciences.* 2020;10(2):487.

[2] Nespoli A, Ogliari E, Leva S, *et al.* Day-ahead photovoltaic forecasting: A comparison of the most effective techniques. *Energies.* 2019;12(9):1621.

[3] Kurani A, Doshi P, Vakharia A, *et al.* A comprehensive comparative study of artificial neural network (ANN) and support vector machines (SVM) on stock forecasting. *Annals of Data Science.* 2023;10(1):183–208.

[4] Ren Y, Suganthan P N, and Srikanth N. Ensemble methods for wind and solar power forecasting—A state-of-the-art review. *Renewable and Sustainable Energy Reviews.* 2015;50:82–91.

[5] Sansana J, Joswiak M N, Castillo I, *et al.* Recent trends on hybrid modeling for Industry 4.0. *Computers & Chemical Engineering.* 2021;151:107365.

[6] Yang D, Kleissl J, Gueymard C A, *et al.* History and trends in solar irradiance and PV power forecasting: A preliminary assessment and review using text mining. *Solar Energy.* 2018;168:60–101. Advances in Solar Resource Assessment and Forecasting.

[7]   Ogliari E, Dolara A, Manzolini G, *et al.* Physical and hybrid methods comparison for the day ahead PV output power forecast. *Renewable Energy.* 2017;113:11–21.

[8]   Mathonsi T and van Zyl T L. A statistics and deep learning hybrid method for multivariate time series forecasting and mortality modeling. *Forecasting.* 2022;4(1):1–25.

[9]   Hajirahimi Z and Khashei M. Hybrid structures in time series modeling and forecasting: A review. *Engineering Applications of Artificial Intelligence.* 2019;86:83–106.

[10] Gupta A, Gupta K, and Saroha S. A review and evaluation of solar forecasting technologies. *Materials Today: Proceedings.* 2021;47:2420–2425.

[11] Fan G F, Wei X, Li Y T, *et al.* Forecasting electricity consumption using a novel hybrid model. *Sustainable Cities and Society.* 2020;61:102320.

[12] Slater L J, Arnal L, Boucher M A, *et al.* Hybrid forecasting: blending climate predictions with AI models. *Hydrology and Earth System Sciences.* 2023;27(9):1865–1889.

[13] Brester C, Kallio-Myers V, Lindfors A V, *et al.* Evaluating neural network models in site-specific solar PV forecasting using numerical weather prediction data and weather observations. *Renewable Energy.* 2023;207:266–274.

[14] Alrashidi M and Rahman S. Short-term photovoltaic power production forecasting based on novel hybrid data-driven models. *Journal of Big Data.* 2023;10(1):26.

[15] Massucco S, Mosaico G, Saviozzi M, *et al.* A hybrid technique for day-ahead PV generation forecasting using clear-sky models or ensemble of artificial neural networks according to a decision tree approach. *Energies.* 2019;12(7):1298.

[16] Belmahdi B, Louzazni M, Bouardi AE. A hybrid ARIMA–ANN method to forecast daily global solar radiation in three different cities in Morocco. *The European Physical Journal Plus.* 2020;135(11):925.

[17] Sina L B, Secco C A, Blazevic M, *et al.* Hybrid forecasting methods—A systematic review. *Electronics.* 2023;12(9):2019.

[18] Ardabili S, Mosavi A, Várkonyi-Kóczy A R. Advances in machine learning modeling reviewing hybrid and ensemble methods. In: Várkonyi-Kóczy A R, editor. *Engineering for Sustainable Future.* Cham: Springer; 2020. pp. 215–227.

[19] Bendiek P, Taha A, Abbasi Q H, *et al.* Solar irradiance forecasting using a data-driven algorithm and contextual optimisation. *Applied Sciences.* 2022;12(1):134.

[20] Dietterich T G. Ensemble methods in machine learning. In: *Multiple Classifier Systems.* Berlin, Heidelberg: Springer; 2000. pp. 1–15.

[21] Lee D S and Son S Y. Weighted average ensemble-based PV forecasting in a limited environment with missing data of PV power. *Sustainability.* 2024;16(10):4069.

[22] Abedinia O and Bagheri M. Execution of synthetic Bayesian model average for solar energy forecasting. *IET Renewable Power Generation.* 2022;16(6):1134–1147.

[23]   Bayrak G, Küçüker A, and Yılmaz A. Deep learning-based multi-model ensemble method for classification of PQDs in a hydrogen energy-based microgrid using modified weighted majority algorithm. *International Journal of Hydrogen Energy*. 2023;48(18):6824–6836. The 5th International Conference on Alternative Fuel, Energy & Environment (ICAFEE 2021).

[24]   Mienye I D and Sun Y. A survey of ensemble learning: Concepts, algorithms, applications, and prospects. *IEEE Access*. 2022;10:99129–99149.

[25]   Ali W and Malebary S. Particle swarm optimization-based feature weighting for improving intelligent phishing website detection. *IEEE Access*. 2020;8:116766–116780.

[26]   Galar M, Fernandez A, Barrenechea E, *et al.* A review on ensembles for the class imbalance problem: bagging-, boosting-, and hybrid-based approaches. *IEEE Transactions on Systems, Man, and Cybernetics, Part C (Applications and Reviews)*. 2012;42(4):463–484.

[27]   Breiman L. Bagging predictors. *Machine Learning*. 1996;24(2):123–140.

[28]   Choi S and Hur J. An ensemble learner-based bagging model using past output data for photovoltaic forecasting. *Energies*. 2020;13(6):1438.

[29]   Friedman J H. Greedy function approximation: A gradient boosting machine. *The Annals of Statistics*. 2001;29(5):1189–1232.

[30]   Yang X, Wang S, Peng Y, *et al.* Short-term photovoltaic power prediction with similar-day integrated by BP-AdaBoost based on the Grey-Markov model. *Electric Power Systems Research*. 2023;215:108966.

[31]   Wolpert D H. Stacked generalization. *Neural Networks*. 1992;5(2):241–259.

[32]   Chu W, Li L, Reyzin L, *et al.* Contextual bandits with linear payoff functions. In: Gordon G, Dunson D, Dudík M, ed. *Proceedings of the Fourteenth International Conference on Artificial Intelligence and Statistics (vol. 15 of Proceedings of Machine Learning Research)*. Fort Lauderdale, FL, USA: PMLR; 2011. pp. 208–214.

[33]   Bifet A and Gavaldà R. Learning from time-changing data with adaptive windowing. In: *Proceedings of the 2007 SIAM International Conference on Data Mining*; 2007. pp. 443–448.

[34]   Wu Y K, Chen C R, and Rahman H A. A Novel Hybrid Model for Short-Term Forecasting in PV Power Generation. *International Journal of Photoenergy*. 2014;2014:1–9.

[35]   Quan D M, Ogliari E, Grimaccia F, *et al.* Hybrid model for hourly forecast of photovoltaic and wind power. In: *2013 IEEE International Conference on Fuzzy Systems (FUZZ-IEEE)*; 2013. pp. 1–6.

[36]   Ahmed R, Sreeram V, Togneri R, *et al.* Computationally expedient Photovoltaic power Forecasting: A LSTM ensemble method augmented with adaptive weighting and data segmentation technique. *Energy Conversion and Management*. 2022;258:115563.

[37]   Chalal L, Saadane A, and Rachid A. Unified environment for real time control of hybrid energy system using digital twin and IoT approach. *Sensors*. 2023;23(12):5646.

[38]   Arcos-Vargas A, Cansino J M, and Román-Collado R. Economic and environmental analysis of a residential PV system: A profitable contribution to the Paris agreement. *Renewable and Sustainable Energy Reviews*. 2018;94: 1024–1035.

[39]   Bertin C, Cros S, Saint-Antonin L, *et al.* Prediction of optical communication link availability: real-time observation of cloud patterns using a ground-based thermal infrared camera. *Proceedings of SPIE* 2015;9641:96410A.

[40]   Simonyan K and Zisserman A. *Very deep convolutional networks for large-scale image recognition.* arXiv:14091556. 2014.

[41]   Ineichen P and Perez R. A new airmass independent formulation for the Linke turbidity coefficient. *Solar Energy*. 2002;73(3):151–157.

Chapter 7
Probabilistic PV forecasting

Martin János Mayer[1] and Sándor Baran[2]

Probabilistic photovoltaic (PV) forecasting involves predicting PV power output while accounting for the inherent uncertainty in various conditions. Unlike deterministic forecasts that provide a single-point estimate, probabilistic forecasts offer a range of possible outcomes along with associated probabilities. The main topics covered in this chapter are the utilization of uncertainty information, workflows to create probabilistic PV forecasts, ensemble methods, parametric and nonparametric postprocessing techniques, and verification of probabilistic forecasts. The explicit quantification of the uncertainty of PV predictions can aid decision-making in grid operation and energy system planning, where uncertainty management is a key consideration. Probabilistic PV power forecasts can be created following several different workflows combining irradiance-to-power conversion and postprocessing steps. Ensemble methods combine forecasts from different sources, models, or initial conditions, treating them as equally probable outcomes to provide a notion of uncertainty. Statistical postprocessing involves various data-driven models to create calibrated probabilistic forecasts by fitting parametric distributions or directly estimating the quantiles of the expected PV output. The verification of probabilistic forecasts goes beyond using scoring rules for overall evaluation by also providing tools for the individual assessment of the reliability and sharpness of the forecasts.

7.1 Introduction

Probabilistic forecasting goes beyond deterministic one in that it not only predicts the future value of the quantity of interest (a weather variable such as wind speed, global horizontal irradiance (GHI), and PV power) but also quantifies the uncertainty of the prediction. As an illustration, consider Figure 7.1, where besides the observed PV power and the corresponding deterministic forecasts, probabilistic forecasts in the form of predictive probability density functions (PDFs) are also depicted. In the early hours, the deterministic forecast overestimates the produced PV power, and the

[1]Department of Energy Engineering, Faculty of Mechanical Engineering, Budapest University of Technology and Economics, Hungary
[2]Department of Applied Mathematics and Probability Theory, Faculty of Informatics, University of Debrecen, Hungary

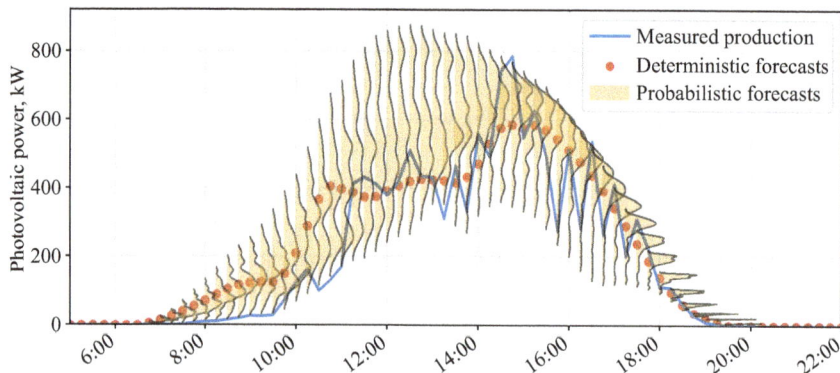

Figure 7.1   Illustration of deterministic and probabilistic PV power forecasts

forecast uncertainty is relatively large, becoming even larger during the peak hours. In contrast, by the end of the day, the predictions are rather accurate, and the corresponding predictive distributions are very sharp (narrow). Thus, a more complete picture is obtained, allowing for estimation of the prediction reliability.

The purpose of this chapter is to provide a comprehensive overview of various aspects of probabilistic forecasting related to PV power. First, Section 7.2 discusses how the uncertainty quantification of probabilistic forecasts can be utilized in practical applications. Then, the specific characteristics of solar energy that are relevant for probabilistic forecasting are introduced in Section 7.3. The technical descriptions and the notable examples of the main probabilistic forecasting methods, divided into the ensemble forecasting and distributional regression categories, are discussed in Sections 7.4 and 7.5, respectively. The verification of probabilistic forecasts is fundamentally different from that of deterministic forecasts; therefore, Section 7.6 presents an overview of the recommended metrics and techniques required for an insight evaluation. Finally, Section 7.7 summarizes the chapter.

7.2   Utilization of probabilistic forecasts

Probabilistic forecasts provide more nuanced information about the expected outcomes; still, many practical applications of PV power forecasting require only point forecasts [1]. Probabilistic forecasting, however, has several advantages even in these applications. The most obvious motivation is the improved accuracy that is often experienced when a deterministic forecasting workflow is extended with probabilistic methods. For instance, the mean of the 50 members of the European Centre for Medium-Range Weather Forecasts (ECMWF) ensemble GHI forecasts was found to be more accurate than the control forecast (which is roughly equivalent to the high-resolution (HRES) deterministic forecast of ECMWF) [2]. Similarly, the mean PV forecasts of a calibrated ensemble of model chains had a lower root mean square error (RMSE) compared to any of the individual model chains [3].

Another advantage of probabilistic forecasts is universality. Deterministic forecasts should always be optimized under a directive that is expressed as a statistical functional, reflecting the error function that the forecasts are intended to minimize [4]. For example, if the forecast is evaluated using the mean square error, the directive is to issue the mean of the underlying predictive distribution as a point forecast. Since different directives are optimized by statistically different deterministic forecasts, a forecast that is calibrated for one metric will be suboptimal for others [5]. This may also lead to verification problems, since evaluating a point forecast with metrics other than the one it was optimized for is theoretically flawed [6], which also calls into question the universal use of the root mean square error skill score for the intercomparability of deterministic forecasts [7]. Whereas different deterministic forecasts must be fitted for different directives, a single probabilistic forecast optimized under a proper scoring rule is more universal, since it can be summarized into a wide range of deterministic forecasts (see Figure 7.2). Yang and Kleissl [8] presented four PV power forecast error penalty cost functions from China, each representing a different error metric and thus minimized by a different statistical functional. In the empirical analysis, a single set of ensemble forecasts was summarized into four different sets of point forecasts, and the lowest costs are always achieved with the forecasts created using the functional consistent with the penalty.

From the perspective of PV plant owners, probabilistic forecasting can also increase revenues by allowing the stochastic optimization of their bids on the day-ahead and intraday markets [9]. Visser *et al.* [10] proposed a market bidding strategy that is implemented by a multistage stochastic optimization method based on probabilistic PV forecasts and imbalance price scenarios. The probabilistic PV power forecasts for the intraday and day-ahead time horizons, along with the generated scenarios are shown in Figure 7.3 for a sample day. The results indicate that the revenues achieved by the PV plant are higher for probabilistic forecasts that have a lower continuous ranked probability score (CRPS) (see also Section 7.6), although

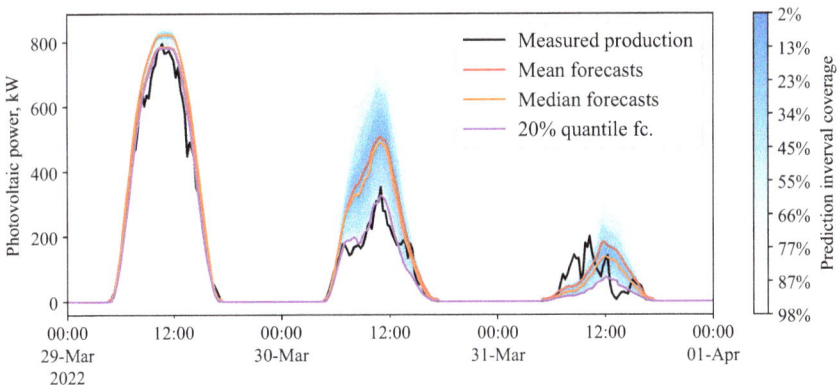

Figure 7.2   Sample probabilistic PV power forecasts and three deterministic forecasts created by summarizing the predictive distributions with mean, median, and 20% quantile functionals

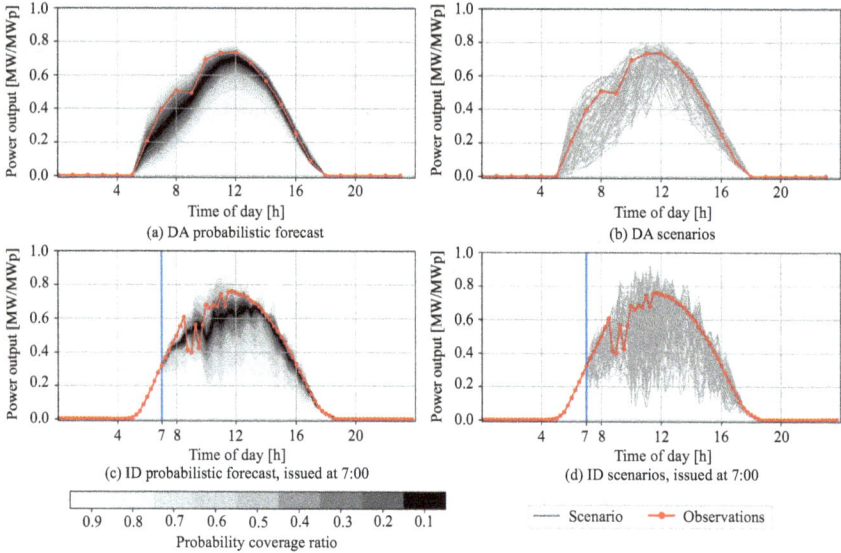

Figure 7.3   Example of the day-ahead (a,b) and intraday (c,d) probabilistic forecasts and subsequent scenarios of the expected PV power output [10]

the relationship is not strict. The study on colocated wind and PV plants in [11] used quantiles with probability levels, depending on the hourly downward and upward balancing costs as optimal bids, finding that more accurate and sharper forecasts reduce balancing costs.

System operators can also utilize probabilistic renewable energy forecasts by conditioning reserve allocation on the expected forecast uncertainty. A solar-conditioned requirement estimation method for the flexible ramp product, the newest operating reserve type of the California Independent System Operator, was proposed in [12]. In this method, the median and the 50% prediction interval width of probabilistic GHI forecasts are further postprocessed into probabilistic net load forecasts, on which the reserve allocation is based. The conditional reserve sizing can reduce ramping reserve oversupply by up to 25% without increasing the risk of reserve shortage [13]. The benefits of dynamic reserve allocation are expected to increase with higher PV penetrations [14].

Optimization of energy management in microgrids and energy storage systems can also benefit from probabilistic solar forecasts. Ramahatana *et al.* [15] described the use of probabilistic forecasting for a microgrid, consisting of a net-zero energy building, a PV system, Li-ion batteries, and a grid connection. The day-ahead schedule of the battery energy storage is optimized using a probabilistic dynamic programming method, relying on probabilistic forecasts of PV power generation. The results highlight the importance of probabilistic forecasting, since even the worst probabilistic forecasts outperformed the deterministic forecasts in this application; whereas the operating cost reduction achieved by the best probabilistic method over the deterministic reference reaches 38%. Finally, for further potential use cases, interested readers are referred to a review of the utilization methods of probabilistic solar forecasting in power systems [16].

7.3   Probabilistic PV power forecasting workflow

This section starts with presenting the possible workflows for creating probabilistic PV power forecasts, followed by a discussion of irradiance-to-power conversion methods, and finally introduces clear sky modeling, as a potential technique for estimating the physically possible range of solar forecasts and deseasonalizing the datasets.

7.3.1   PV power forecasting workflows

Solar forecasting is a general term that covers the prediction of both GHI and PV power output. Accordingly, the papers published in the literature can also be divided into two groups, depending on whether they forecast GHI or PV power, with more papers falling into the first group due to the better public availability of high-quality GHI observation datasets. Fortunately, advances in the solar irradiance forecasting domain also contribute to the development of PV power forecasting. In contrast, even though PV power could also be directly forecast from historical PV data, a two-step framework where GHI forecasts are issued first and then converted to PV power, generally provides better accuracy on time horizons longer than half an hour. The rationale behind this is that the main GHI forecasting methods, such as numerical weather prediction (NWP) and satellite-based forecasting, are able to capture the spatiotemporal nature of the weather evolution, which is not possible in direct time series forecasting of PV power [17]. However, most of the postprocessing methods developed for GHI can also be applied to PV power forecasts without major modifications by simply using the PV power as the predictand instead of the GHI [18].

The two-step framework of PV power forecasting, consisting of GHI forecasting and irradiance-to-power conversion, is often extended by a third step, namely data-driven postprocessing, which aims to improve the accuracy of the forecasts and modify them to suit the requirements better. The comprehensive review by Yang and van der Meer [19] classified the postprocessing tasks into four categories: deterministic-to-deterministic (D2D), probabilistic-to-deterministic (P2D), deterministic-to-probabilistic (D2P), and probabilistic-to-probabilistic (P2P) postprocessing. Among these, methods belonging to the D2P and P2P categories can be used to create or improve probabilistic forecasts. The categories are further divided into thinking tools that reflect the goal accomplished by applying postprocessing. The three thinking tools belonging to the D2P category are the analogue ensemble, the method of dressing, and the probabilistic regression, while the two thinking tools in the P2P category are calibrating ensemble forecasts and combining probabilistic forecasts. The concept of the analogue ensemble is explained in Section 7.4. Probabilistic regressions and the calibration of ensemble forecasts can be performed with similar methods, where the main difference lies in the nature of the input data; these methods are introduced in Section 7.5.

Considering that probabilistic forecasts can be created from either deterministic or probabilistic inputs by postprocessing, many different probabilistic PV power forecasting workflows can be constructed. The postprocessing can be performed either on GHI as the second step or on the PV power as the last step of the workflow. The basic GHI-based workflows that generate probabilistic PV power forecasts are as follows:

- deterministic GHI forecasts → irradiance-to-power conversion → D2P postprocessing,
- deterministic GHI forecasts → D2P postprocessing → irradiance-to-power conversion,
- probabilistic (ensemble) GHI forecasts → irradiance-to-power conversion → P2P postprocessing,
- probabilistic (ensemble) GHI forecasts → P2P postprocessing → irradiance-to-power conversion,

but this list is nonexhaustive since further workflows can still be constructed by adding further postprocessing steps, e.g., a P2P calibration after irradiance-to-power conversion in the second and fourth options. Data-driven methods also provide an opportunity to merge the irradiance-to-power conversion and postprocessing steps to directly create calibrated probabilistic PV power forecasts. Moreover, probabilistic irradiance-to-power conversion methods have also been developed recently [3], further increasing the number of possibilities. Recent studies suggest that performing postprocessing on the PV data results in more accurate forecasts [18,20], but this general trend can be overridden by historical data availability and other specific factors. It is recommended to experiment with different workflows and methods to create the best possible forecasts under the circumstances at hand.

7.3.2   Irradiance-to-power conversion

The methods for converting GHI and supplementary weather variables to PV power output, also called solar power curve modeling, can be categorized into (1) data-driven, (2) physics-based, and (3) hybrid approaches [21].

1.   The data-driven approach covers the data-driven methods that use regression to map the relationship between the predictors, including the GHI forecasts and other arbitrary variables, and PV power output. Due to the nonlinearity of this relationship, the regression is typically performed by machine learning (ML) models. The selection and hyperparameter tuning of the ML model, as well as the chosen predictors and the length of the training data, all have significant effects on the accuracy of the resulting PV power forecasts [22]. The main advantage of the regression approach is flexibility: On the one hand, any seemingly relevant variables can be added to the predictors to assess their effect on the accuracy, and on the other hand, the resulting PV power forecasts can be easily tailored to fit different directives by adjusting the loss function [1]. However,

a drawback of regression is the need for at least a full year of training data for decent performance, which limits its applicability in new PV installations.

2.   The physical approach relies on the so-called model chain, which consists of a successive set of component models representing different stages of energy conversion. The main steps of a model chain are the separation of beam and diffuse irradiance, the transposition of the horizontal irradiance to the tilted module plane, and the estimation of the temperature, efficiency, and power output of the PV module. More detailed model chains, aiming for improved accuracy, may also account for the reflection, soiling, shading, and masking losses, the electric losses of the inverters, cables, and other components, and the degradation of the solar cells. The component models in the main steps can be selected from the tens (or even hundreds) of different models published in the literature, and they can be combined in all possible ways, resulting in hundreds of thousands of different model chains. The selection of the component models has been found to largely affect the errors of PV power forecasts [23]. Physical model chains require knowledge of the design parameters of the PV plant, and even though more information generally allows more detailed modeling, the location, nominal power, and module orientation are already sufficient for good accuracy [24].

3.   The hybrid approaches involve the combined use of physical and data-driven models, leveraging the benefits of both methods [25]. Hybrid models can be constructed in many different ways [26], but the overarching idea is similar: Physical models can be used to provide an initial good guess and to filter out physically impossible forecasts, while ML can correct the forecasts and account for such predictors and effects that cannot be directly considered in the physical modeling.

Irradiance-to-power conversion is traditionally seen as a deterministic procedure. However, the inaccuracies of the conversion models and the increasing need for uncertainty quantification called for their extension into the probability space. In the statistical approach, this can be easily achieved by merging the regression model with a D2P or P2P distributional regression method (see Section 7.5) to directly calculate probabilistic PV power forecasts from GHI-related inputs. Having said that, it must be noted that the important predictors of a deterministic irradiance-to-power conversion model are those that provide information on the expected value of the PV power output. In contrast, good probabilistic forecasting also requires predictors that carry information related to the uncertainty of the forecasts.

Creating a probabilistic physical model chain is a more challenging task. The majority of the component models of a model chain are deterministic; the few exceptions are confined to the probabilistic implementations and irradiance separation [27] and transposition [28] models. To that end, for the model chain as a whole, a better approach is to use an ensemble of physical model chains, where each member consists of different, randomly selected component models [3]. However, since the inaccuracy of the irradiance-to-power conversion is only a small contributor to the total uncertainty of PV power forecasting, the model chain ensemble is highly underdispersed and can provide reliable forecasts only after data-driven calibration, placing this method into the hybrid category. A general description of ensemble forecasting, including further details on this approach, can be found in Section 7.4.

7.3.3   Clear sky irradiance and power

Due to the revolution and rotation of Earth, the apparent path of the Sun in the sky follows a double-seasonal pattern with yearly and diurnal cycles [29]. The zenith and azimuth angles describing the position of the Sun can be precisely calculated using a solar position algorithm based on the time and the geographical coordinates of the location of interest [30]. Accompanied by the declination, these three angles are commonly used as predictors in regression methods to incorporate time information in a periodic way [25].

Consequently, the physically possible ranges of solar irradiance and PV power are time-dependent and follow a similar yearly and diurnal periodicity. The factors affecting the GHI on Earth's surface can be classified into three categories: (1) solar geometry, (2) transmission through the cloud-free atmosphere, and (3) the effects of clouds.

1.   The irradiance at the top of Earth's atmosphere is called extraterrestrial irradiance, and it can be calculated by correcting the solar constant for the annual variation in the Earth–Sun distance [31].
2.   The GHI under cloudless sky conditions, called clear sky irradiance (CSI), can be estimated by clear sky models that account for the attenuation of the radiation in the atmosphere but do not cover the effect of clouds. Simple clear sky models only consider the effect of the air mass, whereas the more complex models also account for the atmospheric composition using variables such as aerosol optical depth and precipitable water vapor content [32]. The inaccuracies of these variables also introduced errors to the CSI estimations [33].
3.   Absorption and scattering by clouds typically decrease the GHI; however, under special circumstances called cloud enhancement, they can even increase it, resulting in higher GHI observations than the CSI [34]. Cloud enhancement events are typically confined to a small area and last up to a few minutes [35]; therefore, they are only relevant in high temporal resolution forecasting but barely present in lower-resolution (15-min or hourly) data [36].

The CSI can be converted into clear sky PV power using a physical model chain, as described in Section 7.3.2. The selection and inaccuracies of the clear sky model, the atmospheric variables, and the model chain all contribute to the errors of the clear sky PV power estimation [37]. Consequently, whereas the typical range of PV power is between 0 and the clear sky PV power, the latter is not a hard limit. The power output of PV plants may also be limited by the nominal AC power of the solar inverters. The inverter sizing factor, which is defined as the ratio of the total DC peak power of the PV modules to the nominal AC power of the inverter, is one of the most important design parameters of a PV plant [38]. A sizing factor higher than 1 indicates that the inverters are undersized compared to the PV modules, which is a common design practice to save on investment costs and improve part-load efficiency [39]. Undersized inverters clip the PV power output at the nominal AC inverter power, posing a strong upper limit to the possible PV power production. Higher inverter sizing factors increase the time under which inverter clipping can occur (i.e., when the clear sky PV power calculated without clipping is higher than nominal AC power), thus making it more important to consider the inverter power limit in the forecasting model. The reducing costs of PV modules shift the design optimum toward higher inverter sizing factors, a trend that is expected to continue in the future [40].

CSI is also commonly used to deseasonalize (i.e., to remove the trend components) the GHI time-series data before fitting a regression method [41]. The clear sky index (k_{cs}) is the ratio of the GHI to the CSI, ideally ranging between 0 and 1, depending on the cloud conditions. In this case, the regression models can be trained by using the clear sky index instead of the GHI as the predictand. Afterward, the trained models can be used to predict the clear sky index, which can be converted back to GHI by multiplying it with the forecast CSI [42]. Similarly, a clear sky index for PV production (k_{PV}) can also be constructed as the ratio of the actual and clear sky PV power [43] and used in the same way as the GHI clear sky index.

7.4 Ensemble forecasting

An ensemble is a set of similar entities that are used together to achieve a goal. Ensembles can be created and utilized in many different ways to perform various tasks. For example, in the ensemble learning subfield of ML, multiple models are trained and used together to create more accurate deterministic predictions. In probabilistic forecasting, ensembles are used to quantify the uncertainty of forecasts through the difference of the individual members, i.e., the spread of the ensemble. This section presents ensemble techniques specifically in the context of probabilistic forecasting.

GHI forecasts can be obtained as outputs of NWP models, describing the dynamical and physical behavior of the atmosphere with the help of nonlinear partial differential equations. However, the solutions of these equations strongly depend on the initial conditions and other uncertainties related to the NWP process, and a single model output (point forecast) does not carry information about forecast uncertainty. Nowadays, the standard approach in NWP is to run the atmospheric models several times with different initial conditions and/or model parametrizations, resulting in a forecast ensemble. Usually, one has a control forecast and several other ensemble members that are obtained using slightly different initial conditions and slightly changed model physics. These perturbed ensemble members are statistically indistinguishable and, hence, considered exchangeable. The use of ensemble forecasts allows the estimation of the probability distribution of the future value of the (weather) quantity at hand, which opens the door for probabilistic forecasting [44], where the probability of various events and forecast uncertainty can also be estimated.

One of the leading services providing ensemble weather forecasts is the independent intergovernmental ECMWF, which pioneers the operational implementation of ensemble prediction systems (EPSs) [45]. The most recent CY49R1 configuration of the ECMWF Integrated Forecast System (IFS) [46] generates 51-member (1 control and 50 perturbed members) medium-range (up to 15 days ahead) and 101-member (1 control and 100 perturbed members) subseasonal range (up to 46 days ahead) ensemble forecasts both at vertical resolution of 137 model levels and horizontal resolutions of 9 km and 36 km, respectively.

In the last couple of years, NWP models got strong competitors in data-driven, ML-based approaches such as Huawei's Pangu-Weather model [47], NVIDIA's FourCastNet [48], or Google's GraphCast [49]. The accuracy of deterministic forecasts provided by these AI-based models is comparable to the control forecast of the ECMWF IFS at several magnitudes lower computational cost. The ECMWF has also launched its Artificial Intelligence/Integrated Forecasting System (AIFS) model [50], which provided first deterministic, then, since June 2024, data-driven 50-member ensemble forecasts (AIFS-CRPS) [51], which is now run operationally, while in December 2024, Google introduced the GenCast [52] AI ensemble model.

Ensemble NWP is categorized as a dynamical ensemble since the forecast values of ensemble members are determined by the dynamical projection of the perturbed initial conditions into the future. Another category of ensemble methods is the model ensemble, also called the poor man's ensemble, where the ensemble members are created by different models. A notable example of this approach in PV power forecasting is the ensemble of model chains proposed in [3] and briefly introduced in Section 7.3.2. The concepts of ensemble NWP and ensemble of model chains are visualized in Figure 7.4. In the model chain ensemble, the same input weather data are fed into the model chains with different component models, providing different estimations for the PV power generation. The investigations of [3] revealed that the component models could be best selected randomly instead of relying on their deterministic modeling accuracy, and the optimal number of ensemble members is around 50 and 100. The ensemble of model chains can also be paired with ensemble NWP [2]. In this case, the weather forecasts missed by each ensemble NWP member are converted to PV power by one or more different model chains. The model chain ensemble alone is less effective than an ensemble NWP, but their combinations outperform both individual approaches in terms of reliability and scores.

Ensemble forecasts can also be created from historical forecasts and observation datasets using the analogue ensemble (AnEn) method, which relies on the similarity of weather patterns [53]. In this, the forecast values of the selected predictors are compared to their past values over a long historical dataset to select the M most similar timestamps, and the past observations for the same timestamps are used to form an M-member ensemble. The predictors on which the similarity is based are the GHI, direct normal irradiance, cloud cover, and 2m temperature forecasts, supplemented by the calculated solar elevation and azimuth angles. A fast similarity search algorithm, such as Mueen's algorithm for similarity search recommended in [54], is crucial for the resource-effective utilization of AnEn. An advantage of AnEn over the dynamical NWP ensemble, owing to its data-driven nature, is that it can provide reliable forecasts without further calibration. However, after a further calibration step (see Section 7.5 for suitable methods), the ensemble NWP of the ECMWF consistently outperformed AnEn for probabilistic GHI forecasting at all tested stations in [55].

A model ensemble can also be created from the outputs of multiple ML models. Yagli *et al.* [56] used an ensemble of 20 ML models for 1-h ahead clear sky index forecasting based on the 12 most recent clear sky index values. As a follow-up,

(a)

Time

• Analysis → control forecast
• Perturbed initial conditions → ensemble members

Climatology

(b)

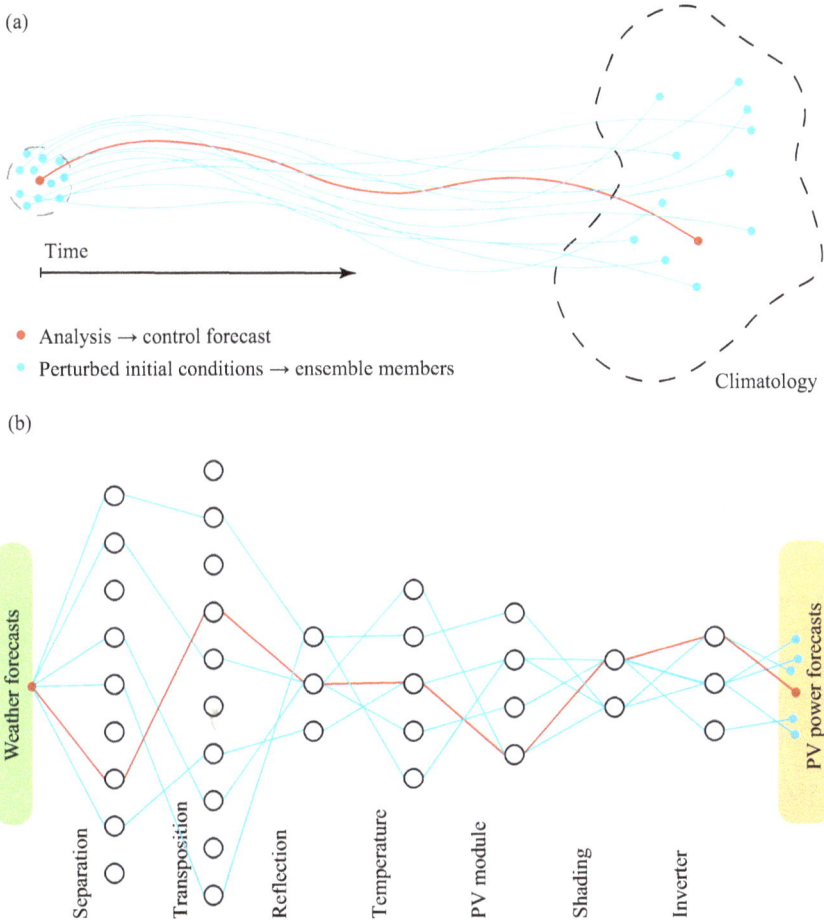

Figure 7.4   *Concepts of the (a) ensemble NWP and (b) ensemble of model chains [2]*

Yagli *et al.* [57] created ensemble GHI forecasts from the GHI data of the 81 surrounding satellite pixels of the location of interest using a dropout neural network (NN) with Monte Carlo sampling. It can be seen that an ensemble can be created from almost everything, but it only works well when the spread of the ensemble correlates with the uncertainty of the forecasts.

Ensemble forecasts might show bias and tend to be underdispersed, i.e., the spread of the ensemble is smaller than the observed errors, since they can only cover some but not all sources of uncertainty [58,59]. Empirical evidence for this can be seen in [2], where ensemble NWP, an ensemble of model chains, and their combination are compared. It is widely accepted that GHI forecasting contributes more to PV forecasting errors than irradiance-to-power conversion, and accordingly, the ensemble of model chains was found to be the most underdispersed, followed by the ensemble NWP, whereas their combination that takes into account both sources of uncertainty was the least underdispersed. Even so, since good reliability is a prerequisite of a decent probabilistic forecast, ensemble forecasts should always be calibrated, which can be performed by the various postprocessing methods presented in Section 7.5. The predictors of the calibration method can either be the summary statistics of the ensemble (e.g., the mean and standard deviation), or the forecast values of all ensemble members [60]. In the latter case, care must be taken with ensemble NWP, in which the members are randomly initialized in each model run, and thus the members with the same number are unrelated between the different model runs, and their characteristics are not consistent over a long dataset. To that end, the ensemble NWP members should always be sorted first before using them as predictors [2].

7.5   Statistical postprocessing

In the last decades, several approaches have been developed for statistical post-processing of forecasts; for a recent overview, see [61]. Among them, parametric methods provide full predictive distributions in the form of parametric probability laws with parameters depending on the raw ensemble forecasts, whereas nonparametric approaches are distribution-free and approximate the predictive distribution by estimating its quantiles. In the following sections, these two approaches will be referred to as distributional and quantile forecasts, respectively.

7.5.1   Parametric methods

A simple and efficient parametric postprocessing approach is the nonhomogeneous regression or ensemble model output statistics (EMOS) proposed by Gneiting *et al.* [62]. The EMOS method models the conditional distribution of the future value Y of the studied quantity at a given location and time, provided the corresponding ensemble forecast $f_1, f_2, \ldots, f_K$ with a given forecast horizon using a single parametric law, where the parameters are connected to the ensemble prediction via appropriate, usually affine, link functions. The choice of the parametric distribution family and the link functions is a crucial step in EMOS modeling, whose selection should reflect the properties of the quantity at hand to be predicted. For instance, the predictive distribution for temperature and sea-level pressure suggested in [62] is Gaussian with mean μ and variance σ^2 expressed as

$$\mu = a_0 + a_1 f_1 + \cdots + a_K f_K \quad \text{and} \quad \sigma^2 = b_0 + b_1 S^2, \tag{7.1}$$

where S^2 denotes the ensemble variance. Following the optimum score principle of [58], model parameters $a_0, a_1, \ldots, a_K, b_0, b_1 \in \mathbb{R}$, $b_0, b_1 \geq 0$, are estimated by optimizing the mean value of a verification score (usually the CRPS or the logarithmic score, see Section 7.6) over the training data comprising past forecast-observation pairs. Note that when the forecast ensemble consists of groups of exchangeable members, forecasts within a given group should get the same EMOS weights [63]. Thus, for instance, in the case of the 51-member ECMWF ensemble, the expression of the mean μ of the Gaussian EMOS model given in (7.1) reduces to

$$\mu = a_0 + a_1 f_{\mathrm{CTRL}} + a_2 \bar{f}_{\mathrm{ENS}},$$

where f_{CTRL} denotes the control forecasts, and $\bar{f}_{\mathrm{ENS}}$ is the mean of the 50 perturbed ensemble members $f_{\mathrm{ENS},1}, f_{\mathrm{ENS},2}, \ldots, f_{\mathrm{ENS},50}$.

By now, EMOS models are available for a wide range of weather quantities, such as wind speed [64–67], precipitation accumulation [68,69], or various heat indices [70], utilizing a broad range of distribution families. Focusing on solar irradiance, to address non-negativity, a common approach is to consider distributions left-truncated at zero (see, e.g., [71,72]). However, these models require the exact specification of the period within a day when positive irradiance can be observed, which strongly depends on the location and day of the year. To circumvent this deficiency, Schultz *et al.* [73] consider EMOS models based on logistic and Gaussian distributions left-censored from below at zero, which might assign a positive probability to the event of zero irradiance (point mass at zero). They suggest linking the location parameter μ and scale parameter σ of these censored laws to the ensemble members as

$$\mu = \alpha_0 + \alpha_1 f_1 + \cdots + \alpha_K f_K + \nu p_0 \quad \text{and} \quad \sigma = \exp\left(\beta_0 + \beta_1 \log S^2\right), \quad (7.2)$$

where p_0 is the proportion of ensemble members predicting zero irradiance, that is

$$p_0 := \frac{1}{K} \sum_{k=1}^{K} \mathbb{I}_{\{f_k = 0\}}.$$

As an alternative, Horat *et al.* [20] investigate a left-censored Gaussian EMOS model for GHI with location and scale parameters given as in (7.1).

Similar ideas can be applied to the building of EMOS models for the photovoltaic (PV) power output. However, in this case, the maximal capacity of the solar plants introduces a new constraint on the set of possible parametric distribution families. As a first step, in this direction, in [20], an EMOS model utilizing a doubly censored Gaussian distribution with point masses at both zero and the maximal capacity of the given plant is considered with location μ and scale σ connected to the PV power ensemble forecasts according to (7.1).

To include additional predictor variables, in [74], a boosted version of EMOS is proposed, automatically selecting the most important input features, thus preventing model overfitting.

A more complex parametric postprocessing approach is the Bayesian model averaging (BMA) introduced by Raftery *et al.* [75]. The BMA predictive distribution of the quantity Y of interest is a weighted mixture with PDFs

$$p(y|f_1, \ldots, f_K; \theta_1, \ldots, \theta_K) := \sum_{k=1}^{K} \omega_k g\left(y|f_k, \theta_k\right), \quad (7.3)$$

where $g\left(y|f_k, \theta_k\right)$ is the component PDF from a parametric family corresponding to the kth ensemble member f_k with parameter (vector) θ_k to be estimated, and ω_k is the corresponding weight satisfying $\omega_k \geq 0$, $k = 1, 2, \ldots, K$, and $\sum_{k=1}^{K} \omega_k = 1$. A component PDF can be interpreted as the conditional PDF of Y provided that the corresponding forecast is the best and that the weights are based on the relative performance of the members of the ensemble over the training data. Such mixture models can be beneficial when the observations indicate a multimodal character that cannot be handled by a single PDF considered in EMOS modeling. BMA weights and model parameters are usually estimated using linear regression and the maximum-likelihood method, where the maximum of the likelihood function is found by the expectation–maximization algorithm. Similar to the EMOS, the choice of the parametric family in BMA modeling is determined by the properties of the quantity to be predicted. For instance, in [75], the predictive PDF for temperature is given as a Gaussian mixture, where the kth component has mean $\alpha_k + \beta_k f_k$ and variance σ^2, that is

$$\sum_{k=1}^{K} \omega_k \, \varphi \left(\frac{y - \alpha_k - \beta_k f_k}{\sigma} \right), \tag{7.4}$$

with φ denoting the PDF of the standard Gaussian law. Again, to account for the existence of groups of exchangeable ensemble members, the same weights and parameters within a given group are suggested [76]. In this way, for the 51-member ECMWF ensemble, (7.4) takes the following form:

$$(1 - \omega)\varphi \left(\frac{y - \alpha_{\text{CTRL}} - \beta_{\text{CTRL}} f_{\text{CTRL}}}{\sigma} \right) + \frac{\omega}{50} \sum_{k=1}^{50} \varphi \left(\frac{y - \alpha_{\text{ENS}} - \beta_{\text{ENS}} f_{\text{ENS},k}}{\sigma} \right),$$

where $\omega \in [0, 1]$, so one has just six unknown parameters to be estimated.

BMA models have also been developed for calibrating wind speed using mixtures of gamma [77] and truncated normal distributions [78], wind direction utilizing a mixture of von Mises distributions [79], or Box–Cox transformed water levels based on a doubly truncated Gaussian mixture [80]. Furthermore, in the case of solar irradiance ensemble forecasts, Aryaputera *et al.* [81] suggested a BMA model mixing skewed Gaussian distributions and compared its forecast skill with the Gaussian BMA and the Gaussian and skewed Gaussian EMOS models. Finally, to handle the discrete-continuous nature of precipitation accumulation, Sloughter *et al.* [82] considered a BMA model, where the individual components contain a point mass at zero to address no precipitation, whereas positive precipitation accumulation is modeled by a gamma distribution.

The same idea is followed in [83], where two BMA models for calibrating PV power forecasts are proposed. For this quantity, on the one hand, the predictive distribution has to be concentrated to the interval $[0, P_{\text{max}}]$, where P_{max} denotes the AC rating of the PV plant. On the other hand, a point mass should be assigned to the event that the plant is clipped, that is, the power output is restricted to AC rating. Clipping is usually performed slightly under P_{max}; in [83], a threshold λP_{max} with $\lambda = 0.995$ is suggested. Given an ensemble member f_k, the probability $\text{P}(y \geq \lambda P_{\text{max}} | f_k)$ of clipping is specified by logistic regression, while the amount of power subject to not clipping is modeled either with a doubly truncated Gaussian distribution or with a beta distribution, both restricted to the $[0, P_{\text{max}}]$ interval.

Recently, ML-based methods for statistical postprocessing of ensemble forecasts have gained more and more popularity. As a direct extension of the EMOS framework, Rasp and Lerch [84] proposed the distributional regression network (DRN) approach, where instead of fixed link functions, the parameters of the predictive distribution are connected to the model predictors via an NN. The advantage of DRN modeling is that it allows an easy extension of the set of predictors by arbitrary problem-related quantities such as forecasts of related weather quantities or location-related information such as geographical coordinates, altitude, or land use; moreover, it can more easily capture the possible nonlinear relationships between the predictors (input features of the NN) and the parameters of the predictive distribution (NN outputs). To model 2m temperature, Rasp and Lerch [84] used a Gaussian predictive distribution and a fully connected feed-forward NN with two output neurons corresponding to the mean and the standard deviation. Similar to the standard approach in EMOS modeling, the NN is trained by optimizing the CRPS as a loss function. Besides summary statistics (mean and standard deviation) of the ensemble forecasts of the target 2m temperature, summary statistics of ensemble predictions of other weather quantities (e.g., surface pressure, sensible and latent heat flux, U wind (zonal), and V wind (meridional) at various levels) are considered as input features together with station-specific information, which were included via embedding. In a case study based on ensemble forecasts for surface observation stations in Germany, the DRN approach outperformed the benchmark Gaussian EMOS and boosted EMOS methods.

Similar to EMOS, the DRN approach has already been tested for several different weather variables such as wind gust [60] or precipitation accumulation [85,86], and the NN-based models consistently outperformed their EMOS counterparts utilizing the same predictive distribution (truncated logistic for wind gust and censored and shifted gamma for precipitation accumulation). However, one should also mention that compared to the handful of parameters in EMOS models, the NNs used in DRN approaches, which are usually fully connected and contain just one or two hidden layers, have a much larger number of network weights to be estimated. Hence, DRN models require far more training data than the corresponding EMOS methods.

Restricting again our attention to solar irradiance, in [87] and [20], GHI ensemble forecasts were calibrated with the help of DRN models based on a Gaussian distribution left-censored at zero. However, in [87], a short (31-day) rolling training window was applied, and the input features of the corresponding EMOS model (7.2) were extended just by the forecast horizon, whereas in [20], deterministic forecasts of 2m temperature and 10m wind speed were also utilized, and the model training was based on a fixed dataset spanning over 29 months.

Finally, [20] also investigated the skill of DRN postprocessing of PV power output ensemble forecasts, where the predictive distribution is the same doubly censored Gaussian, as in the corresponding EMOS model, and the CRPS is applied as the loss function. However, in this case, the advantage of NN-based postprocessing over the simpler EMOS method is rather small, as the available additional predictors do not convey much important information about the PV power to be predicted.

7.5.2   Nonparametric approaches

The nonparametric methods typically estimate discrete quantiles of the predictive probability law instead of the parameters of a predefined distribution function. The predictive cumulative distribution function (CDF) F can also be represented by its τ-quantiles $q_\tau(F) := F^{-1}(\tau) := \inf\{y : F(y) \geq \tau\}$. The essence of quantile regression (QR) is the pinball or quantile loss, defined as

$$\rho_\tau(u) := \begin{cases} u\tau, & \text{if } u \geq 0, \\ u(\tau - 1), & \text{if } u < 0, \end{cases} \tag{7.5}$$

an asymmetric loss function that is minimized by $q_\tau(F)$. The simplest realization of QR is the linear quantile regression (LQR), where the forecasts for the quantile q_τ are calculated as a linear combination of the predictors. If LQR is used for the calibration of ensemble forecasts, the predictors are the K ensemble members, and the calibrated quantile is calculated as

$$q_\tau = \beta_0 + \sum_{k=1}^{K} \beta_k f_k, \tag{7.6}$$

where $\beta_0, \beta_1, \ldots, \beta_K \in \mathbb{R}$ are the regression coefficients. In the traditional deterministic linear regression, the coefficients are most commonly estimated using the least squares method by minimizing the sum of squared errors. In contrast, the coefficients of the LQR are fitted by minimizing the sum of pinball loss as

$$\min \sum_{i=1}^{N} \rho_\tau \left(x_i - q_{\tau,i} \right), \tag{7.7}$$

where $q_{\tau,i}$ is the forecast quantile, and x_i is the observation in the ith sample, and N is the number of training samples. The weighted sum of the quantile score for all quantiles is equal to half of the CRPS [88], which suggests that a quantile forecast estimated by QR is also optimal in terms of the CRPS.

Despite the simplicity of LQR, it was effectively applied for the calibration of model chain ensemble in [3] and ECMWF ensemble NWP in [2], achieving good reliability even though the raw ensemble forecasts were seriously underdispersed. The reliability diagram (see Section 7.6 for description) of the raw and calibrated ensembles in Figure 7.5 clearly shows the effectiveness of this linear method for creating reliable forecasts. However, linearity imposes a significant limitation on the flexibility of QR; therefore, improved performance can be expected by using a nonlinear regression model. Theoretically, any nonlinear regression method whose parameters can be fitted to minimize the pinball loss function can be used for quantile regression.

Quantile regression neural networks (QRNN) use an NN, most commonly a multilayer perceptron (MLP), to provide a nonlinear mapping between the predictors and the forecast quantiles [89]. The pinball loss function, however, is not differentiable everywhere, which may lead to convergence problems in the gradient-based

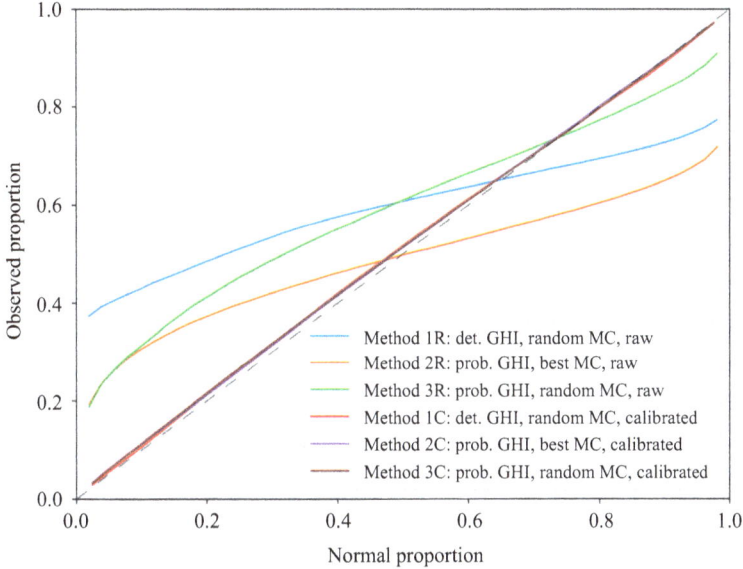

Figure 7.5   *The calibration of PV power forecasts with LQR on a reliability diagram [2]*

optimization methods commonly used in the training of NNs. A remedy proposed in [90] is to use the quantile Huber loss, defined as

$$\rho_{h,\tau(u)} := \begin{cases} h(u)\tau, & \text{if } u \geq 0, \\ h(u)(\tau - 1), & \text{if } u < 0, \end{cases} \tag{7.8}$$

where $h(u)$ is the Huber norm, which combines the L_2 and L_1 norm to absolute values below and above a small ε threshold (e.g., 10^{-8}), respectively, as

$$h(u) := \begin{cases} \frac{u^2}{2\varepsilon}, & \text{if } |u| \leq \varepsilon, \\ |u| - \frac{\varepsilon}{2}, & \text{if } |u| > \varepsilon. \end{cases} \tag{7.9}$$

Many QR methods, including the LQR, can only approximate a single quantile at once; therefore, individual models must be created and trained for each quantile. In such cases, forecasting the whole predictive distribution with high resolution, e.g., forecasting all equidistant quantiles from 1% to 99% with 1% or 2% steps, highly increases computational requirements. In this case, one can opt for creating interval forecasts that represent the uncertainty in the form of central prediction intervals. A central prediction interval with $(1 - \alpha)100\%$ coverage probability is defined by the $q_{\alpha/2}$ and $q_{1-\alpha/2}$ quantiles. However, one must note that interval forecasts carry less information about the overall uncertainty compared to distributional forecasts. Alternatively, QRNN is also capable of including multiple neurons in the output layer with different activation functions. If the activation functions are pinball loss functions with different τ values, the QRNN is able to estimate multiple quantiles (possibly even all quantiles required to define the distribution with the desired resolution) at once with a single model, eliminating the need to initialize and train multiple models.

A notable variant of QRNN is the Bernstein quantile network (BQN) proposed by Bremnes [91]. In BQN, the quantile functions are modeled as linear combinations of Bernstein basis polynomials and the coefficients are estimated from the predictors by an NN. The objective function of the training is the average of the pinball loss for a set of quantile levels. The method was further adjusted in [60] by constraining the coefficients to be nondecreasing, which implies that the quantile function is monotonically increasing. This was implemented by estimating the increments between the coefficients rather than the coefficients themself as the output of the NN using a softplus activation function. A monotonically increasing quantile function eliminates the problem of quantile crossing, which occurs when the forecast value of a lower quantile is higher than that of a higher quantile, i.e., $q_{\tau_1} > q_{\tau_2}$ for $\tau_1 < \tau_2$. This violates that a CDF must be, by definition, monotonically increasing. The BQN has been applied for the D2P postprocessing of solar forecasts in [92]. The inputs are the solar zenith angle and the forecasts of EMCWF's deterministic high-resolution (HRES) model for the GHI (in the form of a clear sky index), 2m temperature, surface pressure, and relative humidity. The BQN is found to outperform a DRN that is used to fit a truncated normal distribution for the same data in terms of calibration and also achieves lower scores, which can be attributed to the BQN's higher flexibility in the shape of the distribution.

A more universal solution to avoid quantile crossing in QRNN has been proposed in [93]. The idea of the noncrossing quantile regression neural network (NCQRNN) is to extend the QRNN with an additional hidden layer before the output layer that ensures a nondecreasing mapping between the outputs of the previous layer and the nodes of the output layer, where each node calculates a quantile. This approach has no requirements on the network structure before the noncrossing and output layer. Therefore, it can be integrated with any type of NN without limitation. The effectiveness of the NCQRNN was demonstrated for the calibration of the ECMWF ensemble GHI forecasts, and it was not only able to fully eliminate quantile crossing but outperformed all 12 benchmark models, including the aforementioned AnEn, EMOS, LQR, BQN, and the QRF presented below [93]. A technical detail that one should be aware of is that in [93], the postprocessing was performed on the clear sky index instead of the GHI itself.

Another important type of QR that is worth mentioning is the quantile regression forest (QRF), which is a probabilistic extension of the random forest ML model [94]. The QRF is based on the discovery that random forecasts provide information about the full conditional distribution of the predictand, which allows the calculation of the conditional quantiles. The first utilization of QRF for postprocessing was for the calibration of surface temperature and wind speed ensemble forecasts, where it performed better than the EMOS [95]. In the context of solar forecasting, the QRF was applied in [96] for the calibration of the ECMWF ensemble GHI forecasts, together with five EMOS and two QRNN variants, simple LQR, and LQR with L_1-penalty (lasso regression). On average, the QRF performed better than EMOS but worse than other QR methods in the comparison for the seven sites of the Surface

Radiation Budget Network (SURFRAD) in the United States. The probabilistic fore-casts obtained by the 10 methods were also combined by simple quantile averaging (SQA) and constrained quantile regression averaging (CQRA), resulting in improved scores compared to the individual methods [96].

7.6   Verification of probabilistic forecasts

As argued by Gneiting *et al.* [97], forecast evaluation should be based on the idea of "maximizing the sharpness of the predictive distributions subject to calibration." Sharpness describes the concentration of the probabilistic forecast, while calibra-tion refers to the statistical consistency between the forecasts and the materialized observations or events. Hence, sharpness is the property of the forecast only, whereas calibration is the joint property of the forecasts and the corresponding observations.

Calibration of probabilistic predictions can be easily assessed visually with the help of the verification rank histograms of ensemble forecasts (see Section 9.7.1 of [98]), and their counterparts for forecasts issued in the form of full predictive dis-tributions (see Section 7.5.1), the probability integral transform (PIT) histograms (see Section 9.5.4 of [98]). The verification rank is the rank of the verifying observation with respect to the corresponding forecast ensemble, while the PIT is the value of the predictive CDF evaluated at the verifying observation with possible randomization at the points of discontinuity. For a properly calibrated K-member forecast ensemble, the verification rank is uniformly distributed over integers $\{1, 2, \ldots, K + 1\}$, while calibrated predictive distributions result in standard uniform PIT values. The shapes of the rank and PIT histograms reflect the source of the lack of calibration: U-shapes and hump-shapes refer to under- and overdispersion, whereas biased forecasts result in triangular shapes.

A powerful graphical tool for assessing forecast calibration is the reliability dia-gram. For dichotomous events, it displays the graph of the observed frequency of the given event against the binned forecast frequencies; see Section 9.4.4 of [98] for details and Figure 7.5 for an example. Reliability diagrams can also be cre-ated for the quantile representation of continuous forecast variables by plotting the observed proportion (i.e., the proportion of the samples where the observation is smaller than the forecast value of the given quantile) as a function of the nominal proportion of the quantile for all available quantiles [99]. In the ideal case, this graph should lie on the main diagonal of the unit square.

The study of the average width and coverage of $(1 - \alpha)100\%$, $\alpha \in]0, 1[$, central prediction intervals also provides information about the sharpness and calibration of probabilistic forecasts, respectively. Sharpness can be quantified by the prediction interval average width (PIAW) metric, where a smaller PIAW indicates sharper fore-casts. In this context, by coverage, we mean the proportion of verifying observations located within the corresponding central prediction intervals, called prediction inter-val coverage probability (PICP), which for a calibrated forecast should be around $(1 - \alpha)100\%$. That said, the PICP alone is insufficient to evaluate calibration since it can take its perfect value even if both quantiles defining the prediction interval are biased in the same direction and the observations outside the interval are distributed highly unevenly between the two sides. Level α is often chosen to match the nominal $(K-1)/(K+1)100\%$ coverage of the range of the K-member ensemble (96.08% for the 50-member operational ECMWF ensemble), which allows a fair comparison of the raw predictions with distributional forecasts discussed in Section 7.5.

Calibration and sharpness can also be addressed simultaneously with the help of proper scoring rules [58]. A scoring rule is a loss function that assigns numerical values to forecast–observation pairs, which is called proper if the expected score is optimal when the distribution of the observation matches the issued predictive distribution and strictly proper when this optimum is unique.

Probably the most popular strictly proper scoring rules are the logarithmic (or ignorance) score, which is the negative logarithm of the predictive PDF evaluated at the verifying observation [100], and the continuous ranked probability score (CRPS; see Section 9.1 of [98]). Given a probabilistic forecast issued in the form of a predictive CDF F and a corresponding observation $x \in \mathbb{R}$, the CRPS is defined as

$$\mathrm{CRPS}(F, x) := \int_{-\infty}^{\infty} \left[F(y) - \mathbb{I}_{\{y \geq x\}} \right]^2 \mathrm{d}y, \tag{7.10}$$

where $\mathbb{I}_H$ denotes the indicator function of a set H. Note that both scores are negatively oriented (the smaller the better), and the CRPS can be expressed in the same units as the observation. Furthermore, the CRPS has a closed form for a large number of probability distributions, a comprehensive summary can be found in [101]. Finally, to calculate the CRPS of an ensemble forecast, one should consider the empirical CDF of the prediction.

Quantile forecasts can be evaluated with the help of the quantile score (QS; see Section 9.6.1 of [98]), which is a proper scoring rule defined as follows:

$$\mathrm{QS}_\tau(F, x) := \rho_\tau\left(x - q_\tau(F)\right), \tag{7.11}$$

where $\rho_\tau(x)$ is the pinball loss function as defined in (7.5). The CRPS is twice the integral of the QS for all quantiles [102].

To investigate the forecast skill in terms of predicting probabilities of dichotomous events, one can also consider the Brier score (BS; see Section 9.4.2 of [98]). For instance, given a predictive CDF F, the BS corresponding to the event that the observed quantity x exceeds a given threshold y is defined as follows:

$$\mathrm{BS}(F, x; y) := \left[F(y) - \mathbb{I}_{\{y \geq x\}} \right]^2. \tag{7.12}$$

Note that the CRPS is the integral of the BS over all possible thresholds.

A proper score that can be used for the evaluation of interval forecasts of central prediction interval forecasts with $(1 - \alpha)100\%$ is the interval score IS [58], which is defined as

$$\mathrm{IS}(L, U; x) := (U - L) + \frac{2}{\alpha}(L - x)\mathbb{I}_{\{x < L\}} + \frac{2}{\alpha}(x - U)\mathbb{I}_{\{x > Us\}}, \tag{7.13}$$

where $L = q_{\alpha/2}$ is the lower and $U = q_{1-\alpha/2}$ is the upper and upper quantile bounding the prediction interval.

The CRPS, QS, and BS admit a very useful additive decomposition into terms representing reliability (calibration), resolution, and uncertainty, for the details see [103], [104], and [105], respectively.

Finally, a unified verification practice could enhance the comparability of the methods presented in different papers, and thus help to identify the most valuable contributions. The CRPS seems to be the most commonly used proper scoring rule in probabilistic solar forecasting, and it can be calculated for parametric distributional, ensemble, and quantile forecasts. However, the CRPS cannot be calculated for interval forecasts, thus, from this perspective, creating high-resolution quantile forecasts is more recommended than only estimating a single prediction interval. Overall, the CRPS is encouraged to be included, preferably with its decomposition, in the verification of all probabilistic GHI or PV power forecasts.

7.7 Summary and conclusions

Owing to the additional information it contains compared to deterministic forecasts, probabilistic forecasting is gaining increasing importance and research interest over the years. Probabilistic forecasts can be created in many different ways, but reliable probabilistic forecasting always requires a data-driven postprocessing step that is able to match the predictive density of the forecasts to the distribution of the observations. That said, incorporating physical knowledge, most commonly in the form of an NWP or model chain ensemble whose spread is well correlated to the uncertainty of the forecasts, is generally beneficial for improving probabilistic PV power forecast accuracy.

Despite the ever-growing literature of probabilistic forecasting, there is still a need for significant research to develop a comprehensive understanding, especially in such specific fields as PV power forecasting. On the one hand, there is still room for the development of new methods that, at least in some respects, perform better than the existing well-established solutions. On the other hand, practical applications are still dominantly rely on deterministic forecasts; therefore, studies on the potential utilization and added value of probabilistic PV power forecasts are also important for reaching the maturity of this field.

Acknowledgments

The preparation of this chapter was supported by the National Research, Development and Innovation Fund under the project numbers OTKA-FK 142702 and OTKA-K 142849.

References

[1] Yang D, Liu B, Zhang H, *et al.* A second tutorial review of the solar power curve: Applications in energy meteorology. *Advances in Atmospheric Sciences*. 2025;42:269–296.

[2]   Mayer MJ and Yang D. Pairing ensemble numerical weather prediction with ensemble physical model chain for probabilistic photovoltaic power forecasting. *Renewable and Sustainable Energy Reviews*. 2023;175:113171.

[3]   Mayer MJ and Yang D. Probabilistic photovoltaic power forecasting using a calibrated ensemble of model chains. *Renewable and Sustainable Energy Reviews*. 2022;168:112821.

[4]   Gneiting T. Making and evaluating point forecasts. *Journal of the American Statistical Association*. 2011;106:746–762.

[5]   Mayer MJ and Yang D. Calibration of deterministic NWP forecasts and its impact on verification. *International Journal of Forecasting*. 2023;39: 981–991.

[6]   Kolassa S. Why the "best" point forecast depends on the error or accuracy measure. *International Journal of Forecasting*. 2020;36:208–211.

[7]   Mayer MJ and Yang D. Potential root mean square error skill score. *Journal of Renewable and Sustainable Energy*. 2024;16:016501.

[8]   Yang D and Kleissl J. Summarizing ensemble NWP forecasts for grid operators: Consistency, elicitability, and economic value. *International Journal of Forecasting*. 2023;39:1640–1654.

[9]   Renkema Y, Brinkel N, and Alskaif T. Conformal prediction for stochastic decision-making of PV power in electricity markets. *Electric Power Systems Research*. 2024;234:110750.

[10]   Visser LR, AlSkaif TA, Khurram A, *et al.* Probabilistic solar power forecasting: An economic and technical evaluation of an optimal market bidding strategy. *Applied Energy*. 2024;370.

[11]   Lindberg O, Lingfors D, Arnqvist J, *et al.* Day-ahead probabilistic forecasting at a co-located wind and solar power park in Sweden: Trading and forecast verification. *Advances in Applied Energy*. 2023;9:100120.

[12]   Hobbs BF, Zhang J, Hamann HF, *et al.* Using probabilistic solar power forecasts to inform flexible ramp product procurement for the California ISO. *Solar Energy Advances*. 2022;2:100024.

[13]   Li B, Feng C, Siebenschuh C, *et al.* Sizing ramping reserve using probabilistic solar forecasts: A data-driven method. *Applied Energy*. 2022;313:118812.

[14]   Wang Q, Tuohy A, Ortega-Vazquez M, *et al.* Quantifying the value of probabilistic forecasting for power system operation planning. *Applied Energy*. 2023;343:121254.

[15]   Ramahatana F, Le Gal La Salle J, Lauret P, *et al.* A more efficient microgrid operation through the integration of probabilistic solar forecasts. *Sustainable Energy, Grids and Networks*. 2022;31:100783.

[16]   Li B and Zhang J. A review on the integration of probabilistic solar forecasting in power systems. Solar Energy. 2020;210:68–86.

[17]   Yang D and Kleissl J. *Solar Irradiance and Photovoltaic Power Forecasting*. Boca Raton: CRC Press; 2024.

[18]   Mayer MJ and Yang D. Optimal place to apply post-processing in the deterministic photovoltaic power forecasting workflow. *Applied Energy*. 2024;371:123681.

[19]   Yang D and van der Meer D. Post-processing in solar forecasting: Ten overarching thinking tools. *Renewable and Sustainable Energy Reviews.* 2021;140:110735.

[20]   Horat N, Klerings S, and Lerch S. Improving model chain approaches for probabilistic solar energy forecasting through post-processing and machine learning. *Advances in Atmospheric Sciences.* 2025;42:297–312.

[21]   Yang D, Xia X, and Mayer MJ. A tutorial review of the solar power curve: Regressions, model chains, and their hybridization and probabilistic extensions. *Advances in Atmospheric Sciences.* 2024;41:1023–1067.

[22]   Markovics D and Mayer MJ. Comparison of machine learning methods for photovoltaic power forecasting based on numerical weather prediction. *Renewable and Sustainable Energy Reviews.* 2022;161:112364.

[23]   Mayer MJ and Gróf G. Extensive comparison of physical models for photovoltaic power forecasting. *Applied Energy.* 2021;283:116239.

[24]   Mayer MJ. Influence of design data availability on the accuracy of physical photovoltaic power forecasts. *Solar Energy.* 2021;227:532–540.

[25]   Mayer MJ. Benefits of physical and machine learning hybridization for photovoltaic power forecasting. *Renewable and Sustainable Energy Reviews.* 2022;168:112772.

[26]   de Oliveira Santos L, AlSkaif T, Barroso GC, *et al.* Photovoltaic power estimation and forecast models integrating physics and machine learning: A review on hybrid techniques. *Solar Energy.* 2024;284:113044.

[27]   Yang D and Gueymard CA. Ensemble model output statistics for the separation of direct and diffuse components from 1-min global irradiance. *Solar Energy.* 2020;208:591–603.

[28]   Quan H and Yang D. Probabilistic solar irradiance transposition models. *Renewable and Sustainable Energy Reviews.* 2020;125:109814.

[29]   Yang D, Wang W, Gueymard CA, *et al.* A review of solar forecasting, its dependence on atmospheric sciences and implications for grid integration: Towards carbon neutrality. *Renewable and Sustainable Energy Reviews.* 2022;161:112348.

[30]   Reda I and Andreas A. Solar position algorithm for solar radiation applications. *Solar Energy.* 2004;76:577–589.

[31]   Duffie JA and Beckman WA. *Solar Engineering of Thermal Processes* Hoboken, NJ: Wiley; 2013.

[32]   Sun X, Bright JM, Gueymard CA, *et al.* Worldwide performance assessment of 75 global clear-sky irradiance models using Principal Component Analysis. *Renewable and Sustainable Energy Reviews.* 2019;111:550–570.

[33]   Fu D, Liu M, Yang D, *et al.* Influences of atmospheric reanalysis on the accuracy of clear-sky irradiance estimates: Comparing MERRA-2 and CAMS. *Atmospheric Environment.* 2022;277:119080.

[34]   Gueymard CA. Cloud and albedo enhancement impacts on solar irradiance using high-frequency measurements from thermopile and photodiode radiometers. Part 1: Impacts on global horizontal irradiance. *Solar Energy.* 2017;153:755–765.

[35]   Järvelä M, Lappalainen K, and Valkealahti S. Characteristics of the cloud enhancement phenomenon and PV power plants. *Solar Energy*. 2020;196:137–145.

[36]   Mayer MJ. Effects of the meteorological data resolution and aggregation on the optimal design of photovoltaic power plants. *Energy Conversion and Management*. 2021;241:114313.

[37]   Abreu EFM, Gueymard CA, Canhoto P, *et al.* Performance assessment of clear-sky solar irradiance predictions using state-of-the-art radiation models and input atmospheric data from reanalysis or ground measurements. *Solar Energy*. 2023;252:309–321.

[38]   Notton G, Lazarov V, and Stoyanov L. Optimal sizing of a grid-connected PV system for various PV module technologies and inclinations, inverter efficiency characteristics and locations. *Renewable Energy*. 2010;35: 541–554.

[39]   Chen S, Li P, Brady D, *et al.* Determining the optimum grid-connected photovoltaic inverter size. *Solar Energy*. 2013;87:96–116.

[40]   Mayer MJ and Gróf G. Techno-economic optimization of grid-connected, ground-mounted photovoltaic power plants by genetic algorithm based on a comprehensive mathematical model. *Solar Energy*. 2020;202:210–226.

[41]   Yang D. Choice of clear-sky model in solar forecasting. *Journal of Renewable and Sustainable Energy*. 2020;12:026101.

[42]   Lauret P, Alonso-Suárez R, Le Gal La Salle J, *et al.* Solar forecasts based on the clear sky index or the clearness index: Which is better? *Solar*. 2022;2:432–444.

[43]   Engerer NA and Mills FP. KPV: A clear-sky index for photovoltaics. *Solar Energy*. 2014;105:679–693.

[44]   Gneiting T and Raftery AE. Weather forecasting with ensemble methods. *Science*. 2005;310(5746):248–249.

[45]   Buizza R. Introduction to the special issue on "25 years of ensemble forecasting". *Quarterly Journal of the Royal Meteorological Society*. 2018;145(51):1–11.

[46]   ECMWF. *IFS Documentation CY49R1 – Part V: Ensemble Prediction System.* Reading, UK: ECMWF; 2024.

[47]   Bi K, Xie L, Zhang H, *et al.* Accurate medium-range global weather forecasting with 3D neural networks. *Nature*. 2023;619:533–538.

[48]   Kurth T, Subramanian S, Harrington P, *et al.* FourCastNet: Accelerating global high-resolution weather forecasting using adaptive Fourier neural operators. In: *Proceedings of the Platform for Advanced Scientific Computing Conference. PASC '23.* New York, NY: Association for Computing Machinery; 2023. p. 13.

[49]   Lam R, Sanchez-Gonzalez A, Willson M, *et al.* Learning skillful medium-range global weather forecasting. *Science*. 2023;382(6677):1416–1421.

[50]   Lang S, Alexe M, Chantry M, *et al.* AIFS: a new ECMWF forecasting system. *ECMWF Newsletter*. 2024;178:4–5.

[51]   Lang S, Alexe M, Clare MCA, *et al. AIFS-CRPS: Ensemble forecasting using a model trained with a loss function based on the Continuous Ranked Probability Score.* arXiv:2412.15832; 2024.

[52]   Price I, Sanchez-Gonzalez A, Alet F, *et al.* Probabilistic weather forecasting with machine learning. *Nature.* 2025;637:84–90.

[53]   Alessandrini S, Monache LD, Sperati S, *et al.* An analog ensemble for short-term probabilistic solar power forecast. *Applied Energy.* 2015;157:95–110.

[54]   Yang D and Alessandrini S. An ultra-fast way of searching weather analogs for renewable energy forecasting. *Solar Energy.* 2019;185:255–261.

[55]   Yang D, Kong Y, Liu B, *et al.* Comparing calibrated analog and dynamical ensemble solar forecasts. *Solar Energy Advances.* 2024;4:100048.

[56]   Yagli GM, Yang D, and Srinivasan D. Ensemble solar forecasting using data-driven models with probabilistic post-processing through GAMLSS. *Solar Energy.* 2020;208:612–622.

[57]   Yagli GM, Yang D, and Srinivasan D. Ensemble solar forecasting and post-processing using dropout neural network and information from neighboring satellite pixels. *Renewable and Sustainable Energy Reviews.* 2022;155:111909.

[58]   Gneiting T and Raftery AE. Strictly proper scoring rules, prediction and estimation. *Journal of the American Statistical Association.* 2007;102(477):359–378.

[59]   Buizza R. Ensemble forecasting and the need for calibration. In: Vannitsem S, Wilks DS, Messner JW, editors. *Statistical Postprocessing of Ensemble Forecasts.* Amsterdam: Elsevier; 2018. pp. 15–48.

[60]   Schulz B and Lerch S. Machine learning methods for postprocessing ensemble forecasts of wind gusts: A systematic comparison. *Monthly Weather Review.* 2022;150(1):235–257.

[61]   Vannitsem S, Bremnes JB, Demaeyer J, *et al.* Statistical postprocessing for weather forecasts – review, challenges and avenues in a big data world. *Bulletin of the American Meteorological Society.* 2021;102(3):E681–E699.

[62]   Gneiting T, Raftery AE, Westveld AH, *et al.* Calibrated probabilistic forecasting using ensemble model output statistics and minimum CRPS estimation. *Monthly Weather Review.* 2005;133(5):1098–1118.

[63]   Gneiting T. Calibration of medium-range weather forecasts. *ECMWF Technical Memorandum.* 2014;719.

[64]   Thorarinsdottir TL and Gneiting T. Probabilistic forecasts of wind speed: Ensemble model output statistics by using heteroscedastic censored regression. *Journal of the Royal Statistical Society: Series A (Statistics in Society).* 2010;173A(2):371–388.

[65]   Lerch S and Thorarinsdottir TL. Comparison of non-homogeneous regression models for probabilistic wind speed forecasting. *Tellus A: Dynamic Meteorology and Oceanography.* 2013;65(1):21206.

[66]   Baran S and Lerch S. Log-normal distribution based EMOS models for probabilistic wind speed forecasting. *Quarterly Journal of the Royal Meteorological Society.* 2015;141:2289–2299.

[67]   Baran S, Szokol P, and Szabó M. Truncated generalized extreme value distribution-based ensemble model output statistics model for calibration of wind speed ensemble forecasts. *Environmetrics*. 2021;32:e2678.

[68]   Scheuerer M. Probabilistic quantitative precipitation forecasting using ensemble model output statistics. *Quarterly Journal of the Royal Meteorological Society*. 2014;140(680):1086–1096.

[69]   Baran S and Nemoda D. Censored and shifted gamma distribution based EMOS model for probabilistic quantitative precipitation forecasting. *Environmetrics*. 2016;27(5):280–292.

[70]   Baran S, Baran Á, Pappenberger F, *et al.* Statistical post-processing of heat index ensemble forecasts: Is there a royal road? *Quarterly Journal of the Royal Meteorological Society*. 2020;146:3416–3434.

[71]   Le Gal La Salle J, Badosa J, David M, *et al.* Added-value of ensemble prediction system on the quality of solar irradiance probabilistic forecasts. *Renewable Energy*. 2020;162:1321–1339.

[72]   Yang D. Ensemble model output statistics as a probabilistic site-adaptation tool for solar irradiance: A revisit. *Journal of Renewable and Sustainable Energy*. 2020;12(3):036101.

[73]   Schulz B, El Ayari M, Lerch S, *et al.* Post-processing numerical weather prediction ensembles for probabilistic solar irradiance forecasting. *Solar Energy*. 2021;220:1016–1031.

[74]   Messner JW, Mayr GJ, Zeileis A. Nonhomogeneous boosting for predictor selection in ensemble postprocessing. *Monthly Weather Review*. 2017;145(1):137–147.

[75]   Raftery AE, Gneiting T, Balabdaoui F, *et al.* Using Bayesian model averaging to calibrate forecast ensembles. *Monthly Weather Review*. 2005;133(5):1155–1174.

[76]   Fraley C, Raftery AE, and Gneiting T. Calibrating multimodel forecast ensembles with exchangeable and missing members using Bayesian model averaging. *Monthly Weather Review*. 2010;138(1):190–202.

[77]   Sloughter JM, Gneiting T, and Raftery AE. Probabilistic wind speed forecasting using ensembles and Bayesian model averaging. *Journal of the American Statistical Association*. 2010;105(489):25–35.

[78]   Baran S. Probabilistic wind speed forecasting using Bayesian model averaging with truncated normal components. *Computational Statistics and Data Analysis*. 2014;75:227–238.

[79]   Bao L, Gneiting T, Grimit EP, *et al.* Bias correction and Bayesian model averaging for ensemble forecasts of surface wind direction. *Monthly Weather Review*. 2010;138(5):1811–1821.

[80]   Baran S, Hemri S, El Ayari M. Statistical postprocessing of water level forecasts using Bayesian model averaging with doubly truncated normal components. *Water Resources Research*. 2019;55(5):3997–4013.

[81]   Aryaputera AW, Verbois H, and Walsh WM. Probabilistic accumulated irradiance forecast for Singapore using ensemble techniques. In: *2016 IEEE*

43rd *Photovoltaic Specialists Conference (PVSC)*. Portland, OR, USA; 2016. pp. 1113–1118.

[82]   Sloughter JM, Raftery AE, Gneiting T, *et al.* Probabilistic quantitative precipitation forecasting using Bayesian model averaging. *Monthly Weather Review*. 2007;135(9):3209–3220.

[83]   Doubleday K, Jascourt S, Kleiber W, *et al.* Probabilistic solar power forecasting using Bayesian model averaging. *IEEE Transactions on Sustainable Energy*. 2021;12(1):325–337.

[84]   Rasp S and Lerch S. Neural networks for postprocessing ensemble weather forecasts. *Monthly Weather Review*. 2018;146(11):3885–3900.

[85]   Ghazvinian M, Zhang Y, Seo DJ, *et al.* A novel hybrid artificial neural network - Parametric scheme for postprocessing medium-range precipitation forecasts. *Advances in Water Resources*. 2021;151:103907.

[86]   Ghazvinian M, Zhang Y, Hamill TM, *et al.* Improving probabilistic quantitative precipitation forecasts using short training data through artificial neural networks. *Journal of Hydrometeorology*. 2022;23(9):1365–1382.

[87]   Baran Á and Baran S. A two-step machine learning approach to statistical post-processing of weather forecasts for power generation. *Quarterly Journal of the Royal Meteorological Society*. 2024;150(759):1029–1047.

[88]   Bröcker J. Evaluating raw ensembles with the continuous ranked probability score. *Quarterly Journal of the Royal Meteorological Society*. 2012;138(667):1611–1617.

[89]   Taylor JW. A quantile regression neural network approach to estimating the conditional density of multiperiod returns. *Journal of Forecasting*. 2000;19:299–311.

[90]   Cannon AJ. Quantile regression neural networks: Implementation in R and application to precipitation downscaling. *Computers and Geosciences*. 2011;37:1277–1284.

[91]   Bremnes JB. Ensemble postprocessing using quantile function regression based on neural networks and Bernstein polynomials. *Monthly Weather Review*. 2020;148:403–414.

[92]   Gneiting T, Lerch S, and Schulz B. Probabilistic solar forecasting: Benchmarks, post-processing, verification. *Solar Energy*. 2023;252:72–80.

[93]   Song M, Yang D, Lerch S, *et al.* Non-crossing quantile regression neural network as a calibration tool for ensemble weather forecasts. *Advances in Atmospheric Sciences*. 2024;41:1417–1437.

[94]   Meinshausen N. Quantile regression forests. *Journal of Machine Learning Research*. 2006;7:983–999.

[95]   Taillardat M, Mestre O, Zamo M, *et al.* Calibrated ensemble forecasts using quantile regression forests and ensemble model output statistics. *Monthly Weather Review*. 2016;144:2375–2393.

[96]   Yang D, Yang G, and Liu B. Combining quantiles of calibrated solar forecasts from ensemble numerical weather prediction. *Renewable Energy*. 2023;189:118993.

[97]   Gneiting T, Balabdaoui F, and Raftery AE. Probabilistic forecasts, calibration and sharpness. *Journal of the Royal Statistical Society: Series B (Statistical Methodology)*. 2007;69B(2):243–268.

[98]   Wilks DS. *Statistical Methods in the Atmospheric Sciences*. 4th ed. Amsterdam: Elsevier; 2020.

[99]   Lauret P, David M, Pinson P. Verification of solar irradiance probabilistic forecasts. Solar Energy. 2019;194:254–271.

[100]   Good IJ. Rational decisions. *Journal of the Royal Statistical Society: Series B (Statistical Methodology)*. 1952;14B(1):107–114.

[101]   Jordan A, Krüger F, and Lerch S. Evaluating probabilistic forecasts with scoringRules. *Journal of Statistical Software*. 2019;90:1–37.

[102]   Gneiting T and Ranjan R. Comparing density forecasts using threshold and quantile-weighted scoring rules. *Journal of Business and Economic Statistics*. 2011;29:411–422.

[103]   Hersbach H. Decomposition of the continuous ranked probability score for ensemble prediction systems. *Weather and Forecasting*. 2000;15:559–570.

[104]   Bentzien S and Friederichs P. Decomposition and graphical portrayal of the quantile score. *Quarterly Journal of the Royal Meteorological Society*. 2014;140(683):1924–1934.

[105]   Murphy AH. A new vector partition of the probability score. *Journal of Applied Meteorology and Climatology*. 1973;12(4):595–600.

Chapter 8

Model optimisation, hyperparameter tuning and performance evaluation in machine learning models for solar PV generation forecast

Hugo Quest[1,2], Christophe Ballif[1,3] and Elham Shirazi[4]

Forecasting solar energy is essential for efficient integration of photovoltaics (PV) into electricity grids and optimally managing renewable energy resources. As PV becomes an increasingly significant component of the energy landscape, accurate predictions of its availability and output are required for efficient integration into energy systems and ensuring a steady supply of electricity. Artificial intelligence (AI)-based PV forecasts are particularly valuable due to their ability to accurately predict PV generation by extracting complex relationships between different variables. However, their performance is heavily influenced by the selection of hyperparameters, such as learning rate, network architecture, and regularisation terms, which significantly affect the model's structure, learning efficiency, and forecasting accuracy. This chapter focuses on the importance of model tuning in AI-driven PV forecasting and explains how selecting the right hyperparameters can enhance forecast accuracy and support more reliable solar energy integration.

8.1 Introduction

The adoption of solar photovoltaics is a critical aspect of the global shift towards carbon-neutral energy systems. Driven by numerous regulations, policies, incentive programmes, technological advancements and environmental concerns, the deployment of PV systems has increased exponentially in recent years [1,2]. However, this rapid expansion can present challenges such as voltage regulation and congestion in the electricity grid if these systems are not integrated efficiently [3–5]. Accurate PV power forecasting is essential for the efficient integration of solar energy into the grid and the optimisation of PV system operations. Key challenges in PV forecasting

[1]École Polytechnique Fédérale de Lausanne (EPFL), Institute of Electrical and Micro Engineering (IEM), Photovoltaics and Thin-Film Electronics Laboratory, Switzerland
[2]3S Swiss Solar Solutions AG, CH-3645 Thun, Switzerland
[3]Swiss Centre for Electronics and Microtechnology (CSEM), Sustainable Energy Centre, Switzerland
[4]Faculty of Engineering Technology, University of Twente, The Netherlands

include managing weather variability, accurately modelling complex non-linear relationships and incorporating diverse data sources such as satellites, sky imagers and on-site measurements [6].

Machine learning (ML) models have proven to be powerful tools for predicting solar energy output due to their ability to analyse large datasets and identify complex patterns [7–11]. PV forecasting models use inputs from different sources, including ground-based sky imagers, satellite images and numerical weather predictions (NWP) [12,13]. Ground-based sky imagers have proven to be highly accurate for short-term forecasts but are expensive to deploy and maintain [14]. Satellite-based images and NWP perform well for long-term (6-h to day-ahead) forecasts but not for short-term, high spatial resolution forecasts [15]. Recent work has therefore focused on using PV data from multiple sites to exploit spatio-temporal relations and improve the forecast accuracy, for example, through graph signal processing (GSP), utilising the fact that dense PV networks can act as virtual or proxy weather stations [16–18]. However, the effectiveness of these models heavily depends on how well they are tuned to specific forecasting tasks [19,20]. In this context, hyperparameters – parameters that govern the learning process – are particularly important. Hyperparameters are preset before model training begins and control the learning process, unlike model parameters, which are learned during training. These hyperparameters can interact in complex, non-linear ways, influencing the behaviour and performance of the model [21–23]. Proper tuning identifies the optimal combination of parameters to maximise predictive accuracy and prevent overfitting or underfitting. Additionally, different datasets and tasks may require tailored tuning approaches to address specific characteristics such as size, noise level and feature distribution, as well as specific goals such as precision, recall or overall accuracy. Proper tuning can significantly reduce errors in predictions, leading to more accurate and reliable solar energy forecasts. Figure 8.1 shows a general ML workflow, from data processing to model optimisation, which forms the focus of this chapter.

Furthermore, tuning AI models for PV forecasting requires a deep understanding of both the models and the unique characteristics of solar energy production. Factors such as weather conditions, geographical location, and seasonal variations

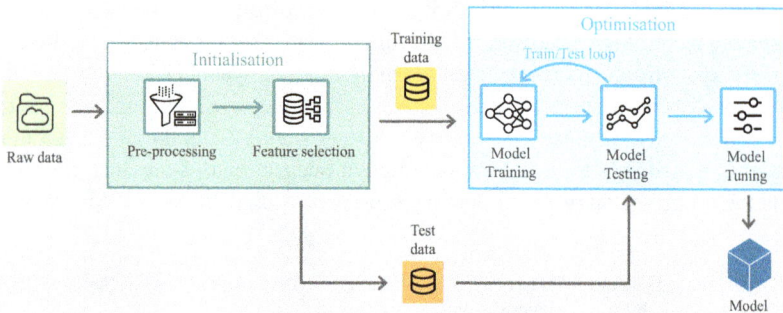

Figure 8.1   General schematic of a ML model workflow. This chapter delves into the model optimisation and tuning (blue box).

play a crucial role in determining solar energy output and must be considered during the tuning process. Advanced techniques, such as cross-validation (CV) and hyperparameter optimisation, are employed to systematically explore different parameter settings and identify the best configuration for the given task [24,25].

By fine-tuning AI models, a higher level of precision can be achieved in solar energy forecasts, which is essential for grid operators and energy planners. Accurate forecasts enable better scheduling of energy resources, reducing the reliance on fossil fuels and minimising the need for costly backup power. Furthermore, improved forecasting helps mitigate the risks associated with the intermittent nature of solar power, thereby enhancing the overall reliability and efficiency of the electricity grid. This chapter is structured as follows: Section 8.2 describes performance evaluation metrics and methods, which form the basis of ML model tuning, as they describe the objectives to minimise and the ways in which to achieve it optimally. Section 8.3 delves into the methods for hyperparameter tuning, detailing how to explore search spaces for effective parameter setting. An overview of key hyperparameters of a few ML models is also provided. Finally, Section 8.4 examines a brief case study of hyperparameter tuning for PV system forecasting, showcasing two methods described in the chapter.

Challenges in hyperparameter tuning for PV power forecasting

- **Computational complexity:** Time-series models, especially neural networks, are resource-intensive, requiring efficient search methods and high-performance computing.
- **Non-stationarity and seasonality:** PV data exhibits seasonal and non-stationary behaviour, which can be addressed using models such as SARIMA or those designed for seasonality.
- **Overfitting:** Poor tuning can lead to overfitting, requiring regularisation, CV, and proper model complexity adjustments.
- **Dynamic nature of PV data:** Changes in PV data over time (e.g., degradation and weather) necessitate continuous re-tuning and learning.

8.2   Performance evaluation

Performance evaluation in ML-based PV forecasting allows the assessment of model efficacy in predicting solar energy output. This process involves various metrics and techniques to measure the accuracy, reliability and robustness of the forecasting models. The choice of evaluation methods, such as CV and time-series split, and evaluation metrics [26,27] are critical, as they are influenced by the model type and the specific forecasting challenges within the PV domain, which will be discussed in this section. By evaluating the performance of the forecasting models, researchers can ensure more accurate and reliable solar energy predictions, leading to more efficient integration of PV into energy systems [28,29].

8.2.1   Evaluation metrics

In the context of tuning ML models for PV power forecasting, evaluation metrics play a crucial role. These metrics provide a quantitative basis for assessing the performance of different models and their configurations, thereby determining the effectiveness of various hyperparameter settings. For classification problems, metrics such as F1 score, recall and precision are commonly used, while for regression problems such as PV power forecasting, metrics such as root mean squared error (RMSE), mean absolute error (MAE), mean bias error (MBE) and the coefficient of determination (R^2) are typically employed.

For instance, RMSE and MAE provide insights into the average magnitude of prediction errors, with RMSE placing more emphasis on larger errors. This can be particularly useful for identifying models that might occasionally produce large deviations. MBE, in contrast, helps to detect any consistent bias in the predictions, indicating whether the model tends to overestimate or underestimate the actual values. The R^2 metric measures the proportion of variance in the dependent variable that is predictable from the independent variables, offering an overall measure of the model's explanatory power.

These evaluation metrics are integral to the process of hyperparameter tuning. By systematically varying the hyperparameters and using these metrics to evaluate the resulting models, the hyperparameter settings that yield the most accurate and reliable forecasts can be determined. This iterative process of tuning and evaluation helps in refining the model to achieve optimal performance.

In this section, different evaluation metrics used in PV power forecasting are described, as well as their significance in the model tuning process.

8.2.1.1   Root mean square error (RMSE)

The RMSE quantifies the difference between the predicted and actual measured values by calculating the square root of the average of the squared differences. In other words, the RMSE measures the average size of the errors between the predicted and observed values. The RMSE is especially valuable because it accounts for individual errors while giving greater emphasis on larger errors, providing a thorough assessment of prediction accuracy. Mathematically, the RMSE is defined as follows:

$$\text{RMSE} = \sqrt{\frac{1}{n}\sum_{i=1}^{n}(X_{\text{pred},i} - x_{\text{obs},i})^2} \tag{8.1}$$

where $X_{\text{pred},i}$ represents the predicted value at time i, $x_{\text{obs},i}$ the observed (actual) value at time i and n the total number of observations. The RMSE provides a single measure of predictive accuracy, with lower values indicating better predictive performance.

8.2.1.2 Normalised root mean square error (nRMSE)

The normalised root mean square error (nRMSE) is the normalised versions of the RMSE, which quantifies the average magnitude of prediction errors. Normalisation makes nRMSE scale-independent, enabling comparison across different datasets.

$$\text{nRMSE} = \frac{\sqrt{\frac{1}{N} \sum_{t=1}^{N} (X_{\text{pred},i} - x_{\text{obs},i})^2}}{\max(x_{\text{obs},i}) - \min(x_{\text{obs},i})} \times 100\% \tag{8.2}$$

where $x_{\text{obs},i}$ is the measured value, $X_{\text{pred},i}$ is the predicted value and N is the number of samples.

8.2.1.3 Mean absolute error (MAE)

The MAE is a straightforward metric that provides insight into the accuracy of predictions by quantifying the average of the absolute differences between the predicted and observed values. Unlike the RMSE, which emphasises larger errors, the MAE gives equal importance to all errors, making it a simpler and more balanced measure of overall prediction accuracy. Mathematically, the MAE is defined as follows:

$$\text{MAE} = \frac{1}{n} \sum_{i=1}^{n} |X_{\text{pred},i} - x_{\text{obs},i}| \tag{8.3}$$

where $X_{\text{pred},i}$ represents the predicted value at time i, $x_{\text{obs},i}$ the observed value at time i and n the total number of observations.

8.2.1.4 Normalised mean absolute error (nMAE)

The normalised mean absolute error (nMAE) is the normalised version of the MAE, which quantifies the average magnitude of prediction errors. Similar to nRMSE, normalisation makes nMAE scale-independent, enabling comparison across different datasets.

$$\text{nMAE} = \frac{\frac{1}{N} \sum_{t=1}^{N} |X_{\text{pred},i-x_{\text{obs},i}}|}{\max(x_{\text{obs},i}) - \min(x_{\text{obs},i})} \times 100\% \tag{8.4}$$

where $x_{\text{obs},i}$ is the measured value, $X_{\text{pred},i}$ is the predicted value and N is number of samples.

8.2.1.5 Mean absolute percentage error (MAPE)

The mean absolute percentage error (MAPE) quantifies the average of the absolute percentage differences between the predicted and observed values. It offers a clear understanding of prediction accuracy by expressing errors as a percentage. The MAPE treats all errors proportionally, relative to the actual observed values. This means each error's impact is scaled according to the size of the actual value, providing a balanced view of prediction accuracy across different magnitudes. This proportional approach makes the MAPE particularly useful for comparing prediction performance across different datasets or time periods. Mathematically, the MAPE is defined as follows:

$$\text{MAPE} = \frac{100\%}{n} \sum_{i=1}^{n} \left| \frac{X_{\text{pred},i} - x_{\text{obs},i}}{x_{\text{obs},i}} \right| \tag{8.5}$$

where $X_{\text{pred},i}$ represents the predicted value at time i, $x_{\text{obs},i}$ the observed value at time i and n the total number of observations.

8.2.1.6   Mean bias error (MBE)

The mean bias error (MBE) quantifies the average difference between the predictions and observations, providing an indication of whether the predicted values are systematically overestimating or underestimating the observed values. Mathematically, the MBE is defined as follows:

$$\text{MBE} = \frac{1}{n} \sum_{i=1}^{n} (X_{\text{pred},i} - x_{\text{obs},i}) \tag{8.6}$$

where $X_{\text{pred},i}$ represents the predicted value at time i, $P_{\text{obs},i}$ the observed value at time i and n the total number of observations. The MBE provides a single measure of predictive bias, with negative values indicating a tendency to underestimate and positive values indicating a tendency to overestimate the actual values.

8.2.1.7   Coefficient of determination (R^2)

The coefficient of determination (R^2) measures the proportion of variance in the observed data that is predictable from the independent variables. It provides insight into the goodness of fit of a model by quantifying how well the predicted values correspond to the actual observed values. An R^2 value of 1 indicates perfect correlation, whereas an R^2 value of 0 suggests that the model fails to explain any of the variability in the response data. Mathematically, R^2 is defined as follows:

$$R^2 = 1 - \frac{\sum_{i=1}^{n} (x_{\text{obs},i} - X_{\text{pred},i})^2}{\sum_{i=1}^{n} (x_{\text{obs},i} - \bar{x}_{\text{obs}})^2} \tag{8.7}$$

where $x_{\text{obs},i}$ represents the observed value at time i, $X_{\text{pred},i}$ the predicted value at time i and $\bar{x}_{\text{obs}}$ the mean of the observed values. The numerator of the fraction calculates the residual sum of squares (RSS), while the denominator calculates the total sum of squares (TSS). The R^2 metric provides a normalised measure of model performance, making it useful for comparing the predictive accuracy of different models or evaluating the same model across different datasets.

What are the metrics evaluating?

- **Regression metrics:**
  - **RMSE and MAE:** describe the average magnitude of prediction errors (with RMSE giving more weight to larger errors).

- **nRMSE and nMAE:** describe the average error. Normalisation makes nRMSE and nMAE scale-independent, enabling comparison across different datasets.
- **MAPE:** evaluates the average percentage error, providing a normalised measure of prediction accuracy relative to the actual values.
- **MBE:** detects consistent bias in predictions, indicating whether models tend to over- or under-estimate actual values.
- **Coefficient of determination** R^2**:** measures an overall indication of the model's explanatory power.

- **Classification metrics:**
  - **Precision:** measures the proportion of correctly predicted positive observations to the total predicted positives.
  - **Recall:** measures the proportion of correctly predicted positive observations to all actual positives.
  - **F1 score:** represents the harmonic mean of precision and recall, balancing both concerns especially in cases of class imbalance.

8.2.2   Evaluation methods

In addition to the evaluation metrics summarised in Table 8.1, selecting appropriate evaluation methods is essential. These methods help ensure that the model's performance is assessed accurately and reliably, providing a robust estimate of how the model will generalise to unseen data. Among these methods, CV techniques are particularly valuable for mitigating overfitting and enabling a comprehensive evaluation by utilising different portions of the data for training and testing. They play a crucial role in the ML model and hyperparameter tuning process, offering a systematic approach to assess model performance across multiple training and testing scenarios

Table 8.1   Summary of the evaluation metrics and their equations, where $X_{pred,i}$ is the predicted value at time i, $x_{obs,i}$ the observed value at time i, n the total number of observations and $\bar{x}_{obs}$ the mean of the observed values.

Evaluation metric	Equation
Root mean squared error	$\mathrm{RMSE} = \sqrt{\frac{1}{n} \sum_{i=1}^{n} (X_{\mathrm{pred},i} - x_{\mathrm{obs},i})^2}$
Mean absolute error	$\mathrm{MAE} = \frac{1}{n} \sum_{i=1}^{n} \lvert X_{\mathrm{pred},i} - x_{\mathrm{obs},i} \rvert$
Mean absolute percentage error	$\mathrm{MAPE} = \frac{100\%}{n} \sum_{i=1}^{n} \left\lvert \frac{X_{\mathrm{pred},i} - x_{\mathrm{obs},i}}{x_{\mathrm{obs},i}} \right\rvert$
Mean bias error	$\mathrm{MBE} = \frac{1}{n} \sum_{i=1}^{n} (X_{\mathrm{pred},i} - x_{\mathrm{obs},i})$
Coefficient of determination	$R^2 = 1 - \frac{\sum_{i=1}^{n} (x_{\mathrm{obs},i} - X_{\mathrm{pred},i})^2}{\sum_{i=1}^{n} (x_{\mathrm{obs},i} - \bar{x}_{\mathrm{obs}})^2}$

(see Figure 8.1 in the optimisation section). By applying CV, it is possible to obtain a more reliable estimate of a model's performance and its sensitivity to various hyper-parameter settings. This, in turn, aids in identifying the optimal hyperparameters that yield the best predictive accuracy and generalisation capability.

CV operates by partitioning the dataset into multiple subsets, known as folds, and iteratively training and testing the model on these different subsets. This technique allows for a comprehensive assessment of the model's performance across various segments of the data, providing a more robust evaluation than a simple train–test split. By systematically rotating through different portions of the dataset as the test set, CV ensures that the model generalises well to unseen data and helps to detect issues related to overfitting or underfitting.

In this section, various CV methods are described, discussing their principles, advantages and limitations.

Overfitting and underfitting?

One of the main challenges in achieving good generalisation is balancing between overfitting and underfitting. Overfitting occurs when a model is too tailored to the training data, capturing noise and specific patterns that do not generalise to new, unseen data. This usually happens with small datasets or highly complex models, leading to poor performance. Underfitting happens when a model is too simplistic and fails to capture underlying patterns in data, often resulting from an overly simple model or insufficient training.

8.2.2.1   Leave-one-out cross-validation

Leave-one-out cross-validation (LOOCV) is a form of CV in which the dataset is split into as many subsets as there are data points. Each data point is treated individually as a separate test set, as seen in Figure 8.2, while the remaining data points form the training set. LOOCV therefore provides an exhaustive evaluation of a model's performance, as it uses all but one data point for training in each iteration. This can be advantageous for smaller datasets, as it maximises the amount of data used for training. As a result, LOOCV tends to have lower bias compared to other CV methods, which is particularly useful when data availability is limited.

However, LOOCV comes with significant computational cost. Since the model must be trained n times, where n is the number of data points, LOOCV can be prohibitively expensive for large datasets. This makes it less practical for high volumes of data. Moreover, each training iteration is also sensitive to outliers, as the exclusion of a single data point can cause a significant variation in the model's performance metric for that iteration. Despite these limitations, LOOCV remains a valuable tool, especially in cases where data is scarce, and the need for thorough model evaluation outweighs the computational cost.

Leave-One-Out Cross-Validation (LOOCV)   K-fold Cross-Validation

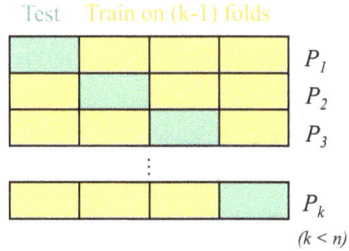

Test   Train on (n-1) points   Test   Train on (k-1) folds

| | | | ... | | | | | P_1 | | | | | | P_1 |

Figure 8.2   Leave-one-out cross-validation and k-fold cross-validation. The performance of each iteration P_i is typically averaged to determine the overall model performance metric.

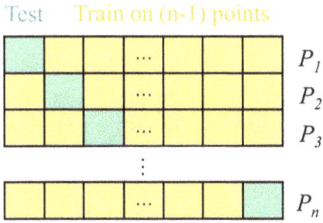

8.2.2.2   K-fold cross-validation

K-fold CV is a widely used and versatile CV method that addresses some of the limitations of LOOCV, particularly its computational inefficiency. In k-fold CV, the dataset is divided into k equal-sized subsets or folds, as shown in Figure 8.2. Unlike LOOCV, which uses only one data point for testing at a time, one of the k folds is used as the validation set, while the remaining $(k-1)$ folds are combined to form the training set [30]. This process is repeated k times, with k typically much smaller than n.

The value of k is typically chosen based on the size of the dataset and the computational resources available. A key advantage of k-fold CV is that it provides a more balanced trade-off between bias and variance compared to LOOCV. By using multiple folds, k-fold CV tends to reduce the variance of the performance estimate while maintaining a relatively low bias. K-fold CV is particularly useful when the dataset is large enough that LOOCV would be computationally expensive but small enough that a simple train–test split might not provide a reliable estimate of model performance. However, as k increases, k-fold CV begins to resemble LOOCV, especially when $k = n$.

Another consideration is the potential for data leakage if the folds are not carefully constructed. In time-series data or data with strong temporal dependencies, it is important to ensure that the training set for each fold does not include future information that would not be available at the time of prediction. In such cases, variants of k-fold CV, such as time-series CV or stratified k-fold CV, can be employed to ensure that the folds are representative of the problem's structure.

Overall, k-fold CV provides a robust and flexible framework for model evaluation, balancing the need for computational efficiency with the goal of obtaining reliable performance estimates. By carefully choosing the value of k and ensuring that the folds are constructed appropriately, k-fold CV can be effectively used in a wide range of ML applications.

Overall performance metric?

Typically, the MAE, RMSE and R^2 are considered as the performance metrics, as described in Section 8.2.1. The final performance measure of the model for both the LOOCV and k-fold CV is the average of the n or k performance measures, respectively. This average is considered a more robust estimate of how the model is expected to perform on unseen data compared to a single train–test split, as it reduces the variability associated with a particular random partitioning of the data.

Comparing LOOCV and k-fold CV

- **LOOCV**: uses n folds (number of data points in the dataset), each with one data point as the test set. It is computationally expensive and can lead to high variance in the performance estimate.
- **K-fold CV**: uses k folds ($k < n$), with each fold used once as the test set. It is more computationally efficient and provides a better balance between bias and variance.

8.2.2.3   Time-series cross-validation

Time-series data presents unique challenges for CV due to its sequential nature, where observations are dependent on previous values. K-fold CV or LOOCV is therefore not always advisable for time series, as they can lead to data leakage, where future information influences past predictions [31–33]. To mitigate this risk, time-series CV employs methods specifically designed to respect the temporal order of data, ensuring that training and testing are conducted in a manner consistent with real-world forecasting scenarios.

Two common approaches to time-series CV are the sliding window and expanding window methods:

- **Sliding window:** In this approach, a fixed-size training set is utilised to predict future outcomes. For instance, for a time series of length n, the training set could consist of the first k observations, used to forecast the $(k + 1)$th observation. After making this prediction, the training window shifts forward by one time step, incorporating the newly available observation while dropping the oldest one, and the process is repeated. This method ensures that the model is continuously trained on the most recent data, making it adaptable to changes in trends or patterns over time. However, a significant drawback is that the fixed-size nature of the training set can limit the model's ability to capture long-term trends if the window is too small.

- **Expanding window:** This approach begins with a small training set that gradually increases in size with each iteration. Starting with the first k observations, the model is trained to predict the $(k + 1)$th step. After the prediction, this observation is added to the training set for the next iteration, and the model is retrained to forecast the next time step. This method allows the model to leverage all previously seen data, potentially leading to better performance as more information is gathered over time. However, as the training set grows, the computational cost also increases, and the model may become less responsive to recent changes in the data.

Both methods ensure that the model is evaluated under conditions that closely resemble real-world forecasting, where only past data is available at the time of prediction. They allow for a more realistic assessment of the model's predictive capability.

8.3   Model optimisation

Achieving optimal performance in PV forecasting models requires more than just selecting the right evaluation method; it demands careful attention to the model's configuration through the process of model optimisation, i.e., hyperparameter tuning. While model performance evaluation helps to diagnose issues such as overfitting and underfitting, optimisation seeks to address these challenges by fine-tuning hyperparameters and therefore the model's architecture and settings. Effective hyperparameter tuning involves systematically searching for the best combination of these settings using methods such as grid search, random search or more advanced techniques like Bayesian optimisation, to ensure that the model not only fits the training data well but also performs robustly on unseen data. This section first explores the strategies and techniques for optimising ML models, then gives a few examples of the most common ML models, which are used in PV power forecast.

8.3.1   Techniques for hyperparameter tuning

Hyperparameter tuning requires the definition of a search space, which can be visualised as an n-dimensional space. In this space, each hyperparameter corresponds to a different dimension, and the range of values that a hyperparameter can have determines the scale of that dimension. This value can be either numerical or categorical. In this search space, every point represents a unique configuration of the model. The optimisation process aims to identify the point or vector in this space that results in the best performance, such as achieving the highest accuracy or the lowest error. Various optimisation algorithms can be used for this purpose such as grid search, random search and Bayesian optimisation [34,35].

8.3.1.1   Grid search optimisation

Grid search is a hyperparameter optimisation technique that systematically explores a specified subset of the hyperparameter space by evaluating every combination of the

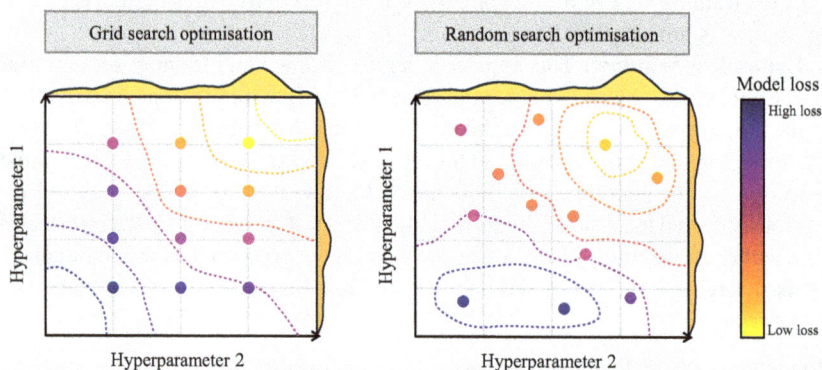

Figure 8.3   Schematic representation of grid search and random search optimisation, with a colour map indicating the model error or loss

provided hyperparameter values to identify the best configuration based on a given evaluation metric, as illustrated in Figure 8.3. It is simple to implement and understand, making it a popular choice for tuning ML models. However, grid search can be computationally intensive and time-consuming, especially when dealing with a large number of hyperparameters or wide ranges of values. Despite its thoroughness and determinism, grid search becomes impractical for large-scale problems due to its exhaustive nature.

Grid search offers a significant advantage in its structured approach, ensuring that every possible combination within the defined search space is evaluated. This guarantees that the global optimum for the hyperparameters, within the given grid, is identified. However, this exhaustive exploration also leads to one of its primary drawbacks: inefficiency, particularly in high-dimensional or continuous hyperparameter spaces. The computational cost escalates exponentially as the number of hyperparameters increases, often making grid search impractical for complex models or large datasets. To mitigate these challenges, one can limit the grid's scope or combine grid search with other techniques, such as random search, to balance thoroughness and computational efficiency.

8.3.1.2   Random search optimisation

Random search is a hyperparameter optimisation technique that involves randomly sampling combinations of hyperparameters from predefined ranges or distributions and evaluating their performance. Unlike grid search, which exhaustively tests every possible combination, random search selects configurations randomly, allowing for a more diverse exploration of the hyperparameter space (see Figure 8.3 for the comparison of both methods). This method is often more efficient and less computationally demanding, as it can find good hyperparameter settings with fewer evaluations, especially when only a few hyperparameters significantly impact performance. Random

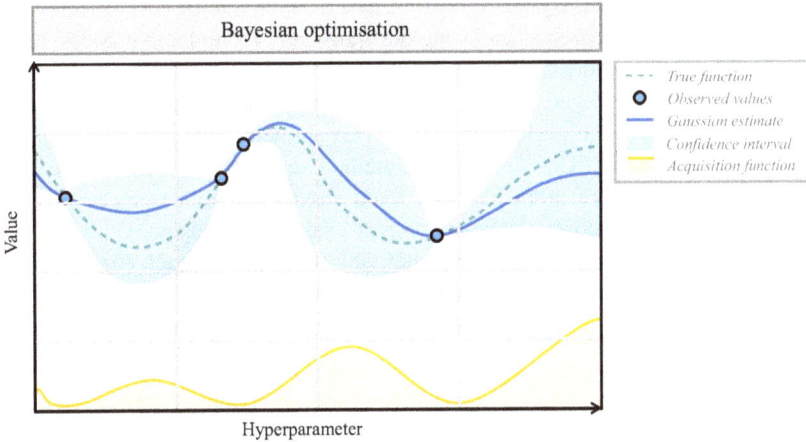

Figure 8.4   *Illustration of the Gaussian process used to predict the objective*
function across a hyperparameter space. The blue curve represents the
Gaussian process mean prediction, with the shaded area indicating the
uncertainty (variance). The blue dots denote observed evaluations of
the objective function, while the acquisition function (orange shaded
curve) guides the selection of the next hyperparameter to evaluate,
aiming to balance exploration and exploitation.

search is particularly useful when dealing with high-dimensional spaces or when the exact relationship between hyperparameters and model performance is unknown.

One of the key advantages of random search is its ability to escape the limitations of grid search, particularly in scenarios where only a subset of hyperparameters has a significant impact on model performance. By not being constrained to a fixed grid, random search can potentially identify optimal or near-optimal configurations more quickly, even in cases where the important hyperparameters are not immediately obvious. This stochastic nature allows random search to efficiently explore larger and more complex hyperparameter spaces, increasing the likelihood of discovering high-performing models with fewer evaluations. However, the randomness also introduces an element of uncertainty, as there is no guarantee that the most effective combination will be sampled, making multiple runs or larger sample sizes sometimes necessary to achieve reliable results.

8.3.1.3   Bayesian optimisation

Unlike grid search and random search, which systematically or randomly explore the hyperparameter space, Bayesian search (or Bayesian optimisation) builds a probabilistic model of the objective function and uses it to select promising hyperparameters to evaluate (see Figure 8.4). This approach balances exploration (searching unexplored areas) and exploitation (refining known good areas), making it more efficient and often faster in finding optimal hyperparameters, especially in complex, high-dimensional spaces. By learning from past evaluations, Bayesian search intelligently navigates the search space using the Gaussian probabilistic process, leading to improved model performance with fewer iterations.

By building a surrogate model, typically using Gaussian processes, Bayesian optimisation predicts the performance of hyperparameter configurations that have not yet been evaluated. This enables the method to focus on areas of the hyperparameter space that are more likely to yield significant improvements in model performance, thereby reducing the number of evaluations needed. This method is particularly advantageous in scenarios where the evaluation of each configuration is computationally expensive, as it prioritises the most promising hyperparameters early in the process, leading to faster and often more reliable optimisation outcomes.

8.3.1.4   Hyperband optimisation

Hyperband is an advanced hyperparameter optimisation algorithm that combines random search with early stopping techniques to efficiently find the best hyperparameters for ML models. It operates by allocating resources (such as computational time or iterations) to a large number of hyperparameter configurations initially and then progressively focusing on the most promising ones. Hyperband performs this through a series of rounds, where each round uses a successively smaller set of configurations with a larger budget per configuration, based on their performance in earlier rounds. This approach balances exploration and exploitation, making Hyperband effective at discovering high-quality hyperparameters with fewer resources and less time compared to exhaustive methods like grid search.

A key innovation is its ability to dynamically allocate resources based on performance, which significantly reduces the computational cost typically associated with hyperparameter optimisation. Hyperband effectively determines how to distribute resources across different configurations, allowing poor-performing configurations to be discarded early. This early stopping mechanism is particularly beneficial when dealing with deep learning models or other computationally intensive algorithms, as it prevents unnecessary evaluation of sub-optimal configurations. Furthermore, Hyperband's flexibility in adjusting the trade-off between the breadth of exploration and the depth of exploitation makes it well-suited for a wide range of ML tasks, from tuning simple models to optimising complex, large-scale systems.

8.3.2   *Hyperparameters of ML models*

Each ML model comes with a set of hyperparameters that influence its learning process and final performance. In this subsection, key hyperparameters of some common ML models, which are used in PV forecasting, will be described, discussing their effect on model performance and how they can be adjusted to optimise model outcomes. A brief introduction to the different ML models is also given for context. Understanding and tuning these hyperparameters is crucial for building effective models tailored to PV forecasting. These models are explained in detail in other chapters.

8.3.2.1 Support vector regression

Support vector regression (SVR) extends the principles of support vector machines (SVM) to regression tasks, which makes it applicable to PV forecasting. Unlike traditional regression models that minimise the error between the predicted and actual values, SVR aims to find a function that fits the data within a specified margin of tolerance, known as the epsilon-insensitive zone. The model tries to maintain a balance between fitting the training data and maintaining model simplicity by minimising the complexity of the function. SVR is particularly effective in scenarios with high-dimensional data or when the relationship between variables is complex and non-linear, making it a versatile choice for tasks like PV forecasting.

One of the key hyperparameters of SVR is the **regularisation parameter** C. This parameter controls the trade-off between maximising the margin and minimising the error. A smaller value of C allows for a wider margin by tolerating more misclassifications, which can lead to better generalisation but potentially lower accuracy. Conversely, a larger value of C penalises misclassifications more heavily, resulting in a narrower margin and potentially higher accuracy on the training data but with an increased risk of overfitting. Another critical hyperparameter is the **kernel function**, which determines how the input data is transformed into a higher-dimensional space to capture complex patterns. The kernel function measures the similarity between data points, and different kernels, such as linear, polynomial, sigmoid and radial basis function (RBF), offering various ways to model the underlying relationships in the data. Each kernel brings its own strengths and suitability depending on the nature of the data and the specific forecasting task at hand. Figure 8.5 shows the result of a SVR model applied to a curve using different kernels.

When using the RBF kernel in SVR, the hyperparameter Gamma (γ) plays a crucial role in shaping the model's decision boundary. The parameter γ controls the influence of each training example; specifically, it determines how much curvature the decision boundary will have. A higher γ value allows the model to create more complex and tightly fitted boundaries by giving more weight to closer points, leading to higher sensitivity to individual data points. In contrast, a lower γ value results in a smoother decision boundary with less curvature, as it gives more weight to distant

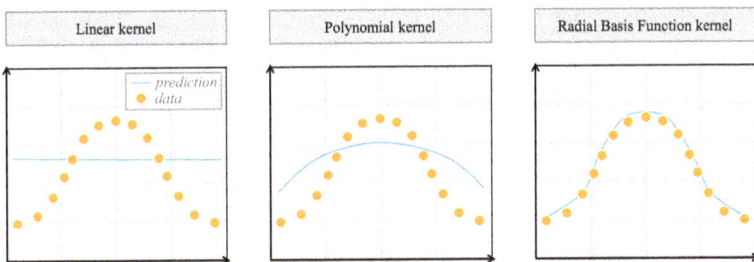

Figure 8.5   Illustrative example of fitting a SVR model on a curve, showing the difference between using linear, polynomial and RBF kernels

points, which can improve generalisation but may underfit the data. The RBF kernel, also known as the Gaussian kernel, is defined as follows:

$$k_{RBF}(\mathbf{x}_1, \mathbf{x}_2) = \exp(-\gamma||\mathbf{x}_1 - \mathbf{x}_2||^2) \tag{8.8}$$

For comparison, other commonly used kernels in SVR include the linear and polynomial kernels:

$$k_{linear}(\mathbf{x}_1, \mathbf{x}_2) = x_1^T \cdot x_2 \tag{8.9}$$

$$k_{Poly}(\mathbf{x}_1, \mathbf{x}_2) = (x_1^T \cdot x_2 + c)^d \tag{8.10}$$

With the RBF kernel, each support vector contributes to the final prediction, with its influence determined by the kernel width parameter γ. The prediction for a new test point u is computed as the weighted sum of the influence of all support vectors:

$$f(x) = \sum_{i=1}^{l} a_i y_i k(\mathbf{x_i}, \mathbf{u}) \tag{8.11}$$

8.3.2.2   K-nearest neighbour

K-nearest neighbours (KNN) model is a simple yet powerful ML model used for both classification and regression tasks. It operates on the principle that similar data points are likely to have similar outcomes. In KNN, predictions are made by identifying the k nearest data points in the training set to a given test point and then using these neighbours to determine the test point's label or value. The model is non-parametric, meaning it makes no assumptions about the underlying data distribution, and its performance is heavily influenced by the choice of k, the distance metric and the method for aggregating the neighbours' contributions. KNN is particularly effective when the data is well-distributed and the decision boundary is not linear, making it a versatile choice for various applications.

The primary hyperparameter in a KNN model is the **number of neighbours**, namely k. A larger value of k results in a less complex model, which may lead to underfitting by smoothing out the decision boundaries and capturing less detail from the training data. Conversely, a smaller value of k creates a more complex model that may overfit the training data by capturing noise and fine details, potentially reducing generalisation to new data. Another important hyperparameter is the **distance metric**, which determines how distances between data points are calculated. Common distance functions include Euclidean, Manhattan and Minkowski distances, each with distinct characteristics:

$$d_{euclidean}(x) = \sqrt{\sum_{i=1}^{n}(x_i - y_i)^2} \tag{8.12}$$

$$d_{manhattan}(x) = \sum_{i=1}^{n}|(x_i - y_i)| \tag{8.13}$$

$$d_{minkowski}(x) = \left(\sum_{i=1}^{n}|(x_i - y_i)|^p\right)^{(1/p)} \tag{8.14}$$

When $p = 1$, the Minkowski distance corresponds to the Manhattan distance, and when $p = 2$, it corresponds to the Euclidean distance. The Manhattan distance, while useful in some cases, is less effective in high-dimensional spaces due to its tendency to underestimate distances, and the Minkowski distance, with its additional tuning parameter p, may introduce further complexity and computational overhead.

8.3.2.3 Artificial neural networks

Artificial neural networks (ANNs) are a class of ML models inspired by the neural structure of the human brain. They are designed to capture complex patterns and relationships in data through multiple layers of interconnected nodes or neurons. ANNs are particularly powerful for tasks such as classification, regression and forecasting, where the relationships between inputs and outputs are intricate and non-linear. Their performance is highly dependent on the careful tuning of several hyperparameters, which defines the network's structure and training process. There are two broad types of the ANN, feed forward neural networks (FFNNs, see Figure 8.6) and recursive neural networks (RNNs), which also include long short-term memory (LSTM), and convolutional neural networks (CNNs). Below are some of the most influential hyperparameters for ANN models (non-exhaustive):

The **number of layers** refers to the depth of the network. In each ANN model, there are at least three layers including the input layer, at least one hidden layer and the output layer. In the case of having two hidden layers or more, the ANN structure is considered deep and it is classified as the deep neural network (DNN). The depth of the network influences its capacity to model complex functions, with deeper networks having the potential to capture more intricate patterns.

The **number of neurons per layer** significantly affects the model's capacity to learn and represent data. Increasing the number of neurons in a layer allows the network to capture more complex patterns and relationships, as it can represent a richer set of features.

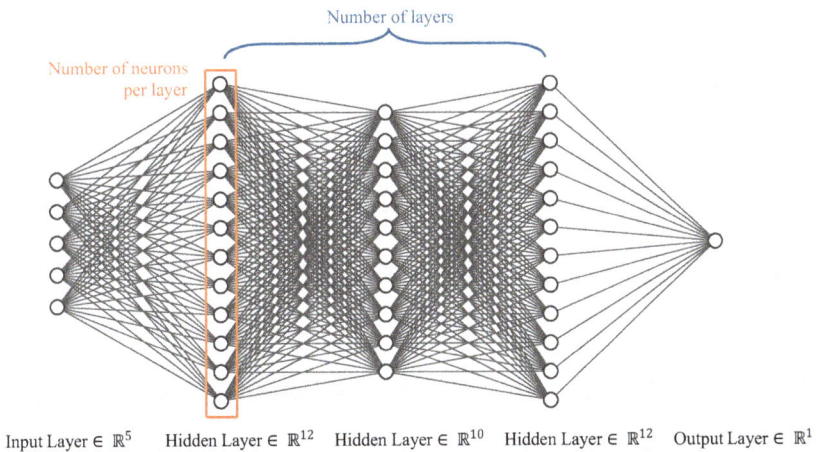

Figure 8.6   Artificial neural network feed forward structure

The **activation function** determines the output of a neuron given an input or a set of inputs. It introduces non-linearity into the model, allowing the network to learn and represent complex patterns in data. Without activation functions, the network would simply perform linear transformations, limiting its ability to model real-world data. The choice of activation function can impact the training speed, convergence and overall performance of the network. Different activation functions have various properties, such as how they handle gradients (affecting issues like vanishing or exploding gradients), and whether they introduce sparsity, which can influence the model's ability to generalise. Common activation functions include sigmoid, hyperbolic tangent (tanh), rectified linear unit (ReLU), leaky ReLU and exponential linear unit (ELU), which are discussed in detail in Chapter 5. Choosing the right activation function depends on the specific problem, network depth and data characteristics. For shallow networks or problems with simple patterns, functions like sigmoid or tanh may suffice, though they can suffer from vanishing gradients in deeper models. In contrast, ReLU and its variants, such as Leaky ReLU or ELU, are preferred for deeper networks because they allow faster training and help maintain gradient flow. For classification tasks, sigmoid is commonly used in the output layer for binary classification, while softmax is ideal for multi-class classification. In regression problems, a linear activation (i.e., no activation) is typically used in the output layer.

The **learning rate** controls the size of the steps the optimiser takes when updating the weights. The learning rate has a significant impact on the model's convergence, training time and overall performance. A higher learning rate may speed up training but risks overshooting the optimal weights, while a lower learning rate ensures more precise convergence but can be slower. Typically, a small positive value such as 0.001 is used as a starting point, but the optimal rate depends on the specific problem, data and model architecture. Various techniques such as learning rate schedules, adaptive optimisers (e.g., Adam or RMSprop) or cyclical learning rates can be employed to dynamically adjust it during training for improved performance.

The **batch size** refers to the number of training samples processed before the model's weights are updated. The batch size affects the trade-off between computational efficiency, model convergence and generalisation. Large batch sizes (e.g., 128 or 256) lead to faster, more stable training but may require more epochs and can risk poorer generalisation. Small batch sizes (e.g., 16 or 32) introduce more stochasticity into the training process, which can help the model generalise better but may also slow down training and cause instability. As the batch size approaches the total number of samples i.e., full-batch gradient descent, the behaviour of the training becomes more deterministic, with each update being based on the full dataset. The optimal batch size depends on the specific model, dataset and hardware used and often seeks a balance between convergence and generalisation.

The **number of epochs** refers to the number of times the entire training dataset is passed through the network. One epoch means that every sample in the training set has been used once to update the model's weights. The number of epochs directly affects the learning process: too few epochs can lead to underfitting, where the model has not learned enough from the data, while too many epochs can result in overfitting, where the model learns the training data too well and performs poorly on new, unseen data. The optimal number of epochs depends on the complexity of the model, the size and variability of the dataset and the learning rate. It is often determined through techniques such as early stopping, where training is halted when performance on a validation set stops improving, helping to prevent overfitting and ensure better generalisation.

The **optimiser** is the algorithm used to adjust the network's weights during training to minimise the loss function. It determines how the model learns from the data by adjusting the direction and magnitude of weight updates during backpropagation. Popular optimisers include stochastic gradient descent (SGD), Adam, and RMSprop, each with different approaches to updating weights and managing learning rates. The choice of optimiser can significantly affect training speed, convergence stability and final model performance. While Adam is widely used due to its robustness and low need for manual tuning, simpler optimisers such as SGD may yield better generalisation in some cases [36]. Selecting an appropriate optimiser often involves experimentation and can be guided by the nature of the data and the specific learning task.

The **dropout rate** is a regularisation technique where a fraction of neurons along with their connections are randomly ignored or 'dropped out' during each training iteration. By forcing the network not to rely on specific neurons, dropout encourages redundancy and robustness, helping the model generalise better to unseen data. A low dropout rate (e.g., 0.1 to 0.3) means that most neurons remain active, allowing the model to learn effectively, especially in simpler models or when the dataset is small. While it can result in faster training, a low dropout rate may not prevent overfitting in more complex models or large datasets. In contrast, a high dropout rate (e.g., 0.5 to 0.7) forces the model to learn more robust features by making it less reliant on specific neurons, which helps improve generalisation and prevent overfitting. However, it can slow down the training process and may lead to undertraining if the dropout rate is too high, as the model has fewer neurons to work with in each iteration. Common values range from 0.2 to 0.5, depending on the network's size and complexity. During testing or validation, dropout is turned off, and all neurons are used, with their outputs scaled appropriately.

Proper **weight initialisation** is crucial for effective training. Proper initialisation helps ensure faster convergence, reduces the risk of getting stuck in local minima, and improves overall model performance. Common methods include random initialisation, where weights are set randomly to break symmetry, and Xavier/Glorot initialisation, which is ideal for sigmoid or tanh activations and ensures that the variance of activations is maintained across layers. For ReLU activation functions, He initialisation is often preferred, as it helps avoid the 'dying ReLU' problem by scaling weights appropriately. LeCun initialisation is suitable for Leaky ReLU, while zero initialisation is generally avoided due to symmetry issues that prevent the network from learning effectively. Orthogonal initialisation can also be used for deep networks, as it helps maintain gradient flow and improve convergence. The choice of initialisation methods is influenced by the network's architecture and the activation function used, and selecting the right one is key to improving training efficiency, preventing gradient-related issues and enhancing model generalisation [37].

Types of hyperparameters?

Hyperparameters can be divided into two main categories: those that control the training process and those that define the model's architecture. Training hyperparameters, including learning rate, batch size and the choice of optimiser, influence how effectively and quickly a neural network learns during training. In contrast, architecture-related hyperparameters, such as the number of hidden layers and the width of each layer, determine the model's capacity to capture complex patterns. These architectural parameters shape the network's structure and play a crucial role in balancing its learning potential with computational demands.

8.3.2.4   Recurrent neural networks

Recurrent neural networks (RNNs) are a class of neural networks designed to process sequential data by maintaining a hidden state that captures information about previous inputs. They are commonly used for tasks where the order of inputs matters, such as time-series prediction, language modelling and sequence classification. RNNs work by applying the same weights recursively over each element in the sequence, allowing them to capture temporal dependencies. In addition to the ANN general hyperparameters described previously, the following are key hyperparameters that influence the performance of RNN models:

Sequence length refers to the number of time steps or data points that the network processes in a given input sequence. In other words, it represents the number of past time steps that the model can 'remember' and use to make predictions. Tuning sequence length is crucial for capturing temporal dependencies effectively. Too short a sequence length may cause the model to miss important long-term dependencies, while too long a sequence can result in higher computational costs and challenges, such as vanishing gradients. Techniques such as padding or bucketing are used to handle variable-length sequences, while strategies like early stopping or CV help assess the performance of different sequence lengths. The optimal sequence length balances the trade-off between computational efficiency and the ability of the model to capture the necessary temporal context and is often determined through experimentation based on the specific dataset and task.

Gradient clipping is a technique used to prevent exploding gradients during training, particularly in deep learning models like RNNs, where the gradients can grow excessively during backpropagation. This can cause the model to become unstable and lead to poor convergence. Gradient clipping works by limiting the magnitude of the gradients to a predefined threshold. There are two common methods for gradient clipping, namely clipnorm and clipvalue. Clipnorm limits the norm of the gradient vector, while clipvalue limits the individual values of the gradients. The choice between clipnorm and clipvalue depends on the model architecture and the specific issues encountered during training. Clipnorm is generally preferred when the overall magnitude of gradients is more important than the individual components. It helps to maintain the relative proportions of the gradient components while keeping the overall scale in check. Clipvalue, in contrast, is useful when you are more concerned about individual gradient components exceeding a threshold, particularly when working with simpler models or datasets where specific weights are sensitive. Both methods aim to stabilise training and improve convergence, especially in deep networks and models with long sequences such as RNNs, where the risk of exploding gradients is higher.

Bidirectionality in RNNs refers to the use of bidirectional RNNs, which process the input sequence in opposite directions, both forward (from past to future) and backward (from future to past). This can improve performance for tasks where future context is as important as past context, such as natural language and some time series forecasts. In the context of PV power forecasting, a bidirectional RNN processes the input time series data in two directions, allowing the network to understand how both preceding and succeeding weather conditions, solar irradiance and other influencing factors contribute to PV output at a given moment. This can lead to more accurate forecasts, especially in scenarios where the influence of certain variables spans across time (e.g., a drop in temperature followed by an increase in cloud cover). However, in real-time forecasting, where future data is not yet available, bidirectional models are typically not used. Instead, they are more applicable in offline training or post-processing scenarios, such as historical analysis or training models with rolling windows of data. Bidirectional models are generally more computationally expensive and not always suitable for tasks where future information is unavailable at prediction time. Choosing to use bidirectionality is a key architectural decision that can significantly impact model performance, depending on the nature of the data and the forecasting task and horizon.

Statefulness in RNNs describes whether the model maintains the state of the RNN across different batches or sequences during training and prediction. This can be useful when sequences are split across batches and the continuity of the sequence matters. A stateful RNN maintains the hidden state between batches, allowing it to remember patterns from previous input sequences that are part of a continuous time series. This is especially useful in PV forecasting, where weather patterns, irradiance levels and temperature trends evolve and carry dependencies that span longer than a single input window. By using statefulness, the model can learn longer-term dependencies without needing to feed excessively long sequences into a single batch. This makes it effective for modelling persistent patterns in solar PV generation, such as daily or seasonal cycles. However, stateful RNNs require careful handling of input data and hidden states to prevent learning from unrelated patterns. In contrast, stateless RNNs reset their state after each batch, treating each sequence independently, which may simplify training, but at the cost of potentially missing long-range temporal dependencies. Choosing between stateful and stateless configurations depends on the specific use case, the characteristics of the dataset and the computational constraints.

8.3.2.5 Long short-term memory

LSTM networks are a specialised type of recurrent neural network designed to capture long-range dependencies and temporal patterns in sequential data. Unlike traditional RNNs, LSTMs are equipped with gating mechanisms that control the flow of information, allowing them to remember information over extended sequences and mitigating issues such as vanishing or exploding gradients. LSTMs are particularly effective for tasks involving time series data, such as PV forecasting, where understanding and predicting future values based on past observations is crucial [38]. In addition to the hyperparameters of ANN and RNN mentioned earlier, the performance of an LSTM model is influenced by several other hyperparameters, including 'hidden state size' and 'cell state initialisation'. These are two distinct yet interrelated concepts in LSTM networks that both influence the ability of the model to learn from sequential data like PV generation time series. Hidden state size sets the space in which the memory is stored, while initialisation decides what memory content you start with.

Cell state initialisation in LSTM networks refers to how the initial memory cell state (C_0) and the hidden state (h_0) are set before the network begins processing a sequence. These two components are fundamental to the ability of LSTM to carry and update information over time. The cell state (C_0) holds long-term memory, while the hidden state (h_0) captures short-term information and contributes to the output at each time step. These states are crucial for the ability of LSTM to store and propagate temporal information across time steps. The most common approach is zero initialisation, which is simple and often effective, particularly for short or independent sequences. However, in tasks like PV power forecasting, where temporal dependencies such as daily or seasonal patterns are important, more complex strategies can be beneficial. For instance, learned initialisation allows the model to optimise the starting states during training, while context-based initialisation uses external features like time of day or weather to set initial values. These approaches can help the model better understand the temporal context right from the beginning of a sequence. Choosing the right initialisation method can enhance learning efficiency and forecasting accuracy, especially in models dealing with long sequences or continuous time-series data.

The **hidden state size** determines the dimensionality of the hidden state vector (h_t) and the cell state (C_t) at each time step. This size is a key hyperparameter that directly influences the ability of the LSTM model to learn and represent complex temporal patterns. A larger hidden state size allows the LSTM to capture more intricate relationships and long-term dependencies in the data, which is particularly valuable in applications such as PV power forecasting, where input features such as solar irradiance, temperature and time of day exhibit non-linear and time-varying interactions. However, increasing the hidden size also raises the number of trainable parameters, which can lead to higher computational costs and a greater risk of overfitting, especially when training data is limited. In contrast, a smaller hidden size may be more efficient and generalise better on simpler tasks, but it might lack the representational power needed for complex patterns. Therefore, choosing an appropriate hidden state size requires balancing model complexity and generalisation performance.

The forget gate in an LSTM controls how much of the previous memory is discarded. The **forget gate bias** can be adjusted to influence the memory retention over time. The forget gate applies a sigmoid activation to a linear combination of the inputs and hidden state, and adding a bias term (the forget bias) shifts this sigmoid function.

A common practice is to initialise the forget bias to a positive value, typically 1 or higher, which biases the gate toward retaining more of the past memory at the start of training. This is particularly useful because it encourages the model to preserve information over longer time periods, helping prevent the early forgetting of useful signals, especially in tasks involving long-term dependencies such as PV power forecasting with daily or seasonal cycles. In contrast, a forget bias of zero or negative values makes the model more likely to discard past information, which can hinder learning in sequential tasks. Choosing an appropriate forget bias is a small but impactful decision in LSTM design, as it stabilises training and can improve the ability of the model to learn temporal patterns, especially when memory persistence is crucial.

8.3.2.6 Convolutional neural networks

CNNs are a specialised class of deep neural networks designed to process structured grid-like data, such as images. CNNs are particularly well-suited for tasks such as image recognition, object detection and other spatial data applications due to their ability to capture spatial hierarchies through convolutional layers (Figure 8.7). In addition to the hyperparameters of ANNs such as number of epochs, learning rate, batch size and dropout rate, the following hyperparameters shape how the network extracts features from the input data, balances computational efficiency and generalises to unseen data:

The **number of filters or kernels** in each convolutional layer defines the depth of the output feature maps (Figures 8.7 and 8.8). It determines how many distinct features the network learns at each convolutional layer. Each filter slides over the input (or feature map from the previous layer), extracting specific patterns such as edges, textures or more complex shapes in deeper layers. A higher number of filters allows the model to capture a wider variety of features, increasing its representational power. For example, early layers may use 32 or 64 filters, while deeper layers may go up to 128, 256 or even 512 filters, depending on the complexity of the task and input data. However, increasing the number of filters also increases the number of trainable parameters and computational cost, which can lead to longer training times

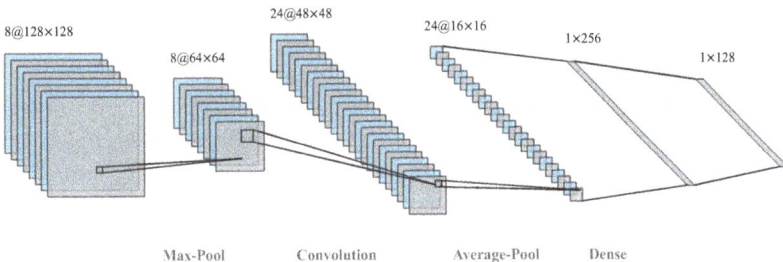

Figure 8.7   Overview of a CNN architecture: the input passes through max-pooling, convolution, average-pooling and dense layers, progressively reducing feature dimensions for classification or regression

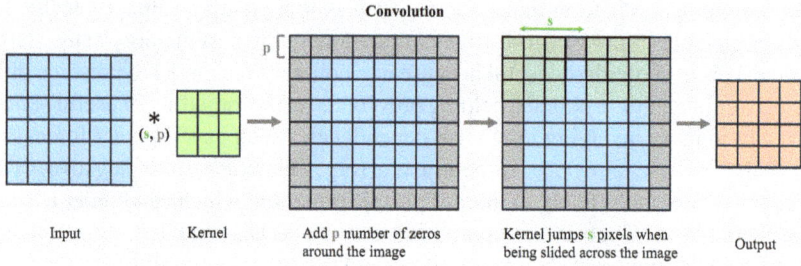

Figure 8.8   *Illustration of the 2D convolution process. The input is padded with zeros (padding p), then a kernel is applied with a defined stride s, resulting in a transformed output feature map.*

and a higher risk of overfitting, especially with limited data. Conversely, too few filters may cause the model to miss important features, reducing its accuracy. In applications such as PV power forecasting using CNNs, especially using images as input e.g., from sky images, a well-chosen number of filters per layer ensures the model effectively captures relevant spatial patterns without unnecessary complexity.

The **filter size or kernel size** defines the dimensions of the sliding window that scans across the input data to extract local features (Figures 8.8 and 8.9(1)). Common filter sizes are 3×3, 5×5 or 7×7, and they determine the receptive field, how much of the input the filter 'sees' at one time. Smaller filters such as 3×3 are widely used because they capture fine-grained local patterns efficiently and can be stacked to cover larger receptive fields. Larger filters such as 5×5 or 7×7 can capture broader patterns in fewer layers but also involve more parameters and computational cost. In applications such as PV power forecasting using spatial inputs (e.g., satellite or sky images), filter size impacts how well the model captures spatial features. A small filter may focus on fine cloud edges, while a larger one might detect broader cloud coverage. Choosing the right filter size is a trade-off between model complexity, computational efficiency and the scale of patterns relevant to the forecasting task. Often, multiple convolutional layers with smaller filters are preferred to gradually build up complex representations while keeping parameter counts manageable. The size of the filters applied during convolution defines the receptive field of the neurons. Smaller filters capture fine details, while larger filters capture broader spatial patterns.

Stride refers to the number of pixels the convolutional filter moves or 'strides' across the input image or feature map during the convolution operation (Figure 8.9(2)). A stride of 1 means the filter moves one pixel at a time, resulting in high-resolution feature maps and preserving more spatial detail. Strides greater than 1 (e.g., stride = 2) reduce the spatial dimensions of the output feature map, leading to down-sampling, which decreases computational load and memory usage. The choice of stride impacts both the resolution of learned features and the efficiency of the model. In PV power forecasting applications using spatial data such as sky images or satellite images, a stride of 1 is often preferred in early layers to capture fine-grained patterns such as cloud edges or irradiance gradients. Higher strides may be used in deeper

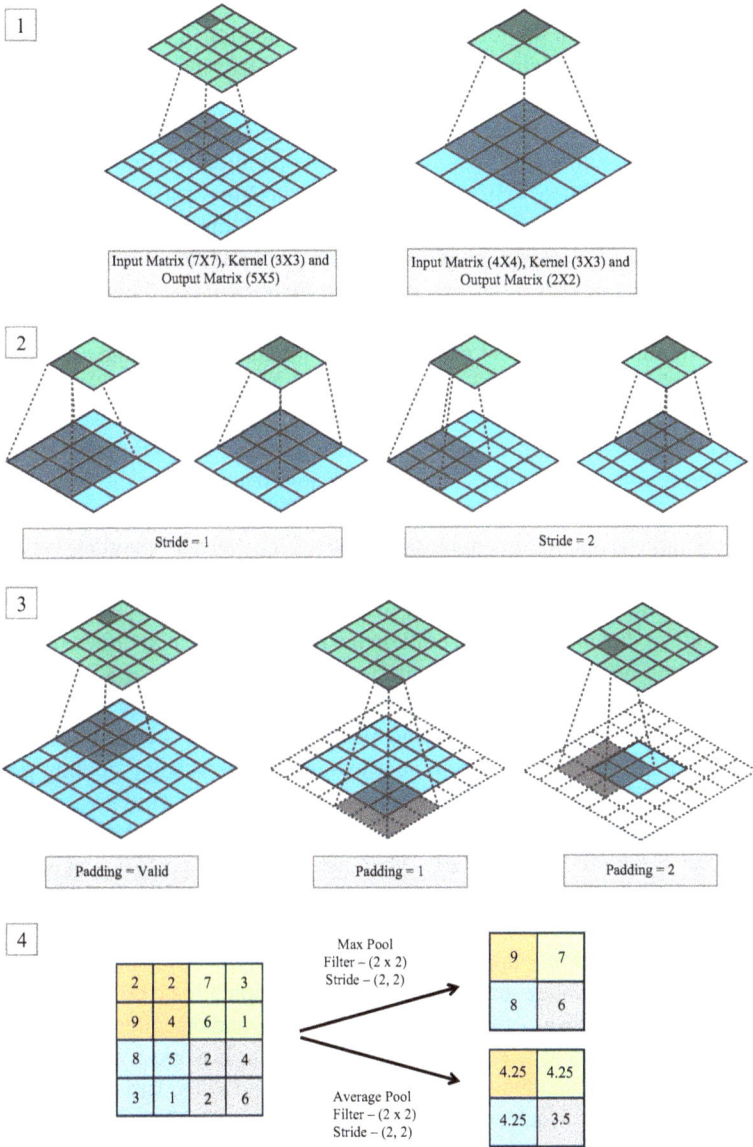

Figure 8.9   Illustration of key operations in convolutional neural networks. (1) Kernel and output size: output size depends on input and kernel dimensions. (2) Stride: increasing stride reduces output resolution by skipping input positions. (3) Padding: 'Valid' uses no padding; padding of 1 or 2 adds zeros to preserve dimensions. (4) Pooling: max pooling selects maximum values; average pooling computes means over 2 × 2 regions with stride (2,2).

layers or combined with pooling to abstract broader spatial trends. However, excessive striding can result in the loss of important information. Therefore, selecting the stride value involves balancing feature detail, model performance and computational efficiency.

Padding refers to the process of adding extra pixels, typically zeros, around the border of the input before applying a convolution operation (Figure 8.9(3)). Padding controls the spatial dimensions of the output feature map and helps manage edge effects during filtering. There are two main types of padding: valid padding and same padding. In the valid padding, which is also called no padding, no additional pixels are added, so the output feature map is smaller than the input. This can lead to the progressive shrinking of feature maps in deeper layers, potentially causing loss of spatial information. In the same padding, zeros are added so that the output has the same spatial dimensions as the input (assuming a stride of 1). This helps preserve edge information and maintain consistency across layers. In PV power forecasting applications that use images as inputs, the same padding is often preferred in early layers to retain full spatial coverage and ensure that features near the edges, such as clouds entering or exiting the field of view, are not lost. Padding also allows for deeper networks without reducing the input too quickly. The choice of padding is therefore important for maintaining spatial detail and supporting effective feature extraction throughout the network.

Pooling is a down-sampling operation that reduces the spatial dimensions of feature maps while retaining the most important information (Figure 8.9(4)). The two most common types are max pooling, which selects the highest value in a local region, and average pooling, which computes the mean. Max pooling is widely used because it effectively captures the most prominent features, while average pooling provides smoother representations. Pooling also introduces a level of spatial invariance, making the model more robust to small shifts or distortions in the input data. In the context of PV power forecasting, particularly when using spatial inputs such as satellite or sky images, pooling helps to distill essential features such as cloud patterns or irradiance gradients while reducing noise and overfitting risk. The pool size and stride determine how aggressively the data is downsampled, typically using a 2×2 window with a stride of 2 to balance feature retention and computational cost.

Batch normalisation is a technique used in CNNs and other deep learning architectures to stabilise and accelerate the training process. It works by normalising the inputs to each layer so that they have a mean of zero and a standard deviation of one, based on the statistics of each mini-batch during training. This helps reduce the problem of internal covariate shift, where the distribution of activations changes during training, requiring the model to continuously adapt. In practice, batch normalisation allows for faster convergence, enables the use of higher learning rates and provides a regularising effect, which can reduce the need for other techniques such as dropout. It is typically applied after the convolution (or fully connected) operation and before the activation function. In the context of PV power forecasting, especially when using deep CNNs to analyse image-based inputs such as sky or satellite images, batch normalisation can improve model generalisation and training stability, particularly when training data is noisy or collected under varying weather conditions.

Fully connected layers are a fundamental component in CNNs, typically placed after the convolutional and pooling layers. In these layers, every neuron is connected to every neuron in the previous layer, enabling the model to combine the extracted features into a final prediction. They act as the decision-making part of the network, learning complex, non-linear combinations of features detected by earlier layers. In PV power forecasting, fully connected layers take the spatial or temporal features learned from inputs and use them to estimate future power outputs. The number of neurons in these layers is a tunable hyperparameter and should match the complexity of the forecasting task. Too few neurons may underfit the data, while too many can lead to overfitting and increased computational cost. These layers are usually followed by an activation function such as ReLU or sigmoid and may be regularised using techniques such as dropout. Fully connected layers are essential for transforming learned patterns into accurate, actionable forecasts at the end of the network.

For a given size of the input (i), kernel (k), padding (p) and stride (s), the size of the output feature map (o) generated is given by:

$$o = \frac{i + 2p - k}{s} + 1 \tag{8.15}$$

8.3.2.7 Decision trees

Decision trees (DTs) are a type of model used for both classification and regression tasks. They work by recursively splitting the input data into subsets based on feature values, resulting in a tree-like structure of decisions. The performance and behaviour of a DT are highly influenced by several key hyperparameters, which control the tree's growth, splitting strategy and complexity. Below are the most important hyperparameters that affect the architecture and training of DTs.

The **criterion** determines the function used to measure the quality of a split at each node, impacting how the tree evaluates potential splits during training, therefore affecting both accuracy and complexity of the model. There are two common options for classification tasks, namely 'gini' and 'entropy'. Gini uses the Gini impurity, which measures the probability of incorrectly classifying a randomly chosen element if it were randomly labelled according to the distribution of labels in the subset. Entropy evaluates how much uncertainty in the data is reduced by the split based on information gain. For regression tasks, there are various options for measuring error and therefore quality, such as squared error, MSE, absolute error, Poisson and Friedman MSE. In PV power forecasting, choosing the right criterion depends on whether the task is classification or regression. For regression-based forecasting, squared error is commonly used to ensure that the model captures detailed variations in power output based on input features.

The **splitter** in a DT controls how the algorithm chooses the feature and threshold to split each node. It defines the strategy used to select the best split and can significantly affect the training time and accuracy of the model. There are two main strategies: best and random. The best strategy considers all features and evaluates all possible thresholds to find the split that most improves the purity of the node. It typically results in more accurate models but can be computationally expensive.

The random strategy, in contrast, selects a random feature and then chooses the best threshold for that feature. While it may be less accurate in a single tree, it increases diversity and is useful in ensemble methods such as random forests, where combining many randomised trees can improve generalisation. In PV power forecasting, the choice of splitter can influence the ability of the model to capture relationships between various variables efficiently. Using 'best' is generally preferred for stand-alone trees to maximise accuracy, while 'random' is practical when building diverse models for ensemble learning.

The **maximum depth** in a DT controls the maximum number of levels the tree can grow (Figure 8.10). It is a crucial parameter for managing the complexity and generalisation of the model. A deep tree (i.e., large max depth) can capture intricate patterns in the training data but runs a high risk of overfitting, especially if the tree grows until all leaves are pure (i.e., 100% correct on the training set). In contrast, a shallow tree (i.e., small max depth) may be too simplistic, leading to underfitting and poor predictive performance. In the context of PV power forecasting, selecting an appropriate max depth ensures that the model can capture relevant non-linear relationships between input features and power output without overreacting to noise or small fluctuations in the training data. It is often tuned through CV, grid search or random search, with typical values ranging from 3 to 20, depending on the dataset size and complexity. Properly constraining max depth helps create a robust and interpretable DT.

The **minimum samples split** specifies the minimum number of samples required to split an internal node. If a node has fewer samples than this value, it becomes a leaf node and will not be further split. This parameter acts as a control mechanism to prevent overfitting by ensuring that splits are made only when there is sufficient data to justify them. A smaller value (e.g., 2) allows the tree to grow deeper and capture more detailed patterns, which can improve training accuracy but may lead to overfitting. A larger value (e.g., 10 or more) results in a more shallow tree, potentially improving generalisation by avoiding splits that rely on small sample sizes and thus reducing variance. In the context of PV power forecasting, setting an appropriate

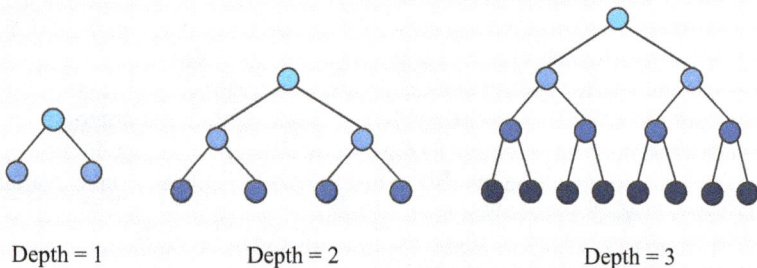

Depth = 1   Depth = 2   Depth = 3

Figure 8.10   Visualisation of decision tree depth. As depth increases from 1 to 3, the tree becomes more complex, allowing finer decision boundaries but also increasing the risk of overfitting

minimum samples split is important for balancing the ability of the model to learn from diverse weather and irradiance patterns while avoiding over-sensitivity to noise or rare conditions in the training data. This parameter is often tuned in combination with others like maximum depth and minimum samples leaf to build a model that performs well on both training and unseen data.

The **minimum samples leaf** defines the minimum number of samples required to be at a leaf node. It ensures that no leaf is created with fewer samples than this threshold, helping to smooth predictions and prevent the model from learning spurious patterns based on very small subsets of data. Using a small value (e.g., 1) allows the tree to fit very closely to the training data, increasing the risk of overfitting. Increasing the value (e.g., to 5 or 10) can make the tree more conservative, reduce variance and improve generalisation, especially in noisy datasets. For PV power forecasting, where input data such as weather variables can be noisy or variable across time and location, setting minimum samples leaf helps ensure that predictions at the leaf nodes are based on sufficiently representative data. This leads to more stable and reliable forecasts, especially under fluctuating or uncertain weather conditions. Like other tree hyperparameters, it is typically tuned via CV to find the best trade-off between bias and variance.

The **maximum features** specifies the maximum number of features to consider when looking for the best split at each node. This parameter controls how much of the input feature space is used during training and can significantly affect both the performance and diversity of the model. There are several common settings for maximum features, such as sqrt where the square root of the total number of features are considered; log2, where the base-2 logarithm of the number of features are considered; an integer, where an exact number of features to consider is defined; a float between 0 and 1, where a fraction of the total features are considered; and none, where all features are considered. Using fewer features per split increases the diversity of splits and reduces overfitting, especially in ensemble models like random forests. However, in single trees, limiting features may reduce accuracy if relevant variables are not considered during splitting. In PV power forecasting, where input features may include temperature, irradiance, time, wind speed and more, adjusting max features helps to balance model complexity and computational cost, especially when working with high-dimensional or correlated data. It is a tunable hyperparameter often optimised during model selection.

The **maximum leaf nodes** sets the maximum number of terminal nodes or leaves that the tree can have. This parameter directly controls the model complexity by limiting how many final decisions or output regions the tree can create. By capping the number of leaf nodes, it helps prevent the tree from growing too complex and overfitting to the training data. A smaller number of leaf nodes leads to a simpler, more generalised model that might miss some nuances in the data (potentially underfitting), while a larger number allows the model to capture more patterns but risks overfitting, especially in the presence of noise. In the context of PV power forecasting, where the relationship between features such as weather conditions and power output can be non-linear and influenced by external factors, setting an appropriate max leaf nodes helps ensure the tree captures meaningful trends without modelling random fluctuations. This parameter is particularly useful when interpretability and generalisation are important and is often tuned using validation techniques to strike a good balance between accuracy and simplicity.

Random state is used to control the randomness involved in certain processes, such as choosing the best split when multiple options are equally good, or initialising random procedures in algorithms like splitter='random'. Setting this parameter ensures ensures reproducibility, meaning that the model will behave the same way every time it is trained, given the same input data and hyperparameters. In PV power forecasting, reproducibility is crucial when comparing different forecasting models or evaluating performance improvements over time. Setting a fixed random state (e.g., random state = 42) can ensure that the same tree structure is generated every time the model is trained, making results deterministic and easier to debug or reproduce.

8.3.2.8   Random forest

Random forest (RF) is an ensemble learning method that constructs a multitude of DTs during training and outputs the mode of their predictions for classification or the mean prediction for regression (Figure 8.11). Since random forest is an ensemble of DTs, they share some hyperparameters such as criterion, splitter, maximum depth, maximum features, maximum leaf nodes, minimum samples split, minimum samples

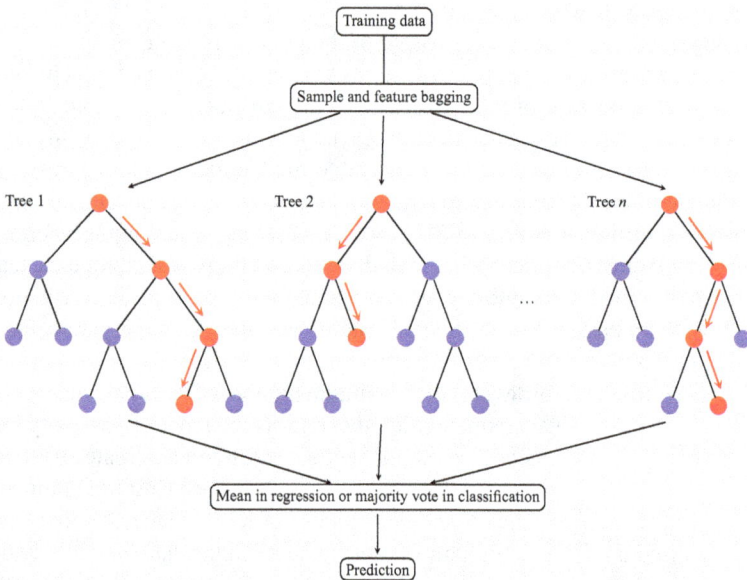

Figure 8.11   Illustration of a random forest model. The training data is used to generate multiple decision trees via sample and feature bagging. The final prediction is obtained by averaging (regression) or voting (classification) the outputs of all trees

leaf and random state. Random forest includes additional hyperparameters, including the total number of trees in the forest and bootstrap.

The total **number of trees in the forest** determines how many trees are grown in the forest. Too few trees may lead to higher variance and overfitting, while more trees lead to better generalisation and stability, reducing the likelihood of overfitting. However, the law of diminishing returns applies, where adding too many trees yields very small improvements but increases training time and memory consumption. A typical starting point is 100 trees, but the optimal number can be found through CV, grid search or random search.

Bootstrap controls whether or not bootstrapping is used to create subsets of the training data for building individual trees in the forest. In case bootstrapping is used, each tree in the random forest is trained on a random subset of the data, sampled with replacement. As a result, each tree sees a slightly different subset of the data, leading to diversity among trees and reducing the risk of overfitting by averaging out predictions. It generally leads to better performance in random forests, especially in noisy or high-dimensional datasets. It also means that some samples may not appear in a tree's training set, and these out-of-bag (OOB) samples can be used for internal validation. In case bootstrapping is not used, each tree is trained on a different subset of the data without replacement. This can lead to lower variance in individual trees since each tree gets a larger portion of the dataset, but it might also reduce the overall diversity of the trees in the ensemble, which can sometimes lead to poorer generalisation. However, this approach might be useful when trying to create less complex models or working with small datasets. In the context of PV power forecasting, where data might be noisy or contain outliers, bootstrapping helps the model generalise better by reducing overfitting to specific instances in the training data. It also provides a way to assess model performance using OOB samples.

General procedure for hyperparameter tuning

1. **Select model and hyperparameters**: Select the ML model and identify the hyperparameters to tune.
2. **Search space**: Determine the range of possible values for each hyperparameter based on prior research, practical experience and initial experiments to narrow down the search space.
3. **Tuning strategy**: Decide between manual tuning, grid search, random search or more advanced methods such as Bayesian optimisation.
4. **Split the dataset and select model evaluation metric**: Divide the dataset into training, validation and test sets to assess the model's performance during tuning. Select and define the model evaluation metrics (e.g., RMSE, MAE and MBE), and consider evaluation methods such as CV or LOOCV to better estimate model performance.
5. **Train and evaluate the model**: For each combination of hyperparameters, train the model and evaluate it using the validation set to avoid overfitting.
6. **Select the best hyperparameters**: Identify the combination of hyperparameters that yields the best performance on the validation set.

7. **Test the final model**: After tuning, retrain the model on the full training set using the best hyperparameters and evaluate it on the test set.

8.4   Hyperparameter tuning case studies in PV forecasting

8.4.1   *Hyperparameter tuning of SVR and KNN: a case study on building integrated PV*

As seen in this chapter, accurate forecasting in PV systems requires careful tuning of ML model hyperparameters. This is particularly true in building-integrated PV (BIPV) installations, where standard PV modelling techniques are not broadly tested or optimised. This section presents a case study that examines the impact of key hyperparameters on the performance of SVR and KNN for a BIPV system, based on a dataset spanning over 6 years of operational data from a single string of panels [39]. In this study, the RBF kernel is employed in the SVR model due to its ability to capture non-linear relationships in the dataset. To evaluate the performance of the PV forecasting models, a K-fold CV approach is used. K-fold CV is chosen because it offers a balance between computational efficiency and performance, as LOOCV is computationally more expensive and may not perform as well in this context [40].

For the SVR model, the hyperparameter C (which controls the trade-off between the margin size and classification error) plays a crucial role in prediction accuracy. Figure 8.12 illustrates the relationship between the RMSE and varying values of C. In this case, higher values of C generally improve the forecast accuracy. For instance, when $C = 1$, the RMSE is 433.47, whereas increasing C to 99 lowers the RMSE to 378.73. As C continues to increase, the RMSE decreases further, reaching a minimum of 370.82 at $C = 1000$. However, the difference between $C = 500$ (which yields an RMSE of 372.46) and $C = 1000$ is minimal, suggesting diminishing returns at higher C values. This indicates that while increasing C improves the model's ability to fit the data, excessively high values offer only marginal improvements. Another critical hyperparameter in SVR is γ, which defines the influence of individual data points. Figure 8.12 also shows the RMSE for different values of γ on a logarithmic scale. The lowest RMSE, 475.64, occurs at $\gamma = 0.000005$. Beyond $\gamma = 0.001$, further increases in γ do not significantly affect the model's accuracy, suggesting that this parameter is less sensitive to fine-tuning at higher values.

For the KNN model, the number of neighbours (k) is a key hyperparameter that directly influences performance. As shown in Figure 8.13, the RMSE decreases as k increases, but not in a strictly linear manner. While small changes in k, such as from 12 to 13 or from 24 to 25, do not result in significant improvements, the overall trend is a reduction in error as k grows. The optimal value of k is found to be 39, where the RMSE reaches its minimum at 376.614. Beyond this point, the RMSE increases, highlighting the importance of choosing an appropriate k value to avoid underfitting or overfitting. A small k can lead to high variance (overfitting), while a large k may cause high bias (underfitting). In this study, tuning k to an optimal value ensures

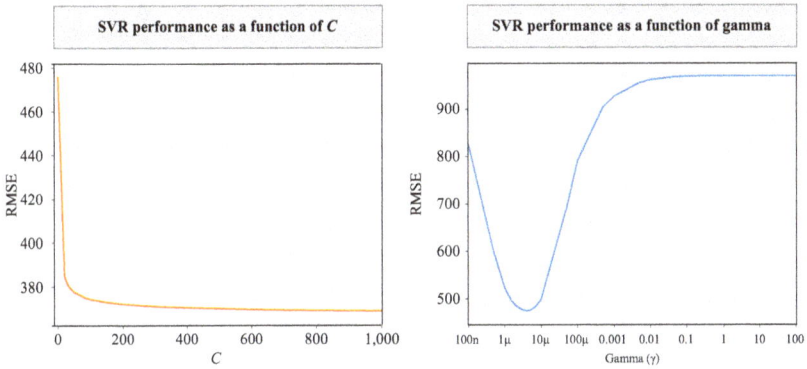

Figure 8.12   SVR model tuning for a case study of a BIPV system, adapted
from [39]. (Left) RMSE of the SVR model as a function of the
hyperparameter C. (Right) Optimisation of the hyperparameter γ

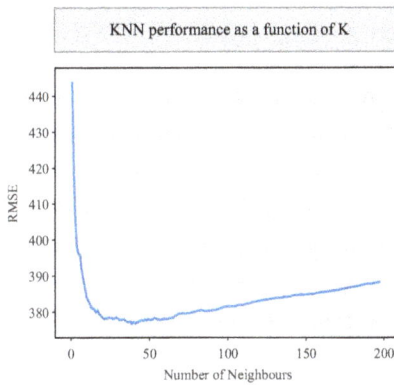

Figure 8.13   KNN model tuning for a case study of a BIPV system, adapted
from [39]. Model RMSE as a function of k number of neighbours

a balance between bias and variance, thereby improving the model's generalisation
capabilities.

8.4.2   Hyperparameter tuning of CNN: a case study on a PV charging station

A CNN model was developed to forecast PV generation from the PV charging station
at the University of Twente, with the hyperparameter configurations detailed in Table
8.2. Hyperparameter tuning was performed using a grid search approach. The dataset
spans a 26-month period, from September 2021 to November 2023, and includes PV

Table 8.2   Search space of CNN hyperparameters

Number of filters	Kernel size	Pool size	Dense unit	Dropout rate	Learning rate	Activation
32, 64, 128	3, 5, 7	2, 3	32, 64	0.2, 0.3	0.001, 0.0005	*tanh, relu*

output data alongside meteorological variables such as global horizontal irradiance (GHI), wind speed, ambient temperature and relative humidity [41].

Due to the large number of hyperparameters associated with CNNs, tuning all of them simultaneously is computationally impractical. To manage this complexity, several hyperparameters were kept constant, and a simplified CNN architecture was adopted to systematically evaluate the impact of key parameters. The architecture, as outlined in Table 8.2, consists of the following layers:

1. **Convolution layer:** Applies a set number of filters with a specified kernel size, using the defined activation function.
2. **Pooling layer:** A max pooling operation with a pool size to reduce spatial dimensions and control overfitting.
3. **Convolution layer:** A second convolutional layer with twice the number of filters as the first, using the same kernel size and activation function.
4. **Pooling layer:** A second max pooling layer with the same pool size as the first.
5. **Flatten layer:** Transforms the multi-dimensional tensor output into a 1D vector, preparing it for the fully connected layers.
6. **Dense layer:** A fully connected layer with dense units and the chosen activation function, capturing high-level representations.
7. **Dropout layer:** Includes a dropout rate to prevent overfitting by randomly deactivating neurons during training.
8. **Output layer:** A final dense layer with a single neuron that outputs the forecasted value.

The CNN models with the best, average and worst performances were identified and evaluated against a persistence forecast baseline, which yielded an nRMSE of 20.01%. Figure 8.14 illustrates the day-ahead PV generation forecasts over a sample period of four days. The corresponding hyperparameter settings and performance metrics are summarised in Table 8.3.

The best-performing CNN model achieved a 50.17% improvement over the persistence benchmark, while even the worst-performing model delivered a 23.58% gain. On average, the CNN-based forecasts improved upon the baseline by 38.48%. These results underscore the effectiveness of CNN architectures in PV forecasting: even sub-optimal configurations outperform traditional benchmarks, confirming the robustness of CNNs for this task.

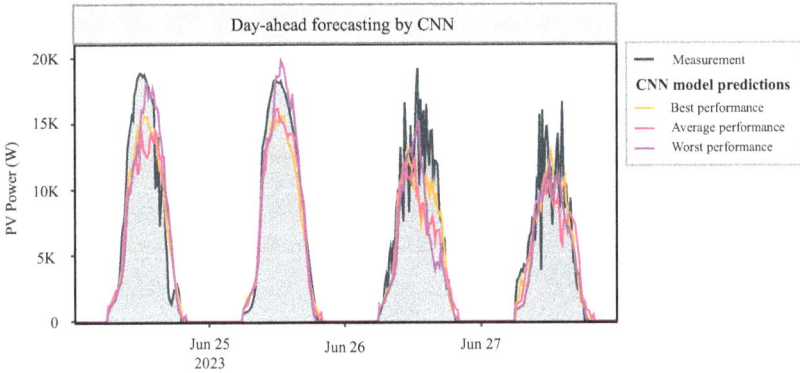

Figure 8.14   Comparison of day-ahead forecasting by best, average and worst performing CNN models

Table 8.3   Tuned CNNs with best, average and worst performances

Hyperparameter	Number of Filters	Kernel size	Pool size	Dense units	Drop rate	Learning rate	Activation	nRMSE (%)
Best CNN	32	5	2	32	0.3	0.0005	tanh	9.97
Average CNN	128	5	3	32	0.3	0.0005	tanh	12.31
Worst CNN	64	7	3	64	0.3	0.001	tanh	15.29

8.5   Conclusion

This chapter explored evaluating and optimising ML models for PV forecasting. Key evaluation metrics such as RMSE, nRMSE, MAE and MAPE were discussed to provide a comprehensive understanding of how model performance is quantified. Additionally, different evaluation methods were examined, including K-fold and LOOCV, which ensure robust and reliable model validation. A major focus was placed on model optimisation strategies, particularly hyperparameter tuning, as a means of improving model performance. Through careful adjustment of key hyperparameters, significant enhancements in forecasting accuracy were demonstrated through two case studies, a BIPV system in Switzerland and a PV charging station in the Netherlands. This process helps prevent common issues such as overfitting, where a model learns the noise in the training data at the cost of generalisation, and underfitting, where it fails to capture the underlying patterns. Overall, model tuning is not only a technical necessity for improving accuracy but also a key step toward understanding the behaviour of 'black-box' ML models in PV forecasting.

References

[1]   Haegel NM, Verlinden P, Victoria M, *et al.* Photovoltaics at multi-terawatt scale: Waiting is not an option. *Science.* 2023;380(6640):39–42.

[2]   SolarPower Europe. *Global Market Outlook For Solar Power 2023-2027*; 2023.

[3]   Ahmadi B and Shirazi E. Optimal allocation of voltage regulations to maximize the hosting capacity of distribution systems. In: *2023 IEEE 50th Photovoltaic Specialists Conference (PVSC)*; 2023. pp. 1–6.

[4]   Ahmadi B, Ceylan O, Ozdemir A. Reinforcement of the distribution grids to improve the hosting capacity of distributed generation: Multi-objective framework. *Electric Power Systems Research*. 2023;217:109120.

[5]   Rabiee A and Mohseni-Bonab SM. Maximizing hosting capacity of renewable energy sources in distribution networks: A multi-objective and scenario-based approach. *Energy*. 2017;120:417–430.

[6]   Shirazi E, Lemmens J, Gofran Chowdhury M, *et al. Cloud Detection for PV Power Forecast based on Colour Components of Sky Images*; 2021. pp. 2389–2391.

[7]   Bazionis IK, Kousounadis-Knousen MA, Georgilakis PS, *et al.* A taxonomy of short-term solar power forecasting: Classifications focused on climatic conditions and input data. *IET Renewable Power Generation*. 2023;17(9): 2411–2432.

[8]   Alcañiz A, Grzebyk D, Ziar H, *et al.* Trends and gaps in photovoltaic power forecasting with machine learning. *Energy Reports*. 2023;9:447–471.

[9]   Barhmi K, Heynen C, Golroodbari S, *et al.* A review of solar forecasting techniques and the role of artificial intelligence. *Solar*. 2024;4(1):99–135.

[10] Yang D, Wang W, Gueymard CA, et al. A review of solar forecasting, its dependence on atmospheric sciences and implications for grid integration: Towards carbon neutrality. *Renewable and Sustainable Energy Reviews*. 2022;161:112348.

[11] Markovics D and Mayer M. Comparison of machine learning methods for photovoltaic power forecasting based on numerical weather prediction. *Renewable and Sustainable Energy Reviews*. 2022;161:112364.

[12] Sarkis R, Oguz I, Psaltis D, *et al.* Intraday solar irradiance forecasting using public cameras. *Solar Energy*. 2024;275:112600.

[13] Scolari E, Sossan F, Haure-Touzé M, *et al.* Local estimation of the global horizontal irradiance using an all-sky camera. *Solar Energy*. 2018;173: 1225–1235.

[14] Shirazi E, Gordon I, Reinders A, *et al.* Sky images for short-term solar irradiance forecast: A comparative study of linear machine learning models. *IEEE Journal of Photovoltaics*. 2024;14(4):691–698.

[15] Antonanzas J, Osorio N, Escobar R, *et al.* Review of photovoltaic power forecasting. *Solar Energy*. 2016;136:78–111.

[16] Carrillo R, Schubnel B, Langou R, *et al.* Dynamic graph machine learning for multi-site solar forecasting. In: *40th European Photovoltaic Solar Energy Conference and Exhibition (EUPVSEC)*, Lisbon, Portugal; 2023.

[17] Simeunović J, Schubnel B, Alet PJ, *et al.* Interpretable temporal-spatial graph attention network for multi-site PV power forecasting. *Applied Energy*. 2022;327:120127.

[18]   Simeunović J, Schubnel B, Alet PJ, *et al.* Spatio-temporal graph neural networks for multi-site PV power forecasting. *IEEE Transactions on Sustainable Energy.* 2022;13(2):1210–1220.

[19]   Parmezan ARS, Souza VMA, and Batista GEAPA. Evaluation of statistical and machine learning models for time series prediction: Identifying the state-of-the-art and the best conditions for the use of each model. *Information Sciences.* 2019;484:302–337.

[20]   Das UK, Tey KS, Seyedmahmoudian M, *et al.* Forecasting of photovoltaic power generation and model optimization: A review. *Renewable and Sustainable Energy Reviews.* 2018;81:912–928.

[21]   Feurer M and Hutter F. Hyperparameter optimization. In: Hutter F, Kotthoff L, Vanschoren J, editors. Automated Machine Learning: Methods, Systems, Challenges. Cham: Springer; 2019. pp. 3–33.

[22]   Bischl B, Binder M, Lang M, *et al.* Hyperparameter optimization: Foundations, algorithms, best practices, and open challenges. *WIREs Data Mining and Knowledge Discovery.* 2023;13(2):e1484.

[23]   Yang L and Shami A. On hyperparameter optimization of machine learning algorithms: Theory and practice. *Neurocomputing.* 2020;415:295–316.

[24]   Belle V and Papantonis I. Principles and practice of explainable machine learning. *Frontiers in Big Data.* 2021;4.

[25]   Probst P, Boulesteix AL, and Bischl B. Tunability: importance of hyperparameters of machine learning algorithms. *Journal of Machine Learning Research.* 2019;20(53):1–32.

[26]   Makridakis S, Spiliotis E, and Assimakopoulos V. Statistical and machine learning forecasting methods: Concerns and ways forward. *PLoS One.* 2018;13(3):e0194889.

[27]   Zhang W, Li Q, and He Q. Application of machine learning methods in photovoltaic output power prediction: A review. Journal of Renewable and Sustainable Energy. 2022;14(2):022701.

[28]   Voyant C, Notton G, Kalogirou S, *et al.* Machine learning methods for solar radiation forecasting: A review. *Renewable Energy.* 2017;105:569–582.

[29]   Bojer CS. Understanding machine learning-based forecasting methods: A decomposition framework and research opportunities. *International Journal of Forecasting.* 2022;38(4):1555–1561.

[30]   Stone M. Cross-Validatory Choice and Assessment of Statistical Predictions. *Journal of the Royal Statistical Society: Series B (Methodological).* 2018;36(2):111–133.

[31]   Bergmeir C, Hyndman RJ, and Koo B. A note on the validity of cross-validation for evaluating autoregressive time series prediction. *Computational Statistics & Data Analysis.* 2018;120:70–83.

[32]   Bergmeir C and Benítez JM. On the use of cross-validation for time series predictor evaluation. *Information Sciences.* 2012;191:192–213.

[33]   Sharadga H, Hajimirza S, and Balog RS. Time series forecasting of solar power generation for large-scale photovoltaic plants. *Renewable Energy.* 2020;150:797–807.

[34]   Andradóttir S. Chapter 20 An Overview of Simulation Optimization via Random Search. In: Henderson SG, Nelson BL, editors. *Simulation. Handbooks in Operations Research and Management Science*, vol. 13. Amsterdam: Elsevier; 2006. pp. 617–631.

[35]   Fayed HA and Atiya AF. Speed up grid-search for parameter selection of support vector machines. *Applied Soft Computing*. 2019;80:202–210.

[36]   Zhou P, Feng J, Ma C, *et al.* Towards theoretically understanding why SGD generalizes better than ADAM in deep learning. In: *Proceedings of the 34th International Conference on Neural Information Processing Systems. NIPS '20*. Red Hook, NY, USA: Curran Associates Inc.; 2020.

[37]   Narkhede MV, Bartakke PP, and Sutaone MS. A review on weight initialization strategies for neural networks. *Artificial Intelligence Review*. 2022;55:291–322.

[38]   Sardarabadi A, Ardakani AH, Matrone S, *et al.* Multi-temporal PV power prediction using long short-term memory and wavelet packet decomposition. *Energy and AI*. 2025; p. 100540.

[39]   Quest H and Shirazi E. Model tuning and performance evaluation in machine learning models for PV power forecasting: case study of a BIPV system in Switzerland. In: *2024 IEEE 52nd Photovoltaic Specialists Conference (PVSC)*; 2024. pp. 438–442.

[40]   Fonseca-Delgado R and Gómez-Gil P. An assessment of ten-fold and Monte Carlo cross validations for time series forecasting. In: *2013 10th International Conference on Electrical Engineering, Computing Science and Automatic Control (CCE)*; 2013. pp. 215–220.

[41]   The Royal Netherlands Meteorological Institute. Available from: https://www.knmi.nl/home.

Chapter 9
Sky imager based solar photovoltaic forecast

Khadija Barhmi[1], Elham Shirazi[2], Sara Mirbagheri Golroodbari[1] and Wilfried van Sark[1]

The ever-increasing photovoltaic (PV) system deployment is an essential part of the global transition toward a fully renewable energy-based energy system. However, the inherent intermittency and variability of solar irradiance lead to fluctuations in PV power generation. This may present significant challenges to the stability and reliability of electrical power grids, especially at high PV penetration. Accurate forecasting of PV generation is essential to mitigate these challenges, as it will allow grid operators and other stakeholders to mitigate any grid instability at the lowest cost. This chapter will present so-called nowcasting of solar irradiation and derived PV power based on image sequences from sky imagers, also denoted as all-sky imagers (ASIs). Forecasting methods will be presented based on artificial intelligence and machine learning methods, as these have been developed in recent years and have been shown to be able to forecast solar irradiance with high accuracy. In particular, for ASIs, such high accuracy is demonstrated at high spatial and temporal resolutions.

9.1 Introduction

The global transition toward renewable energy systems is accelerating, driven by the urgent need to reduce greenhouse gas emissions and combat climate change. Among various renewable energy technologies, solar PV and wind energy are expected to play a leading role in future energy systems. Several studies, including so-called 100% renewable energy scenarios, have shown that the rapid growth of PV and wind energy must continue over the coming decades to meet climate goals [1].

However, both PV and wind energy are inherently weather-dependent. In the case of PV systems, the variability of solar irradiance, caused mainly by cloud movements, introduces significant challenges to maintaining grid stability and ensuring a reliable energy supply. Rapid fluctuations in PV power output can disrupt the balance between supply and demand, affect electricity market dynamics, and increase the need for reserve power and grid flexibility [2].

[1]Faculty of Geosciences, Copernicus Institute of Sustainable Development, Utrecht University, The Netherlands
[2]Faculty of Engineering Technology, University of Twente, The Netherlands

To address these challenges, accurate short-term forecasting of solar irradiance and PV generation has become crucial. Forecasts with lead times ranging from a few minutes to a few hours can support real-time grid operation, enable proactive energy management, and reduce costs associated with imbalance penalties. The choice of forecasting approach depends largely on the time horizon and data availability. While long-term forecasts (days to weeks ahead) typically rely on numerical weather prediction (NWP) models, short-term and very short-term forecasts benefit more from real-time observational data, such as those from ground-based instruments [3].

Among ground-based tools, ASIs have received increasing attention. ASIs provide high-resolution hemispherical images of the sky, enabling detailed observation of cloud dynamics on a local scale [4]. Their ability to capture the temporal and spatial evolution of clouds makes them particularly valuable for very short-term forecasting, where traditional methods often fail due to their coarse resolution or low update frequency [5]. An example of a typical ASI setup is shown in Figure 9.1, illustrating the imaging capability at different temporal resolutions.

In recent years, artificial intelligence (AI) has emerged as a powerful enabler for enhancing the performance of solar forecasting models. Machine learning (ML) and deep learning (DL) techniques can model complex, nonlinear relationships and process large volumes of heterogeneous data, including sky images, satellite data, and meteorological measurements [3,6]. These AI-based models provide higher accuracy, adaptability, and robustness than traditional statistical or physics-based methods, especially for localized and rapidly changing weather conditions, see also other chapters in this book.

This chapter focuses on short-term solar power forecasting using AI techniques in combination with sky images from ASIs. The aim is to explore how DL models can extract meaningful patterns from sky imagery to predict cloud evolution,

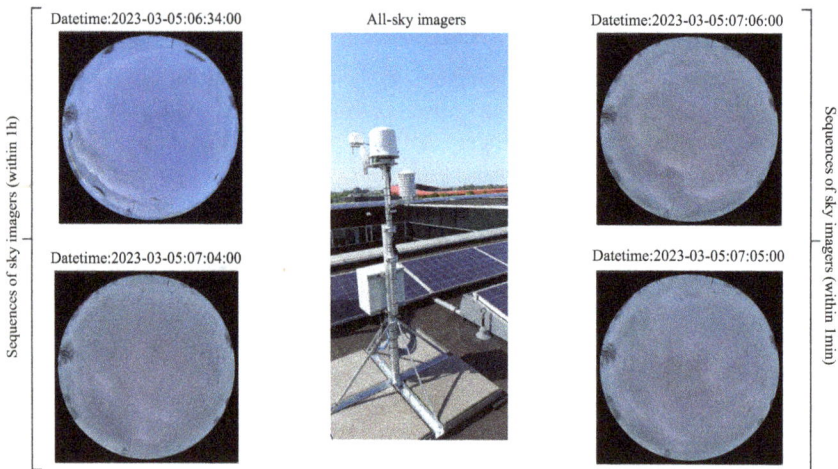

Figure 9.1   A typical all-sky imager installation (middle), and images captured by the camera, at 1 h (left) and 1 min (right) time interval

solar irradiance, and ultimately PV power output. The subsequent sections address cloud detection, irradiance forecasting, and PV power prediction, followed by data fusion strategies that combine satellite and physical models and a discussion on key challenges and future research directions.

9.2   Solar PV forecasting pipeline

AI models have become increasingly valuable for improving short-term solar PV forecasting, particularly when combined with high-resolution sky images from ASIs. These images capture detailed cloud patterns and sky conditions in real time, which AI algorithms can analyze to detect clouds, forecast solar irradiance, and forecast PV power output. This section presents the key components of this forecasting pipeline (see also Figure 9.2): cloud detection, irradiance prediction, and PV output forecasting, highlighting how AI enhances accuracy across each stage.

A typical PV forecasting pipeline using all-sky imagers consists of the following stages, see also Figure 9.2:

1.   Acquisition of sky images using a hemispherical camera;
2.   Cloud detection and classification from image data;
3.   Cloud motion estimation using techniques such as optical flow or convolutional neural network (CNN) and/or recurrent neural network (RNN) architectures;
4.   Irradiance forecasting based on cloud movement and clear-sky models;
5.   Conversion of predicted irradiance into PV power output.

Training the AI models typically involves supervised learning using datasets of timestamped sky images and corresponding irradiance measurements from pyranometers, see also previous section. The pipeline generally includes:

1.   **Data preprocessing:** Resize images, normalize pixel values, and apply data augmentation (e.g., rotation, flipping) to improve generalization;
2.   **Loss functions:** Common choices include mean absolute error (MAE) and mean squared error (MSE), defined, respectively, as follows:

$$\text{MAE} = \frac{1}{N} \sum_{i=1}^{N} |I_{\text{pred},i} - I_{\text{true},i}| \tag{9.1}$$

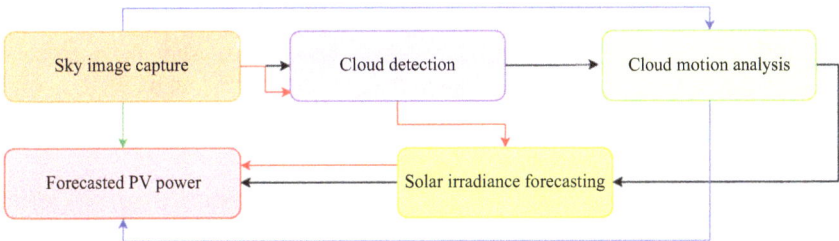

Figure 9.2   Overview of short-term solar PV forecasting using sky imagers

$$\text{MSE} = \frac{1}{N} \sum_{i=1}^{N} (I_{\text{pred},i} - I_{\text{true},i})^2 \tag{9.2}$$

3. **Training:** Models are trained using optimizers, such as stochastic gradient descent (SGD) or adaptive moment estimation (Adam) [7], with regularization techniques like dropout and early stopping to prevent overfitting.

Evaluation metrics used to assess model performance include mean absolute percentage error (MAPE), root mean squared error (RMSE), coefficient of determination R^2, forecast skill (FS). The benchmark model used in FS determination is usually a persistence forecast but can also be smart persistence [8].

$$\text{MAPE} = \frac{1}{N} \sum_{i=1}^{N} \left| \frac{I_{\text{pred},i} - I_{\text{true},i}}{I_{\text{true},i}} \right| \times 100 \tag{9.3}$$

$$\text{RMSE} = \sqrt{\frac{1}{N} \sum_{i=1}^{N} (I_{\text{pred},i} - I_{\text{true},i})^2} \tag{9.4}$$

$$R^2 = 1 - \frac{\sum_{i=1}^{N} (I_{\text{true},i} - I_{\text{pred},i})^2}{\sum_{i=1}^{N} (I_{\text{true},i} - \overline{I_{\text{true}}})^2} \tag{9.5}$$

$$\text{FS} = 1 - \frac{\text{Error}_{\text{forecast}}}{\text{Error}_{\text{reference}}} \tag{9.6}$$

9.2.1 Cloud detection

To achieve reliable solar forecasting, it is critical to accurately analyze and predict the cloud cover, as clouds have a significant impact on solar radiation [5]. Here, various models and advancements in short-term solar forecasting are explored, specifically focusing on cloud detection using sky imagers [9].

Cloud detection from sky images is a fundamental task in short-term solar forecasting [10]. A common starting point is manual labeling, where experts manually annotate cloud pixels in the images, thereby identifying the *ground truth*, which has been widely employed to train cloud detection algorithms. However, manual labeling can be labor-intensive, time-consuming, and subject to human error, particularly when handling a large number of images [11]. Therefore, there is a need to explore automatic manual labeling models to overcome these limitations and improve the efficiency of training datasets [12].

Traditional cloud detection techniques typically rely on thresholding approaches using color ratios in RGB images. For instance, the blue-to-red ratio (BRR) is often employed, as clouds generally reflect more blue light than the clear sky:

$$BRR = \frac{I_B}{I_R} \tag{9.7}$$

where I_B and I_R are the blue and red channel intensities, respectively [13]. Pixels exceeding a predefined threshold T are classified as cloud regions. However, this threshold can vary depending on weather conditions and must often be dynamically adjusted [14].

In addition, alternative color-based indices have been proposed. For example, Shirazi *et al.* [15] introduced a method using red–blue ratios and other color components for efficient real-time cloud detection, particularly suitable for low-computation scenarios.

Beyond thresholding, ML models have enabled more robust cloud detection by learning from spectral and texture features. Classical ML approaches, such as support vector machines (SVMs) and decision trees (DTs), have been trained on hand-crafted features to identify cloud patterns. More recently, CNNs have outperformed traditional methods by learning hierarchical features directly from raw image data [16]. A flowchart summarizing these cloud detection models is shown in Figure 9.3.

Cloud classification is another important step for forecasting accuracy. Different cloud types have distinct optical properties that influence irradiance in different ways. For instance, cumulus clouds may allow partial sunlight to pass through, while stratus clouds can block it entirely. Schmidt *et al.* [17] developed a cloud classification algorithm using a support vector classifier (SVC), trained on 600 labeled sky images, to identify nine cloud types (e.g., cumulus, stratocumulus, and cirrocumulus). Their method utilized 16 texture-based features, including metrics from the gray-level co-occurrence matrix (GLCM), such as angular second moment and correlation [18], achieving a classification accuracy of 92%. Recently, vision transformers (ViTs) have been explored for this task due to their ability to model long-range spatial dependencies and varying texture scales more effectively than CNNs [19].

The next critical component of cloud analysis is predicting cloud movement. Moving clouds cause rapid fluctuations in solar irradiance, making their tracking

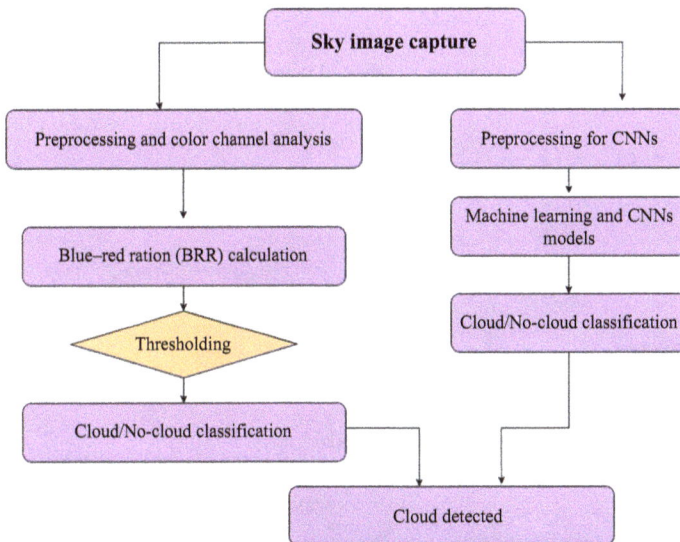

Figure 9.3   Flowchart of cloud detection models for sky imagers

vital for short-term forecasting. Cloud motion estimation traditionally uses physical models such as *optical flow*, which detects pixel intensity changes across sequential images. The optical flow equation is expressed as follows:

$$\frac{\partial I}{\partial t} + u\frac{\partial I}{\partial x} + v\frac{\partial I}{\partial y} = 0 \tag{9.8}$$

where u and v represent horizontal and vertical motion components [20,21]. Another common approach is *block matching*, which divides images into small blocks and tracks their displacement across frames:

$$\mathbf{d} = \arg\min_{\mathbf{d}'} \sum_{(x,y)\in B} |I_t(x,y) - I_{t+\Delta t}(x + d_x, y + d_y)| \tag{9.9}$$

where B is a block in the image, and d_x, d_y are displacements in x, y directions [22].

AI models have further advanced cloud motion prediction. CNN–RNN hybrids, where CNNs extract spatial cloud features and RNNs (especially long short-term memory (LSTM) models) capture temporal dynamics, are widely used for short-term movement prediction [23]. Figure 9.4 illustrates a typical AI-based cloud motion prediction pipeline.

Optical flow can also be predicted directly using deep learning. Encoder–decoder architectures such as FlowNet and PWC-Net are trained to estimate flow vectors from image sequences [24], bypassing the need for explicit image differencing. More recently, ViTs have been adopted to model global dependencies between cloud regions in sequential images, offering promising improvements in cloud tracking accuracy [25].

The influence of cloud cover on irradiance is often modeled using cloud cover fraction C_f, a modification factor M_c, and clear-sky irradiance I_{clear}, as follows:

$$I_{solar} = I_{clear} \times (1 - C_f) \times M_c \tag{9.10}$$

Alternatively, the irradiance can be expressed using a cloud transmission factor T_{cloud}, where $T_{cloud} = M_c$:

$$I_{solar} = I_{clear} \times (1 - C_f) \times T_{cloud} \tag{9.11}$$

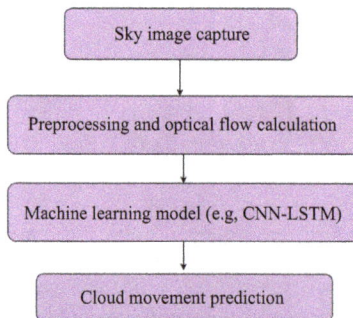

Figure 9.4   Flowchart of cloud movement prediction

These models quantify how cloud density and type affect the solar radiation received by PV panels, directly impacting power generation. Accurate and dynamic estimation of C_f and T_{cloud} is thus crucial for real-time PV forecasting [26]. The practice of using oktas to quantify cloud cover, in which 0 oktas reflects a clear sky and 8 oktas reflects a completely overcast sky, is insufficiently accurate for irradiance estimation. Cloud cover and transmission determined from, e.g., ASIs provide a more accurate irradiance estimation.

Effective cloud detection, classification, and motion prediction are indispensable components of AI-based PV forecasting pipelines. While traditional methods offer interpretable and computationally efficient approaches, AI models, especially those incorporating deep learning and transformer-based architectures, provide the adaptability and precision required for operational forecasting in dynamic environments. Future improvements may lie in integrating sky imagers with external data sources such as satellite observations and NWP, thereby enhancing spatial coverage and temporal accuracy.

9.2.2 Solar irradiance forecast

Solar irradiance prediction is essential for forecasting PV power generation, particularly in short-term applications where fluctuations in cloud cover and movement play a significant role, such as congestion management in low-voltage distribution grids. Traditional physical models rely on atmospheric parameters, empirical formulas, and radiative transfer equations to estimate solar irradiance. However, recent advances in AI have introduced sophisticated models capable of learning complex, nonlinear relationships from sky-imager data. AI-based approaches have demonstrated notable improvements in prediction accuracy and computational efficiency, particularly in capturing the temporal and spatial dynamics of clouds.

Physical models for solar irradiance prediction use meteorological and radiative transfer principles to estimate the irradiance reaching the Earth's surface. One common approach is the clear-sky model, which calculates irradiance under cloud-free conditions based on factors such as the solar zenith angle θ_z, atmospheric optical depth τ, and extraterrestrial irradiance I_0 [27]. Clear-sky irradiance I_{clear} can be expressed as follows:

$$I_{clear} = I_0 \times \exp\left(-\frac{\tau}{\cos(\theta_z)}\right) \tag{9.12}$$

where $I_0 \approx 1361 \text{ W/m}^2$ is the solar constant [28].

In the presence of clouds, additional factors such as light scattering and absorption must be considered. This is typically addressed using a cloud modification factor M_c, which adjusts irradiance based on cloud cover fraction C_f and optical properties [29]. The resulting solar irradiance under cloudy conditions is then expressed using (9.10).

While physical models are useful under clear or stable conditions, they often struggle to capture dynamic cloud behavior, prompting the adoption of AI models for short-term irradiance forecasting. Recent developments in AI have led to a variety of models for solar irradiance prediction using sky image data. Among the most commonly used approaches are artificial neural networks (ANNs), CNNs, RNNs, and hybrid CNN–RNN architectures, see also previous chapters for more details:

1. **ANNs** were among the first AI models applied to solar forecasting, leveraging historical weather and sky image features. They can model basic nonlinear relationships but may underperform on complex image data [30].
2. **CNNs** are highly effective in extracting spatial features from sky images, including cloud shape, density, and texture, making them suitable for irradiance prediction based on cloud analysis [31].
3. **RNNs**, especially LSTM networks, are used to model temporal dependencies in sequential sky images. They help capture cloud movement and transitions, enhancing the prediction of irradiance variability [32].
4. **Hybrid CNN–RNN models** combine spatial and temporal analysis. CNN layers process spatial features from sky images, while LSTM layers track temporal dynamics across frames [33].

These AI models are capable of learning subtle variations in cloud features and their effects on irradiance. For example, thin cirrus clouds may not appear visibly dense but still reduce irradiance through scattering, which CNNs can learn to detect. Recent work by Hendrikx *et al.* [34] demonstrated the effectiveness of LSTM networks for irradiance forecasting, reporting a ramp FS score of 0.39 in sunny conditions and 0.25 in partly cloudy scenarios. A comparative overview of model performance across cloudy, partly cloudy, and sunny conditions is shown in Figure 9.5.

Transformer-based models such as ViTs have also shown promise in irradiance prediction, particularly due to their ability to model long-range spatial dependencies in cloud structures [25].

In summary, AI-based models, particularly deep learning architectures that integrate spatial and temporal information, offer a powerful solution for solar irradiance prediction. Their ability to adapt to cloud variability and weather conditions makes them well-suited for deployment in PV forecasting. The next section explores how these irradiance forecasts can be further improved by integrating sky imagers with other data sources, such as satellite and NWP models.

9.2.3   Solar PV forecast

Forecasting solar PV power output is a critical step in managing solar energy integration, particularly in power grids with high PV penetration. Once the solar irradiance forecast is obtained, whether through physical models, AI-based approaches, or hybrid methods, it must be converted into a forecasted PV power generation. This conversion requires understanding the characteristics of the PV system and the interaction between solar radiation and the PV modules.

Short-term PV forecasting using sky imagers is primarily based on real-time analysis of cloud cover and cloud motion. Sky imagers capture high-frequency images of the sky (typically every 30 s to 5 min), which are processed to identify cloud location, thickness, and dynamics [35]. By tracking cloud movement across consecutive images, the future positions of clouds can be predicted, enabling the prediction of incoming solar radiation at specific times. This irradiance forecast is then used to compute the expected PV output.

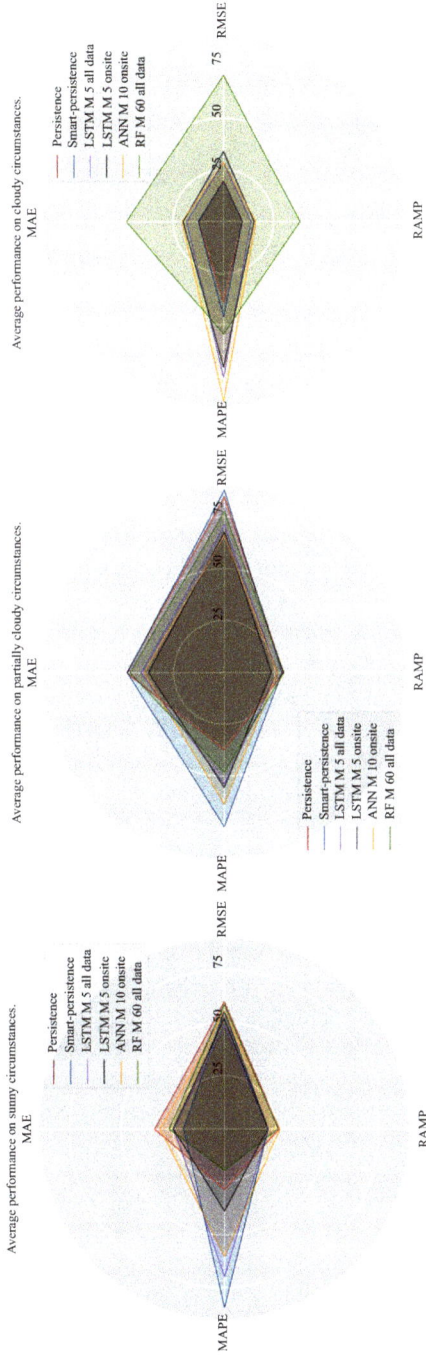

Figure 9.5   Performance comparison of various models on short-term irradiance forecasting across weather conditions, adapted from [34]

The conversion from solar irradiance to PV power is typically modeled using the following physical equation [36]:

$$P_{\mathrm{PV}} = I_{\mathrm{solar}} \times A_{\mathrm{PV}} \times \eta_{\mathrm{PV}} \times PR \tag{9.13}$$

where I_{solar} is the predicted irradiance incident on the PV module (W/m^2), A_{PV} is the surface area of the PV array (m^2), η_{PV} is the nominal efficiency of the PV modules, and PR is the performance ratio accounting for losses (e.g., temperature, dust, and wiring).

While physical models provide a straightforward means of forecasting PV power, their accuracy is often limited under rapidly changing atmospheric conditions, especially in the presence of intermittent clouds. To address this challenge, AI models are increasingly used to improve the mapping from irradiance and sky image features to PV output.

Incorporating AI into PV forecasting allows models to learn complex, nonlinear relationships between input features (e.g., sky image characteristics, time of day, and past irradiance values) and the resulting PV power output. CNNs are commonly used to extract spatial patterns from sky images, such as cloud texture and coverage. LSTM or gated recurrent unit (GRU) layers can then capture the temporal evolution of these features, enabling short-term predictions of PV power. Hybrid CNN–LSTM models have shown improved performance by combining both spatial and temporal learning capabilities.

Recent work by Qu *et al.* [37] demonstrated the effectiveness of using CNN–RNN hybrid architectures for PV power forecasting, achieving improved forecast accuracy by integrating cloud spatial patterns and movement sequences. Moreover, ViT-based models have begun to gain traction in PV forecasting due to their ability to model global spatial dependencies in large-scale sky image datasets [38]. To train these models, datasets typically include time-synchronized sky images and ground-truth PV power measurements collected from inverters or monitoring systems. During training, the model minimizes a loss function, such as MAE or MSE, between predicted and actual PV power values.

Ultimately, PV forecasting models based on sky imagers must balance accuracy with computational speed to be useful in operational environments. High-frequency image acquisition and real-time inference are necessary for grid operators to make informed decisions regarding dispatching, load balancing, and curtailment. The use of AI-enhanced forecasting pipelines, particularly those leveraging hybrid deep learning models and physics-informed architectures, continues to improve short-term forecasting accuracy, especially under dynamic and partially cloudy conditions.

9.3   AI-driven data fusion of sky imager, satellite, and NWP models

Integrating sky imager data with satellite imagery and NWP models has become a powerful approach to improve the accuracy and spatiotemporal coverage of solar irradiance forecasting. While traditional physical models rely heavily on satellite and NWP data, each source brings unique strengths: satellite images offer broad spatial coverage, making it possible to monitor cloud formations beyond the field of view of ground-based sky imagers, while NWP models simulate atmospheric conditions using physical laws, providing forecasts of temperature, humidity, wind, and cloud cover over time [39]. Sky imagers, in contrast, provide high-resolution, real-time observations of local cloud dynamics at the PV site. The combination of these sources allows for more accurate short-term predictions, especially when clouds are moving rapidly. A common technique to combine these data streams is weighted averaging, where irradiance forecasts from each source – sky imagers I_{imager}, satellite-based cloud index CI, and NWP predictions I_{NWP} – are fused as follows:

$$I_{integrated} = w_{imager} \cdot I_{imager} + w_{satellite} \cdot CI + w_{NWP} \cdot I_{NWP} \qquad (9.14)$$

where w_{imager}, $w_{satellite}$, and w_{NWP} are weights determined based on the relative accuracy of each source over different time horizons [2]. However, fixed-weight fusion has limitations in adapting to rapidly changing cloud conditions, which has led to increased use of AI-driven data fusion techniques. Recent advancements in AI have enabled the integration of multisource data using models such as CNNs, RNNs, and attention-based transformers. CNNs are often used to process sky and satellite images in parallel through separate convolutional branches, followed by feature-level fusion to combine local and large-scale cloud information [40]. To capture temporal dependencies, LSTM-based RNNs are employed to integrate sequences from sky imagers, satellite imagery, and NWP forecasts, allowing the model to learn cloud evolution patterns and produce more robust irradiance predictions [41]. Furthermore, attention mechanisms embedded in transformer-based models have shown strong performance by dynamically weighting input sources based on their contextual relevance. For instance, ViTs and hybrid CNN-transformer architectures can prioritize spatial regions within sky and satellite images that are more indicative of future irradiance conditions [25].

Building on this, various hybrid AI models have been developed to exploit the complementary strengths of sky imager, satellite, and NWP data. CNN–LSTM architectures are commonly used for this purpose, where CNN layers extract spatial features from sky images, and LSTM layers incorporate temporal patterns from NWP forecasts. This structure allows the model to blend real-time observations with longer-term meteorological trends, improving its ability to predict short-term irradiance fluctuations [42]. Ensemble models further extend this concept by combining the outputs of separate models trained on each data source. Predictions from CNNs, RNNs, and physically based NWP systems can be aggregated through techniques, such as weighted averaging or stacking to produce a more robust and accurate forecast [43]. Additionally, Physics-Informed Neural Networks (PINNs) offer a powerful extension to data fusion by embedding physical laws – such as radiative transfer or energy balance equations – directly into the neural network training process. Unlike traditional neural networks that rely solely on data, PINNs integrate domain knowledge in the form of partial differential equations, ensuring that model outputs remain consistent with known physics even in sparse or noisy data scenarios [44]. In a recent study by Paletta *et al.* [45], ASI data was combined with a physics-informed transfer learning for irradiance forecasting. In this work, an ML model (convolutional LSTM) that was trained for a large ASI dataset in Paris was subject to transfer learning approaches to adapting the ML model for another data set in California, i.e., the open source SKIPP'D dataset [46]. While the adapted ML model performed well under stable cloud conditions, it faced challenges during fast-evolving cloud dynamics. To address this, the authors proposed supplementing ASI data with real-time inputs from additional sensors or satellite imagery, enhancing the model's spatial awareness and further using the strengths of physics-informed learning in dynamic weather conditions.

A noteworthy example of an sensor network is a network of ASIs in north-west Germany [47]. This network comprises 34 ASIs complemented with weather stations, radiometers, and ceilometers and will cover an area of over $10,000$ km^2. It is able to detect clouds and track cloud movement over large distances. Data fusion with satellite images would allow forecasts of up to 6 h ahead established [48,49]. A similar, albeit smaller network of 10 ASIs is being constructed to allow for solar forecasting for the city of Utrecht. With such networks, stereo-imaging can be employed to infer cloud base heights [50].

A notable application of AI-based data fusion is presented in Kumari and Toshniwal's work [42], where a CNN–LSTM model was used to forecast short-term irradiance in California using sky images and NWP forecasts. The CNN extracted spatial cloud features, while the LSTM captured evolving weather patterns. This hybrid structure enabled accurate irradiance forecasting with a MAPE of approximately 12% under variable conditions. The model's architecture is depicted in Figure 9.6.

The model's performance was validated against ground-truth measurements, showing high accuracy during clear-sky conditions and some deviation under fast-changing clouds. Seasonal performance comparisons, shown in Figure 9.7, illustrate that the CNN–LSTM model consistently outperformed other methods, especially in winter and spring. These seasons exhibited the highest FS improvement over the smart persistence model, while summer and autumn showed lower gains [42].

Another key factor in AI-driven solar forecasting models is the weather dependency across different geographic locations. For example, in the Netherlands, where cloud cover is highly variable due to the maritime climate, forecasting models face unique challenges. A study conducted by Hendrikx *et al.* [34] utilized LSTM networks for short-term solar irradiance forecasting and recorded different levels of accuracy depending on weather conditions. Their model achieved an RMSE of 86 W/m^2 on sunny days, which increased to 109 W/m^2 on partly cloudy days, illustrating the challenges of accurately predicting irradiance under changing cloud conditions. Similarly, annual RMSE values of CNN-based forecasting for three major cities in California (San Francisco, Los Angeles, and San Diego) were found to range between approximately 68 and 90 W/m^2 [42], with the southernmost city having the lowest RMSE. The proposed CNN–LSTM model improved RMSE values between approximately 45 and 60 W/m^2 [42].

In comparison, in arid regions such as Arizona, United States, forecasting models tend to achieve higher accuracy due to more stable weather patterns. For instance, a CNN model applied in Arizona achieved an RMSE of approximately 60 W/m^2, benefiting from predictable cloud movement in the region's low-humidity climate. This stability allows simpler ML models such as CNNs to perform reliably, with fewer recalibrations required than in the Netherlands [42].

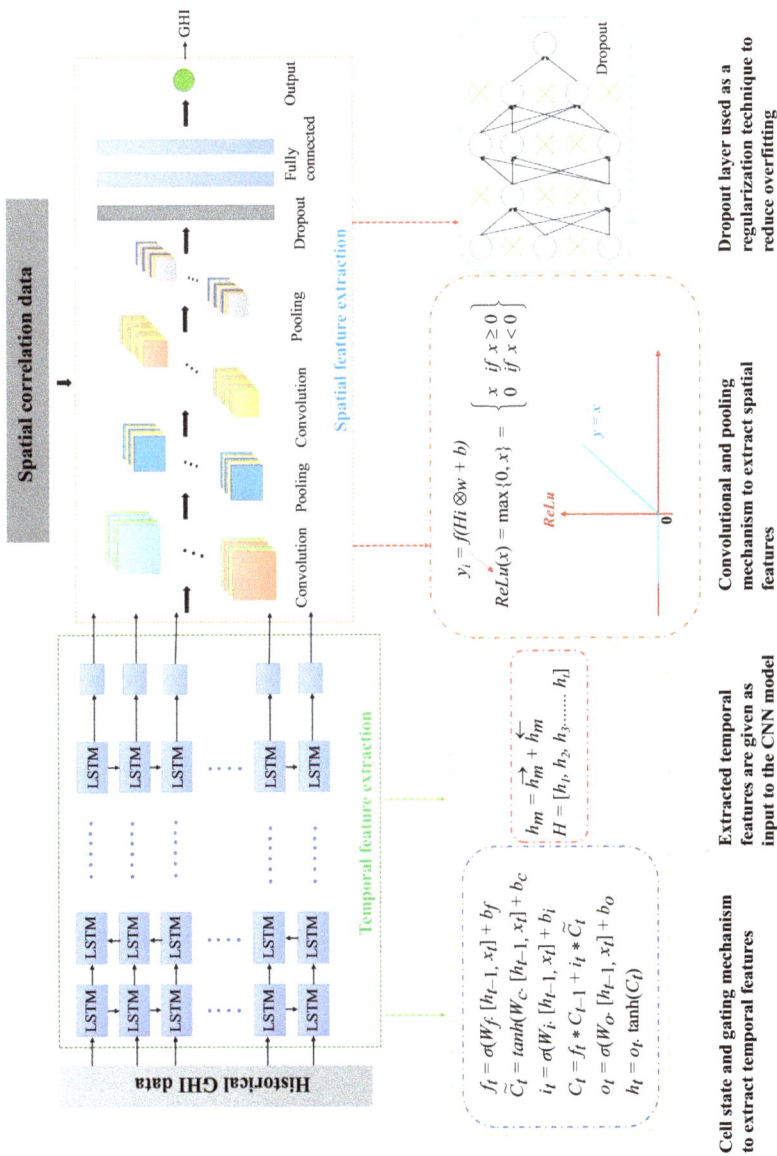

Figure 9.6 Structure of the CNN–LSTM model used for short-term irradiance prediction (adapted from [42])

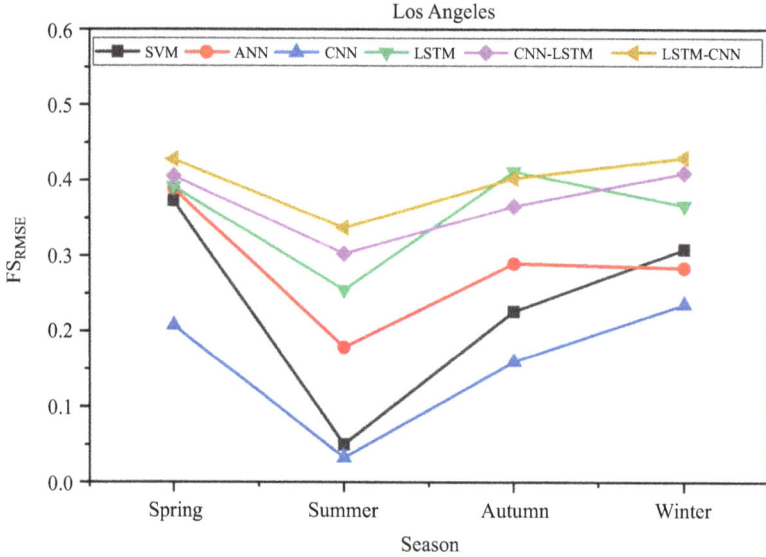

Figure 9.7   Seasonal FS score for various AI models for Los Angeles, CA, United States (adapted from [42])

A further comparison with tropical climates, such as Singapore, highlights additional challenges posed by extreme variability. In Singapore, where high humidity and frequent storms lead to rapid cloud dynamics, a GRU network model achieved an RMSE of 95 W/m². Despite the higher error, the model's ability to handle frequent and abrupt changes in cloud cover underscores the value of RNN-based models in capturing temporal dependencies necessary for tropical weather [51].

These studies emphasize the need for adaptive, region-specific forecasting models. In stable climates, simpler ML models trained with sky images are effective and efficient. In contrast, variable and tropical climates benefit from recurrent architectures, such as LSTM and GRU models, that adjust dynamically to cloud variability. This demonstrates the importance of selecting and training models that are tailored to regional climate characteristics for maximizing PV forecasting accuracy and reliability.

9.4   Case study Netherlands

This case study presents a recently developed AI-based forecasting framework that addresses the challenge of short-term solar irradiance variability, resulting from rapidly evolving cloud conditions. Developed by Barhmi *et al.* [52], the method combines high-resolution sky imagery with computer vision, ML, and filtering techniques. The overall methodology is illustrated in Figure 9.8. The framework integrates cloud cover classification, CNN-based cloud cover estimation, CSI computation, and machine-learning-based GHI forecasting to produce accurate 30-min-ahead

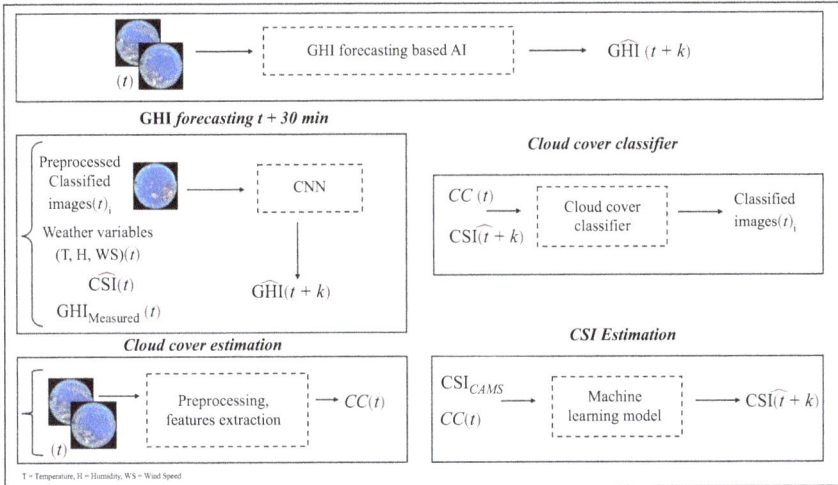

Figure 9.8   Flowchart of the methodology for short-term solar forecasting using sky images, SVM, and CNN (from [52])

predictions at 1-min resolution. The approach is particularly suited for regions with highly variable weather, such as the Netherlands.

Traditional models, such as persistence or NWP, often fail to capture the rapid cloud transitions that significantly impact irradiance within very short time frames. To address this, the proposed framework uses a four-stage pipeline using data from the Cabauw Baseline Surface Radiation Network (BSRN) station, spanning 7 months and over 75,000 sky images.

First, sky images are segmented using the Simple Linear Iterative Clustering (SLIC) algorithm to generate superpixels, which are then classified as cloud or clear-sky regions using red–blue ratio (RBR) and Otsu thresholding. This improves cloud detection accuracy while validated against the Cabauw Experimental Site for Atmospheric Research (CESAR) Nubiscope measurements. Second, a Support Vector Machine (SVM) model predicts the Clear Sky Index (CSI) using the estimated cloud cover and historical Copernicus Atmosphere Monitoring Service (CAMS) CSI data. To enhance temporal consistency and reduce noise, a Kalman filter is applied to the CSI predictions. Third, weather conditions are classified using k-means clustering based on the CSI and cloud cover values. This enables the model to adapt dynamically to different atmospheric regimes by distinguishing between fully clouded, partially clouded, and clear-sky scenarios. Finally, a CNN is trained using downsampled 128×128 RGB sky images and auxiliary features, such as CSI, cloud cover, solar zenith angle, ambient temperature, and humidity. The CNN forecasts GHI for the next 30 min with a 1-minute update interval.

As shown in Figure 9.9, the integrated system demonstrated strong performance across different weather conditions when compared to measured BSRN data. Under clear-sky conditions (March 1 and 2), the forecasted GHI values are closely aligned with actual measurements, showing tight clustering around the perfect fit line with

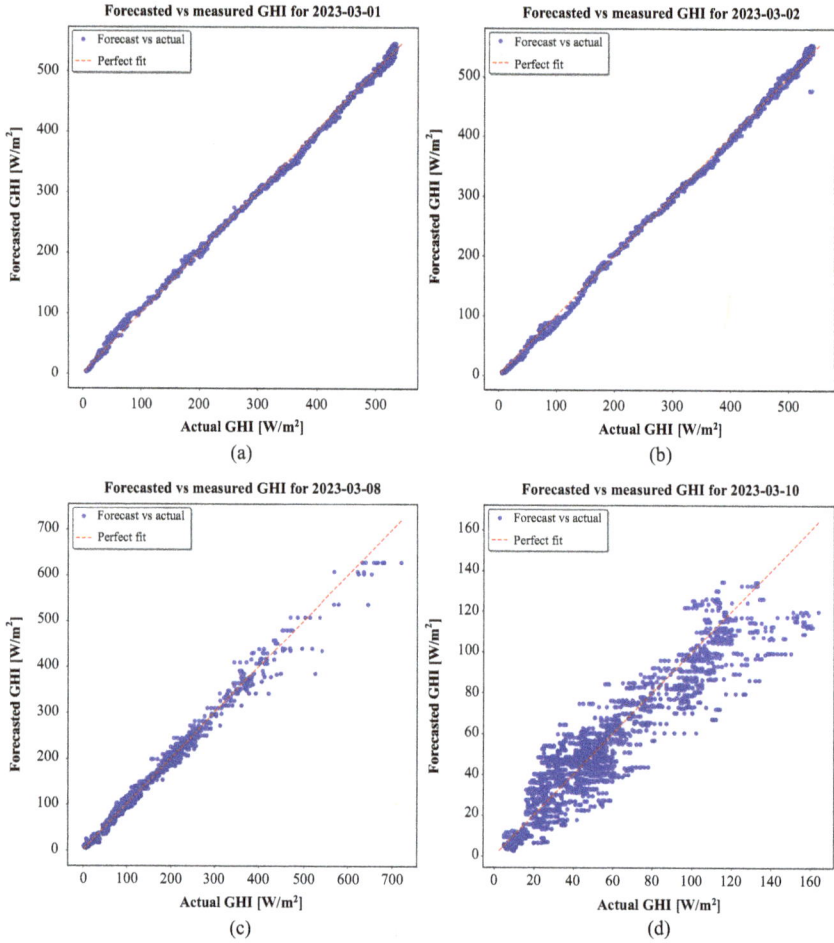

Figure 9.9   GHI forecast evaluation across four representative days in March 2023, covering a range of weather conditions from clear to highly variable (from [52]). (a) March 1, clear-sky conditions; (b) March 2, clear-sky conditions; (c) March 8, variable cloud conditions; and (d) March 10, high variability.

minimal scatter and excellent correlation across the full range of solar irradiance values. However, under variable cloud conditions (March 8), increased scatter is particularly evident at higher GHI values above 400 W/m², where cloud variability introduced greater uncertainty in the forecasts. Despite these challenges, the system shows to have reasonable accuracy across the range of weather conditions tested, with forecast skill improved by up to 22% over measured GHI and by 18% over a smart persistence baseline. Additionally, the application of the Kalman filter reduced the RMSE of CSI predictions by around 20%, especially under overcast and unstable sky conditions.

This multistage, weather-adaptive pipeline highlights the benefits of combining visual information from all-sky imagers with AI models for highly granular solar forecasting. The framework offers a scalable and computationally efficient solution suitable for real-time deployment in smart grids and energy-flexible buildings.

9.5   Challenges and outlook

The implementation of AI-based forecasting systems using sky imagers for PV power prediction has shown promising potential, yet several critical challenges remain, including forecasting accuracy, sky imager hardware limitations, high computational load, spatiotemporal data alignment, data quality and handling, and model generalizability. Here, we discuss these challenges in some more detail. Addressing these challenges is fundamental to reliable adoption of AI-based forecasting in solar energy systems.

9.5.1   Challenge 1: accuracy

A key obstacle in operational settings is maintaining forecasting accuracy under rapidly changing cloud conditions. In a study by Paletta *et al.* [45], an ASI was combined with a physics-informed transfer learning approach to improve solar irradiance forecasting based on an earlier developed ConvLSTM model. While the system achieved high accuracy during clear-sky and stable cloud formations, it struggled during dynamic weather events, where clouds moved quickly or evolved rapidly. To mitigate such limitations, Paletta *et al.* [45] proposed integrating local data from additional sensors or satellite imagery to supplement ASI observations. This multisource approach enhances spatial context and enables the forecasting model to better predict incoming cloud formations.

As shown in Figure 9.10, integrating local data significantly improved the alignment between predicted and actual irradiance values across 10 consecutive days. Here, the smart persistence model (gray line) is compared to ground truth measurements (black line) and four other models, i.e., a local expert model, transfer learning, and adding a few days of local data (Few-shot Learning, FSL), or no local data addition (Zero-shot Learning, ZSL).

The forecast skill improvement is particularly evident during periods of high variability (days 2–3 and 6–7), where the data integration approach shows reduced prediction errors. The corresponding RMSE plots (Figure 9.10) highlight that learning models that do not include local data (e.g., ZSL) exhibit significant error spikes reaching ±20 kW under dynamic conditions, whereas the integrated local data approach maintains more stable performance, with errors typically contained within ±10 kW. This improvement is most pronounced during peak irradiance periods (15–30 kW range) and rapid transitions, where traditional models struggle with the nonlinear dynamics of cloud movement, while the integrated system uses real-time sensor data to adapt more effectively to changing conditions.

To complement this, we present results from a case study based on our own published work using ASI-driven deep learning methods for solar forecasting [34]. In

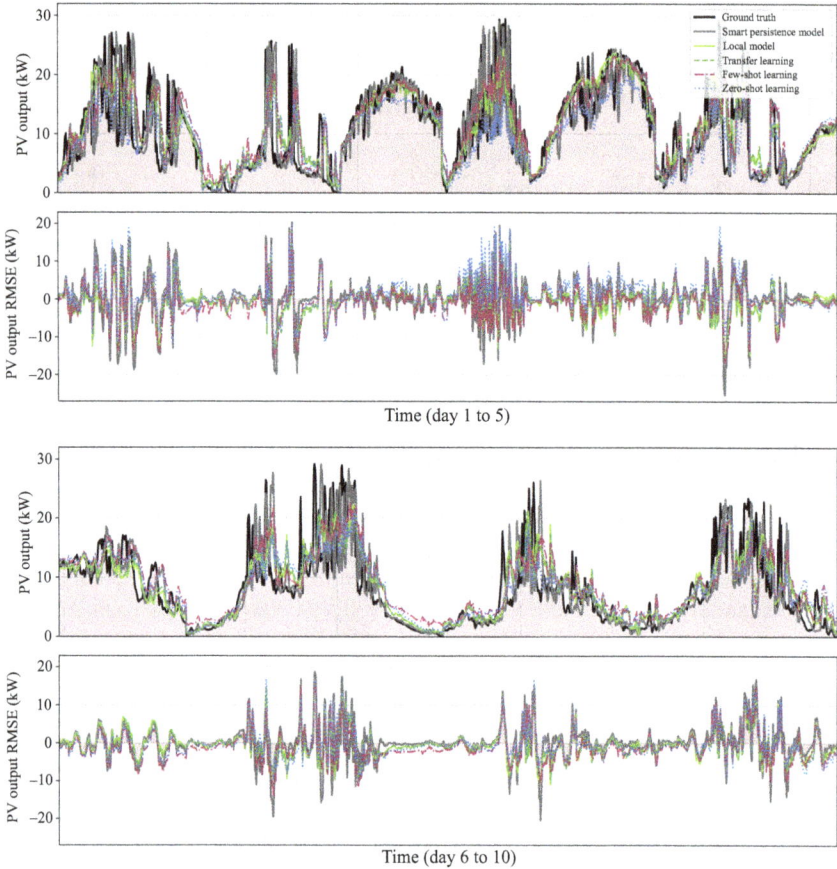

*Figure 9.10   Solar forecasting results for the adapted ConvLSTM model for diverse
data availability contexts, i.e., from no additional to a large quantity
of local data, for a selection of 10 cloudy days. Night time and
missing data are excluded (from [45]).*

our study, we deployed two high-resolution ASI systems at the Plataforma Solar de
Almería (PSA) in southern Spain and developed an LSTM neural network to forecast
GHI up to 20 min ahead. The model was trained using both on-site meteorological
data and sky image-derived features, such as cloud coverage and luminance, and was
benchmarked against baseline models, including persistence, smart persistence, RF,
and ANN. Our results showed that the LSTM model with a short input sequence
(5 min) and access to on-site and image-derived data outperformed all other tested
models. It achieved a FS, based on RSME, of 0.18 and 0.20 under sunny and partially cloudy conditions, respectively, clearly outperforming the persistence-based
baselines and even strong competitors like the SKIPP'D model [46]. In addition, the
LSTM model delivered better ramp-score performance, capturing rapid irradiance
fluctuations more effectively, especially beyond a 10-min horizon.

This case study highlights how combining ASI images with temporal modeling via LSTM networks can significantly improve accuracy in short-term solar forecasting. When real-time visual and meteorological data are used, deep learning models can better adapt to environmental dynamics, providing more reliable forecasts during cloud transitions. These results demonstrate the practical value of hybrid imaging–data-driven approaches in addressing one of the core challenges in solar nowcasting: achieving accurate performance under dynamic sky conditions.

9.5.2   Challenge 2: sky imager device

Beyond accuracy, technical constraints also challenge the deployment of such systems. Sky imagers must operate in variable environmental conditions and are often limited during dawn, dusk, or heavy overcast periods due to poor lighting. Thin or high-altitude clouds, which are optically faint, can go undetected in standard visible-light imagery. This affects the precision of cloud-modified irradiance forecasts, particularly under dynamic conditions. Furthermore, sky imagers alone cannot always capture cloud motion accurately, which is crucial for short-term forecasting. Physically, solar irradiance I_{solar} is typically estimated by modifying clear-sky irradiance models using cloud parameters, but rapid cloud evolution complicates real-time updates to these parameters.

9.5.3   Challenge 3: data quality

The quality and resolution of sky imager data significantly affect model performance. High-resolution images capture detailed cloud structures but increase data size and processing demands. In contrast, low-resolution images may miss important cloud features, leading to degraded forecasts. Additionally, image quality is sensitive to glare, sensor noise, and environmental artifacts. A study by [53] proposed preprocessing techniques such as fisheye removal and contrast enhancement to improve image consistency and model accuracy.

9.5.4   Challenge 4: data handling

Data handling and storage also pose significant barriers. Sky imagers typically capture images every 1–5 minutes, resulting in large volumes of high-frequency data. Processing this in real time is fundamental for nowcasting but requires scalable computational infrastructure. In resource-constrained settings, this limits the feasibility of deploying advanced models. Lightweight neural architectures, model compression techniques, and region-specific optimization may help reduce overhead without significantly compromising accuracy.

9.5.5   Challenge 5: computational load

Deep learning models such as CNNs, LSTMs, and transformer-based architectures have significantly improved the ability to extract spatial and temporal patterns from sky images. However, their deployment introduces practical challenges. High-quality, temporally aligned datasets are needed, and preprocessing steps such as glare removal, cloud masking, and contrast normalization introduce latency and computational overhead. These models are resource-intensive and may not be viable in remote or hardware-limited environments. Cloud or edge computing could mitigate this issue, but trade-offs exist: cloud solutions depend on reliable internet connectivity, while edge devices must balance speed with accuracy.

9.5.6 Challenge 6: spatiotemporal data alignment

Another challenge lies in the alignment between spatial features from sky images and the temporal forecasting frameworks used by AI models. CNNs excel at extracting spatial features, while temporal evolution, key-based for cloud dynamics, is better handled by RNN-based models like LSTM. Hybrid architectures (e.g., CNN-LSTM) attempt to integrate both [42], as shown in Figure 9.11. However, these models often degrade in performance when rapid adaptation to real-time input is required.

9.5.7 Challenge 7: generalizability

AI models trained in one geographical region often struggle to generalize to other locations due to variations in cloud characteristics, movement patterns, and irradiance profiles. This lack of transferability limits broad deployment. Transfer learning and domain adaptation techniques can help by fine-tuning pretrained models on local datasets.

9.5.8 Outlook

The convergence of recent technological advances is positioning AI-enhanced sky imager systems to overcome the limitations that have historically constrained their adoption. ViTs are revolutionizing sky imagery processing by capturing global spatial dependencies across entire images, enabling better tracking of distant cloud formations and significantly improving prediction accuracy during periods of high variability.

Simultaneously, next-generation sky imagers are integrating multispectral sensing with embedded AI processors, creating intelligent edge computing platforms that perform real-time cloud analysis without external dependencies. This addresses key challenges around computational scalability and network reliability by bringing AI inference directly to the point of data collection.

Modern forecasting frameworks are using sophisticated fusion strategies that dynamically combine sky imager data (and sky imager networks) with satellite imagery and numerical weather models. These multimodal approaches automatically adjust data source contributions based on forecast horizons and weather patterns, ensuring optimal performance across different temporal scales.

Perhaps most significantly for practical deployment, recent research has focused intensively on developing lightweight and energy-efficient AI models specifically

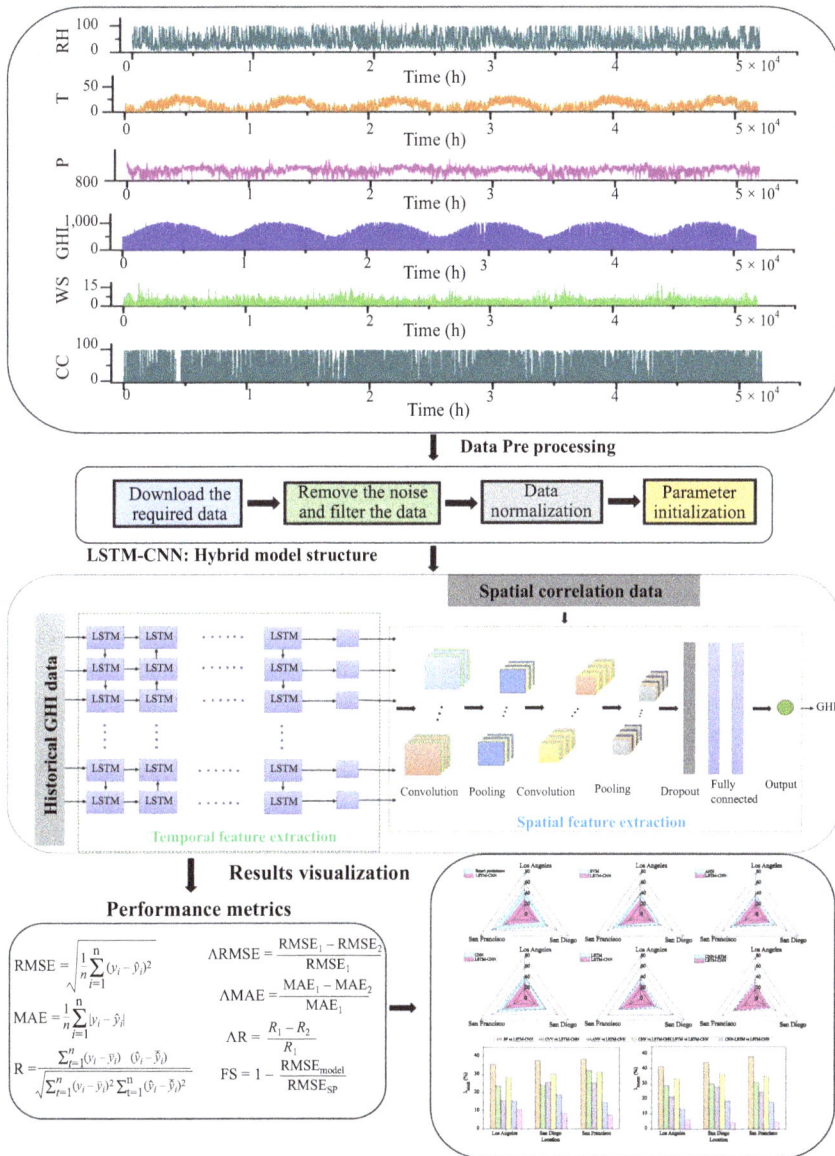

Figure 9.11   Integration of CNN and LSTM layers for processing sky imager data in forecasting models. Demonstrates how combining spatial and temporal data improves short-term solar forecasting accuracy (adapted from [42]).

optimized for embedded hardware deployment in resource-constrained environments. These developments address one of the most persistent barriers to widespread adoption by ensuring that advanced forecasting capabilities can be deployed across diverse geographic locations and economic contexts without requiring extensive computational infrastructure. Transfer learning techniques enable rapid deployment across diverse geographic regions with minimal local data requirements, making advanced forecasting capabilities economically viable in varied contexts.

As these technologies mature and integrate, they are establishing the foundation for highly accurate, scalable solar forecasting systems essential for renewable energy expansion. The field is moving toward a future where AI-enhanced forecasting will be as ubiquitous as current weather monitoring, providing the predictive accuracy necessary for large-scale solar grid integration and more stable renewable energy systems.

9.6   Conclusion

The integration of AI models with sky imager technology marks a significant step forward in short-term solar forecasting, particularly for PV power generation. Sky imagers provide high-resolution, real-time observations of cloud patterns and movements, offering localized insights that conventional forecasting methods often lack. When combined with AI techniques, these systems enable precise detection, tracking, and prediction of cloud dynamics, enhancing the accuracy of solar irradiance forecasts on minute-level timescales.

The analysis presented in this chapter has highlighted the particular strengths of AI architectures in processing sky imagery data. CNNs and LSTM networks have demonstrated accurate capabilities in analyzing both spatial and temporal aspects of sky images, with hybrid CNN–LSTM architectures showing especially promising results across diverse climatic conditions. These systems successfully capture real-time sky conditions while predicting short-term cloud evolution, directly contributing to improved PV power forecasts and enhanced grid stability. The evidence from multiple studies confirms that these AI-enhanced approaches consistently outperform traditional persistence and statistical models, particularly during periods of dynamic weather conditions.

However, this chapter has also identified significant challenges that must be addressed for widespread practical deployment. The comprehensive challenge analysis demonstrates critical limitations in dynamic weather adaptability, hardware and environmental constraints, computational scalability requirements, and cross-regional model transferability. These challenges are interconnected, with computational limitations affecting real-time processing capabilities, while hardware constraints impact data quality and system reliability. Economic viability concerns and the lack of standardization further complicate practical implementation across diverse deployment scenarios.

Despite these challenges, the technological outlook remains highly promising. Emerging AI technologies, particularly ViTs with their ability to model long-range spatial dependencies, are demonstrating superior performance in capturing broad cloud formations across entire sky domains. The concurrent development of next-generation sky imagers equipped with multispectral sensors and embedded AI processors represents a paradigm shift toward real-time, on-site data processing that addresses key operational constraints while reducing dependence on external computational infrastructure.

The evolution toward multimodal forecasting systems that integrate sky imager (network) data with satellite observations and NWP models represents perhaps the most significant opportunity for advancing the field. These comprehensive approaches use the complementary strengths of each data source, providing scalable and reliable PV power forecasts even under highly variable weather conditions. The dynamic fusion of local high-resolution sky imagery with broader spatial and temporal context from satellite and weather models creates reliable forecasting frameworks capable of adapting to diverse operational requirements.

The path forward requires coordinated advances across multiple dimensions. Technical development must focus on lightweight and transferable AI models that enable accurate forecasting in resource-constrained environments while maintaining high performance standards. Domain adaptation techniques will be essential for tailoring predictions to specific geographic regions and climatic conditions with minimal additional data requirements. Simultaneously, the standardization of data formats, communication protocols, and hardware interfaces will be crucial for enabling widespread interoperability and reducing deployment costs.

As these technological advances mature and integrate, AI-enhanced sky imager systems are positioned to comprehensively address the current limitations that have constrained their adoption. The convergence of advanced AI architectures, next-generation hardware capabilities, and sophisticated data fusion approaches is establishing the foundation for highly accurate, scalable, and economically viable solar forecasting systems.

Accurate short-term solar forecasting is becoming increasingly critical as PV installations scale to meet growing renewable energy targets. The systems and approaches analyzed in this chapter provide the predictive capabilities necessary for managing the inherent variability of solar resources while maintaining grid stability and reliability. As AI models and sky imager technologies continue to advance, short-term solar forecasting must and will achieve the accuracy and adaptability required to support a stable, resilient, and fully sustainable energy grid.

References

[1]   Breyer C, Khalili S, Bogdanov D, *et al.* On the history and future of 100% renewable energy systems research. *IEEE Access*. 2022;10:78176–78218.

[2]   Yang D, Kleissl J, Gueymard CA, *et al.* History and trends in solar irradiance and PV power forecasting: A preliminary assessment and review using text mining. *Solar Energy*. 2018;168:60–101.

[3]   Inman RH, Pedro HT, and Coimbra CF. Solar forecasting methods for renewable energy integration. *Progress in Energy and Combustion Science*. 2013;39(6):535–576.

[4]   Urquhart B, Kurtz B, Dahlin E, *et al.* Development of a sky imaging system for short-term solar power forecasting. Atmospheric Measurement Techniques. 2015;8(2):875–890.

[5]   Barbieri F, Rajakaruna S, and Ghosh A. Very short-term photovoltaic power forecasting with cloud modeling: A review. *Renewable and Sustainable Energy Reviews.* 2017;75:242–263.

[6]   Barhmi K, Heynen C, Golroodbari S, *et al.* A review of solar forecasting techniques and the role of artificial intelligence. *Solar.* 2024;4:99–135.

[7]   Kingma DP and Ba J. Adam: A method for stochastic optimization. *Conference Track Proceedings, Proceedings 3rd International Conference on Learning Representations, ICLR 2015*, San Diego, CA, USA, May 7–9; 2015.

[8]   Liu W, Liu Y, Zhou X, *et al.* Use of physics to improve solar forecast: Physics-informed persistence models for simultaneously forecasting GHI, DNI, and DHI. *Solar Energy.* 2021;215:252–265.

[9]   Richardson Jr W, Krishnaswami H, Vega R, *et al.* A low cost, edge computing, all-sky imager for cloud tracking and intra-hour irradiance forecasting. *Sustainability.* 2017;9(4):482.

[10]   Chauvin R, Nou J, Thil S, *et al.* Cloud detection methodology based on a sky-imaging system. *Energy Procedia.* 2015;69:1970–1980.

[11]   Liu Y, Li Q, Li X, *et al.* Leveraging Physical Rules for Weakly Supervised Cloud Detection in Remote Sensing Images. *IEEE Transactions on Geoscience and Remote Sensing.* 2023;61:1–18.

[12]   Skakun S, Wevers J, Brockmann C, *et al.* Cloud Mask Intercomparison eXercise (CMIX): An evaluation of cloud masking algorithms for Landsat 8 and Sentinel-2. *Remote Sensing of Environment.* 2022;274:112990.

[13]   Chow CW, Urquhart B, Lave M, *et al.* Intra-hour forecasting with a total sky imager at the UC San Diego solar energy testbed. *Solar Energy.* 2011;85(11):2881–2893.

[14]   Chu Y, Pedro HT, Nonnenmacher L, *et al.* A smart image-based cloud detection system for intrahour solar irradiance forecasts. *Journal of Atmospheric and Oceanic Technology.* 2014;31(9):1995–2007.

[15]   Shirazi E, Lemmens J, Chowdhury MG, *et al.* Cloud detection for PV power forecast based on colour components of sky images. *2021 IEEE 48th Photovoltaic Specialists Conference (PVSC).* IEEE; 2021. pp. 2389–2391.

[16]   Ye L, Cao Z, and Xiao Y. DeepCloud: Ground-based cloud image categorization using deep convolutional features. *IEEE Transactions on Geoscience and Remote Sensing.* 2017;55(10):5729–5740.

[17]   Schmidt T, Kalisch J, Lorenz E, *et al.* Evaluating the spatio-temporal performance of sky-imager-based solar irradiance analysis and forecasts. *Atmospheric Chemistry and Physics.* 2016;16(5):3399–3412.

[18]   Haralick RM, Shanmugam K, Dinstein IH. Textural features for image classification. *IEEE Transactions on Systems, Man, and Cybernetics.* 1973;(6): 610–621.

[19]   Li X, Qiu B, Cao G, *et al.* A novel method for ground-based cloud image classification using transformer. *Remote Sensing.* 2022;14(16):3978.

[20]   Horn BK and Schunck BG. Determining optical flow. *Artificial Intelligence.* 1981;17(1–3):185–203.

[21]   Lucas BD and Kanade T. An iterative image registration technique with an application to stereo vision. *IJCAI'81: 7th International Joint Conference on Artificial Intelligence*. vol. 2; 1981. pp. 674–679.

[22]   Dazhi Y, Walsh WM, Zibo D, *et al.* Block matching algorithms: Their applications and limitations in solar irradiance forecasting. *Energy Procedia*. 2013;33:335–342.

[23]   Zang H, Liu L, Sun L, *et al.* Short-term global horizontal irradiance forecasting based on a hybrid CNN-LSTM model with spatiotemporal correlations. *Renewable Energy*. 2020;160:26–41.

[24]   Ilg E, Mayer N, Saikia T, *et al.* Flownet 2.0: Evolution of optical flow estimation with deep networks. *Proceedings of the IEEE Conference on Computer Vision and Pattern Recognition*; 2017. pp. 2462–2470.

[25]   Dosovitskiy A, Beyer L, Kolesnikov A, *et al.* An image is worth 16x16 words: Transformers for image recognition at scale. *International Conference on Learning Representations*; 2021.

[26]   Quesada-Ruiz S, Linares-Rodríguez A, Ruiz-Arias J, *et al.* An advanced ANN-based method to estimate hourly solar radiation from multi-spectral MSG imagery. *Solar Energy*. 2015;115:494–504.

[27]   Gueymard C. Critical analysis and performance assessment of clear sky solar irradiance models using theoretical and measured data. *Solar Energy*. 1993;51(2):121–138.

[28]   Iqbal M. *An Introduction to Solar Radiation*. Amsterdam: Elsevier; 2012.

[29]   Vignola F, Michalsky J, and Stoffel T. *Solar and Infrared Radiation Measurements*. Boca Raton: CRC Press; 2019.

[30]   Yadav AK and Chandel SS. Solar radiation prediction using Artificial Neural Network techniques: A review. *Renewable and Sustainable Energy Reviews*. 2014;33:772–781.

[31]   Yang H, Wang L, Huang C, *et al.* 3D-CNN-Based sky image feature extraction for short-term global horizontal irradiance forecasting. *Water*. 2021;13(13):1773.

[32]   Harrou F, Kadri F, and Sun Y. Forecasting of photovoltaic solar power production using LSTM approach. *Advanced Statistical Modeling, Forecasting, and Fault Detection in Renewable Energy Systems*. London: IntechOpen; 2020; p. 3.

[33]   Shi X, Chen Z, Wang H, *et al.* Convolutional LSTM network: A machine learning approach for precipitation nowcasting. *Advances in Neural Information Processing Systems*. 2015;28.

[34]   Hendrikx NY, Barhmi K, Visser LR, *et al.* All sky imaging-based short-term solar irradiance forecasting with Long Short-Term Memory networks. *Solar Energy*. 2024;272:112463.

[35]   Caldas M and Alonso-Suárez R. Very short-term solar irradiance forecast using all-sky imaging and real-time irradiance measurements. *Renewable Energy*. 2019;143:1643–1658.

[36]   Reich NH, Mueller B, Armbruster A, *et al.* Performance ratio revisited: is *PR* >90% realistic? *Progress in Photovoltaics: Research and Applications*. 2012;20(6):717–726.

[37] Qu Y, Xu J, Sun Y, *et al.* A temporal distributed hybrid deep learning model for day-ahead distributed PV power forecasting. *Applied Energy.* 2021;304:117704.

[38] Xu S, Zhang R, Ma H, *et al.* On vision transformer for ultra-short-term forecasting of photovoltaic generation using sky images. *Solar Energy.* 2024;267:112203.

[39] Singla P, Duhan M, and Saroha S. A comprehensive review and analysis of solar forecasting techniques. *Frontiers in Energy.* 2021: 1–37.

[40] Ajith M and Martínez-Ramón M. Deep learning based solar radiation micro forecast by fusion of infrared cloud images and radiation data. *Applied Energy.* 2021;294:117014.

[41] Pérez E, Pérez J, Segarra-Tamarit J, *et al.* A deep learning model for intra-day forecasting of solar irradiance using satellite-based estimations in the vicinity of a PV power plant. *Solar Energy.* 2021;218:652–660.

[42] Kumari P and Toshniwal D. Long short term memory–convolutional neural network based deep hybrid approach for solar irradiance forecasting. *Applied Energy.* 2021;295:117061.

[43] Kleissl J. *Solar Energy Forecasting and Resource Assessment.* Academic Press; 2013.

[44] Raissi M, Perdikaris P, and Karniadakis GE. Physics-informed neural networks: A deep learning framework for solving forward and inverse problems involving nonlinear partial differential equations. *Journal of Computational Physics.* 2019;378:686–707.

[45] Paletta Q, Nie Y, Saint-Drenan YM, *et al.* Improving cross-site generalisability of vision-based solar forecasting models with physics-informed transfer learning. *Energy Conversion and Management.* 2024;309:118398.

[46] Nie Y, Li X, Scott A, *et al.* SKIPP'D: A SKy Images and Photovoltaic Power Generation Dataset for short-term solar forecasting. *Solar Energy.* 2023;255:171–179.

[47] Blum NB, Wilbert S, Nouri B, *et al.* Analyzing spatial variations of cloud attenuation by a network of all-sky imagers. *Remote Sensing.* 2022;14(22).

[48] Blum N, Schmidt T, Nouri B, *et al.* Nowcasting of Irradiance Using a Network of All-Sky-Imagers. *Proceedings EU PVSEC*; 2019. pp. 1403–1409.

[49] Nouri B, Lezaca J, Hammer A, *et al.* Multi-source observations to improve solar forecasting within the Smart4RES project. *Smart4RES Projekt*; 2021.

[50] Nguyen DA and Kleissl J. Stereographic methods for cloud base height determination using two sky imagers. *Solar Energy.* 2014;107:495–509.

[51] Zameer A, Jaffar F, Shahid F, *et al.* Short-term solar energy forecasting: Integrated computational intelligence of LSTMs and GRU. *PLoS One.* 2023;18(10):e0285410.

[52] Barhmi K, Mirbagheri Golroodbari S, Knap W, *et al.* All-sky imager-based solar forecasting: A multi-stage AI approach with Kalman filter post-processing. (*Submitted Renewable Energy, Elsevier*). 2025.

[53] Shirazi E, Gordon I, Reinders A, *et al.* Sky images for short-term solar irradiance forecast: A comparative study of linear machine learning models. *IEEE Journal of Photovoltaics.* 2024;14(4):691–698.

Chapter 10

Solar photovoltaic forecasting for energy system integration and control

Silvana Matrone[1], Amirhossein Heydarian Ardakani[2], Emanuele Ogliari[1], Sonia Leva[1] and Elham Shirazi[2]

Given the variable nature of solar photovoltaics (PV) production, accurate forecasting becomes essential for energy system integration and control. This chapter provides a comprehensive exploration of different aspects associated with PV forecasting for energy system integration and control. In this regard, we first introduce different temporal resolutions in PV forecasting and then provide information on how PV is integrated into the energy system. Another topic discussed in this chapter is the use of PV forecasting to facilitate the integration of PV systems together with other renewable energy sources and energy storage into energy system.

This chapter highlights how PV forecasting techniques can enhance grid flexibility, allowing infrastructure to respond to fluctuations in solar PV generation. It also explores the role of PV forecasting in energy management. By forecasting changes in PV output, system operators can better align energy demand with supply through demand-side measures such as load shifting and demand response, and generation-side strategies, including flexible dispatch and dynamic generation scheduling. Regulatory and policy frameworks also play a crucial role in enabling the integration of PV forecasting into energy systems control. This chapter also reviews policies and incentives that have been implemented to promote the adoption of PV forecasting technologies.

Lastly, to illustrate the practical applications and benefits of PV forecasting, this chapter presents three case studies: a robust unit commitment model using uncertainty sets tailored to solar variability; a coordinated voltage control strategy using forecast-based scheduling of local assets; and an energy management system for an EV charging station, where forecast accuracy significantly impacts operational costs.

Forecasting can also be embedded in a foundational model, a large, pre-trained model that serves as a base for many downstream tasks. Its detailed design and implementation lie beyond the scope of this book and will be addressed in a separate study.

[1]Department of Energy, Politecnico di Milano, Italy
[2]Faculty of Engineering Technology, University of Twente, The Netherlands

10.1   Introduction

The integration of solar PV forecasting has emerged as a key component in optimizing energy production, enhancing grid stability, and improving efficient energy use within modern energy systems. As renewable energy sources such as solar PV become increasingly prominent in global energy portfolios, accurately predicting solar PV generation is crucial in managing and stabilizing energy grids, reducing operational costs, and ensuring a reliable energy supply [1]. Solar PV forecasting involves predicting how much electricity a solar PV system will generate over a specified period in the future. This process is essential because solar energy is inherently variable, affected by weather conditions such as cloud cover, temperature, and the incident angle of sunlight on PV modules. Accurate forecasting helps grid operators and energy producers manage these fluctuations and plan accordingly, allowing for better integration of solar energy into the grid.

10.1.1   *Temporal resolution of PV forecasts*

PV forecasting generally encompasses multiple time horizons, each serving distinct purposes within the energy system. Short-term forecasts, spanning minutes to hours ahead, are essential for real-time grid management and operational decisions, such as balancing supply and demand or adjusting energy storage dispatch. Medium-term forecasts, ranging from days to weeks, support maintenance scheduling, market participation, and resource allocation. These forecasts rely on weather prediction models that simulate future weather conditions [2], such as those provided by the European Centre for Medium-Range Weather Forecasts (ECMWF) [3]. Meanwhile, long-term forecasts of years ahead provide a macroscopic view that informs infrastructure development and policy planning and mostly are termed predictions and used for optimal PV site selection.

The type of forecast primarily used to manage energy systems incorporating PV generation is the short-term forecast. It focuses on predicting PV output over the next few minutes or hours. This is necessary to manage real-time energy supply and demand balance in the grid. Factors such as cloud cover, temperature, and atmospheric conditions can change quickly, so forecasting systems rely on high-resolution weather data, satellite imagery, and real-time monitoring of PV systems to predict rapid changes in solar output. Different time frames call for different techniques to be applied, see also other chapters in the book. For short-term forecasting, models such as numerical weather prediction (NWP), machine learning (ML), and statistical methods are commonly used to predict solar power output within hours to days, each with distinct advantages. Medium-term forecasting (days to weeks) often employs hybrid approaches, combining ML and statistical methods, along with weather patterns and time-series analysis. Long-term forecasting (months to years) relies on climate models, scenario analysis, and geospatial studies to project PV production, helping inform strategic energy planning and investment decisions [4].

Various ML models are frequently used for short-term solar power forecasting due to their ability to manage complex and changing data. Artificial neural networks (ANNs) are particularly favored for their ability to model complex, non-linear relationships between variables such as weather data and solar energy output, even in the presence of noisy or incomplete information [5]. Support vector machines (SVMs) are commonly applied in regression tasks, offering strong accuracy, especially when working with smaller datasets [6]. Random forests (RFs), which are ensemble learning models, improve prediction accuracy by combining multiple decision trees, making them effective for handling large datasets with significant variability [7]. K-Nearest neighbors (KNN), a simple yet efficient technique, forecasts solar output based on similar historical data points. More sophisticated models such as gradient boosting machines (GBMs), including XGBoost and LightGBM, enhance forecasting accuracy by combining weak learners to form stronger predictive models [8]. Finally, recurrent neural networks (RNNs) [9] and long short-term memory (LSTM) [10] networks are specialized for time-series data, identifying patterns in sequential information such as weather changes and thus enhancing prediction precision. These models are discussed in detail in Chapters 3–9.

10.1.2   *PV forecast in combination with other renewables*

Coordinating PV power with other renewable energy sources is crucial for maintaining a reliable and efficient energy supply, especially given the intermittent nature of solar energy. Accurate forecasting of solar and wind facilitates the integration of complementary sources, such as wind power, which can generate electricity when solar output is low. For example, wind energy often peaks at night and during colder seasons, while solar energy is most productive during sunny days. This complementary relationship allows for a more consistent energy flow, as periods of low solar generation can be balanced by wind power production.

Hybrid energy systems, which combine solar, wind, and other renewable sources, create a more resilient energy infrastructure while reducing dependency on fossil fuels and improving the overall efficiency of clean energy utilization [11]. Deploying PV and other renewable technologies in geographically diverse areas mitigates the impact of local weather variations. Pairing PV systems in sunny areas with wind farms in wind-rich regions can balance generation. For instance, Frank *et al.* [12] presented a study on balancing the potential of natural variability and extremes in PV and wind energy production in European countries. Wind energy typically increases during the night and colder seasons, particularly in temperate regions. This is due to weather phenomena such as stronger pressure differences and more dynamic air flows that occur when the sun is not heating the Earth's surface. Coastal regions can experience stronger nighttime winds because the temperature difference between the land and sea creates pressure gradients that drive air movement. In contrast, solar energy production peaks during daytime, with the highest output occurring on sunny days, especially in spring and summer [13]. With accurate PV forecasts, the use of wind energy can be optimized to compensate for low solar generation, thereby enhancing overall grid reliability. Conversely, when solar energy is abundant, wind generation may decline. Seasonally, solar power tends to reach its peak in the summer, while wind is more productive in the winter months. This seasonal and daily complementarity ensures a more consistent energy output throughout the year when both sources are used together. This combination helps hybrid systems reduce the inherent variability and intermittency of renewable energy, offering a more stable power supply. As a result, the need for fossil-fuel backup generation is reduced, and grid reliability is improved, making renewable energy more dependable and efficient [14].

10.1.3 PV forecast for PV integration with energy storage

Integrating PV systems with energy storage technologies, such as batteries, offers an effective solution to the variability and intermittency of solar power. These systems store excess energy generated during peak solar production and discharge it during periods of low generation, such as at night or during overcast conditions. Where geographically feasible, pumped hydro storage can also be employed to take advantage of surplus renewable energy, acting as a large-scale energy reserve [15]. This coupling not only ensures a more stable and continuous power supply but also enhances grid reliability, reduces dependence on fossil fuels, and enables more efficient energy system management.

Introducing PV forecasting into PV-storage systems is essential for optimizing their performance and ensuring effective grid integration. Accurate forecasts help anticipate solar energy production, enabling more precise control of charging and discharging cycles. This allows storage systems to contribute to grid stability through services such as frequency regulation and peak shaving. Forecast-informed energy management also supports demand response (DR) strategies by aligning consumption with expected generation and enables to export of energy based on market prices. Long-term forecasts can guide seasonal storage planning, ensuring excess energy generated under favorable conditions is available during less productive periods.

Various storage technologies complement PV systems in capturing surplus solar energy for later use. Lithium-ion batteries are widely used in residential and commercial settings due to their high energy density, fast response times, and efficiency [16]. For large-scale applications, pumped hydro storage is employed, where excess solar power is used to elevate water, later releasing it to generate electricity during low-output periods [15]. Emerging technologies such as hydrogen storage and compressed air energy storage (CAES) also offer promising solutions. In hydrogen systems, solar energy powers electrolysis to produce hydrogen, which can be stored and later reconverted to electricity or used as fuel [17]. In CAES, surplus energy compresses air, which is released later to drive turbines and generate electricity [18].

The size of lithium-ion batteries has to be calculated based on the needs of the system, considering the generation capacity of the PV system and assessing the daily energy consumption together with other objectives such as cost, self-consumption or self-sufficiency ratio.

In this context, PV forecasting predicts solar energy generation based on weather conditions, enabling optimal battery management. When high solar production is expected, the system can prioritize charging the battery to full capacity before output declines, such as during evening hours or cloudy weather, thereby minimizing energy waste from overproduction [16,19]. Conversely, during predicted periods of low solar output, the system can discharge stored energy to maintain a stable supply. Accurate forecasts also support DR programs by allowing users to schedule energy-intensive activities during peak solar production, reducing grid dependence and lowering electricity costs.

Moreover, forecasting contributes to extending battery lifespan by aligning charging and discharging cycles with expected solar availability. This prevents overcharging or deep discharging, slowing down battery degradation and enhancing long-term performance [20]. From an economic standpoint, in grid-tied systems, PV forecasting enables strategic energy export. By anticipating surplus generation, the system can charge the battery first and schedule exports during peak demand periods, capitalizing on higher feed-in tariffs [21]. Additionally, during times of high grid stress and low solar availability, forecasts can prompt the system to discharge stored energy, helping to stabilize the grid and reduce reliance on backup generation.

10.2   PV forecasting for energy system control

The integration of large PV plants in the electricity grid requires careful attention to various factors, such as grid stability and frequency control, voltage regulation and reactive power management, fault ride-through, protection coordination, congestion management, and compliance with grid codes. For successful integration, PV systems must be effectively interconnected with the transmission and distribution grids, which involves a thorough understanding of their impact on the voltage level of the grid.

A PV plant, designed to supply solar energy to the grid, consists of several key components, including PV panels, inverters, meters, grid connection, and both direct current (DC) and alternating current (AC) cabling. Solar modules generate DC power, while grids operate on AC power; therefore, inverters play a central role in integrating PV systems into the grid by converting the DC power into AC.

In addition to converting DC to AC, inverters ensure steady voltage and frequency despite fluctuations in load, manage reactive power for reactive loads, maintain power quality, and synchronize the PV system with the grid. Transformers are another essential component, adjusting the voltage levels of the grid.

The inherent variability of solar energy poses significant challenges to grid stability and reliability. These variation necessitates careful planning considering diurnal variation. Seasonal variations introduce even more fluctuation during the year, which should be considered in long-term planning of the energy system. Cloud cover, rain, and storms can significantly reduce solar irradiance, leading to abrupt decreases in PV output. For instance, a cloud movement over a PV system can reduce production by as much as 70% in a matter of minutes [22]. Such daily and seasonal trends are quite different from the energy demand profiles that have to be met. Maintaining a balance between energy supply and demand at all times becomes critical. Traditional power systems rely on consistent generation to meet load requirements; they always guarantee to meet the power demand. The sudden drops in solar generation can lead to supply deficits, potentially causing voltage fluctuations and grid instability. To counteract the perceived unpredictability of solar energy caused by weather conditions, utilities may need to maintain additional backup generation resources, often from fossil fuels, which can be ramped up quickly to meet demand when solar output is low. However, this reliance on backup generation can diminish the environmental benefits of solar energy. Hence, accurate solar PV forecasting is required to reduce the requirement for fossil fuel backup generation.

10.2.1   Operation and control of electricity grid

Real-time management of solar PV power integrated with grids is crucial for addressing the variability and intermittency of solar energy. Fluctuations in solar PV generation present challenges to grid operators who must continuously balance supply and demand. PV forecasting helps grid operators to effectively manage these variations in real time and adjust the grid condition accordingly [23].

Power quality is a major concern for both consumers and utilities. Power systems are designed to function at a sinusoidal voltage of a specific amplitude and frequency. Any substantial changes in voltage magnitude, frequency, or waveform purity are classified as power quality issues. Variations in solar radiation lead to changes in the operating points of solar cells, resulting in fluctuations in voltage and current. Since the power system can control voltage but not current, it is essential to keep voltage levels within acceptable limits to maintain power quality and ensure that PV systems connected to distribution networks do not adversely affect the system.

PV systems are connected to the grid through inverters. Therefore, they are commonly referred to as inverter-based resources (IBRs). Smart inverters are capable of providing grid support and voltage regulation. These inverters can adjust their output in real time based on grid conditions, helping to stabilize voltage and reduce harmonics. Additionally, active power control functions can be implemented in inverters to manage output based on grid demand, preventing excessive generation during low-demand periods.

Unlike synchronous generators, IBRs lack rotating components and rely entirely on power electronics. The consequence is that these systems have significantly lower inertia and experience faster frequency changes. This reduced inertia limits the system's ability to slow down frequency deviations during disturbances or supply/demand imbalances. Therefore, with the controllers having less time to respond, the power system is prone to instability [24]. Moreover, the location of IBRs and spatial inertia distributions within the grid affects the power system robustness [25]. A solution to improving the stability of low-inertia grids is using the inverters of PV systems as virtual inertia (VI) devices to mimic the inertial response of synchronous generators [26]. VI devices are commonly applied in two forms: grid-following and grid-forming. Grid-forming inverters regulate the voltage and frequency of the PV inverter to specific setpoints, similar to the operation of synchronous machines. This approach is also known as the virtual synchronous generator (VSG), which aims to replicate the dynamic behavior of synchronous generators and provide virtual inertia. Therefore, the grid can support the integration of a high share of PV systems without compromising its stability [27]. Grid-following inverters, in contrast, synchronize their output with the grid's voltage and frequency to avoid destabilizing the system [28]. In this regard, short-term PV forecasting can enhance inverter controllers to optimize grid synchronization [29].

Grid stability can be achieved through grid-level coordination of PV inverters using PV forecasts over longer time horizons (ranging from several hours to days) [30]. Accurate PV forecasts enable system operators and aggregators to schedule and allocate power reserves for grid-supportive functions, including voltage regulation and congestion mitigation. From a power flow perspective, the influence of PV power injections on the grid can be described using the standard AC power flow equations [31] as follows:

$$P_i^{\mathrm{PV}} - P_i^{\mathrm{L}} = \sum_{j=1}^{N} V_i V_j \left(G_{ij} \cos(\theta_i - \theta_j) + B_{ij} \sin(\theta_i - \theta_j) \right) \tag{10.1}$$

$$Q_i^{\mathrm{PV}} - Q_i^{\mathrm{L}} = \sum_{j=1}^{N} V_i V_j \left(G_{ij} \sin(\theta_i - \theta_j) - B_{ij} \cos(\theta_i - \theta_j) \right) \tag{10.2}$$

In these equations, P_i^{PV} and Q_i^{PV} represent the active and reactive power injected by the PV inverter at bus i, respectively, while P_i^{L} and Q_i^{L} are the corresponding active and reactive power demands (loads) at that same bus. The terms V_i and V_j denote the voltage magnitudes at buses i and j, and θ_i and θ_j are the voltage phase angles at those buses, respectively. The parameters G_{ij} and B_{ij} represent the real and imaginary parts of the admittance matrix connecting buses i and j, respectively. The index N indicates the total number of buses in the network.

When instantaneous power production exceeds instantaneous power consumption, the resulting imbalance causes a net power flow backward through the medium-voltage to low-voltage transformers, meaning that the electric power flows from the low-voltage to the medium-voltage network. Therefore, it is important to determine the limit of the penetration level of PV that can be fed into a power network without causing problems to the power system. Active power curtailment (APC) relies on controlling the active power output of PV inverters, allowing them to slightly reduce PV power output when the grid voltage becomes too high. This approach minimizes power losses compared to complete disconnection but requires accurate forecasting of PV generation. A centralized APC control strategy for active distribution networks has been proposed [32], which relies on consumption and generation profiles of prosumers. To achieve this, PV generation profiles must be forecasted and adjusted based on weather forecasts for each prosumer. These forecasted PV profiles are then used as inputs to the control algorithm, enabling utilities to determine active power setpoints for each prosumer.

Curtailment of PV generation during excess production should be the last resort to maintain grid stability; in this view, an accurate PV forecast is a key player to predict overproduction and avoid curtailment. The forecasted PV power overproduction can be shifted toward other applications. Energy storage systems such as batteries or pumped hydro storage can be employed in these cases (as discussed in detail in Section 10.1.3). These systems store excess solar energy during periods of high production and release it during low-production periods, ensuring a more consistent power supply. PV forecast also makes possible real-time coordination with other energy sources (Section 10.1.2), such as wind or conventional power plants, helping to maintain grid stability.

Another potential solution to address grid stability issues is reactive power control (RPC). Depending on whether reactive power is injected or absorbed, it can either raise or lower voltage levels. PV inverters have a capability curve, similar to synchronous generators, that limits the active and reactive power [33]:

$$Q_i^{PV} \le \sqrt{(S_i^{PV})^2 - (P_i^{PV})^2} \tag{10.3}$$

where Q_i^{PV} is the reactive power output of the PV inverter at bus i, P_i^{PV} is the active power output at the same bus, and S_i^{PV} denotes the apparent power capacity of the inverter. Emarati *et al.* [33] proposed an RPC strategy for PV-storage systems to improve grid stability through voltage regulation in distribution networks. In this context, PV forecasts are used to schedule storage, while overvoltage caused by forecast errors is mitigated through the reactive power capability of PV inverters. Moreover, RPC methods can be used to minimize the active power losses in the grid [34].

Other voltage regulation techniques are static VAR compensators (SVCs) [35] to provide dynamic reactive power support and capacitor banks that can be automatically switched in and out based on voltage conditions [36].

Unit commitment (UC) in the electricity grid is also a challenging problem, particularly the growth in PV penetration. The UC problem aims for scheduling controllable generators over a time horizon while considering the physical characteristics of generating units. UC optimization models incorporate PV uncertainty using stochastic and robust optimization techniques. These models rely on probabilistic forecasts to describe the PV uncertainty. In this context, Ref. [37] presents a computationally efficient approach that combines forecasting and optimization for intraday UC (IUC). Their method includes two stages. The first stage generates dynamic PV forecasts, while the second stage applies a UC model to produce generator schedules that are both reliable and efficient. Similarly, Ref. [38] integrates PV output prediction into a stochastic UC problem, considering fluctuations in solar radiation. In this study, a feedforward NN is used to forecast PV generation, which is then used in a stochastic UC model aimed at reducing operational costs and improving voltage stability. By using a probabilistic operation plan based on these forecasts, they achieve stable system operation across different PV scenarios while maintaining the grid voltage stability.

10.2.2   Energy management systems

In energy systems that include multiple renewable energy sources and storage technologies, effective coordination of energy flows is critical, especially in standalone or off-grid configurations. A central component enabling this coordination is the energy management system (EMS). In this regard, PV forecasting enhances EMS functionality by providing accurate predictions of solar generation. These forecasts allow the EMS to make informed decisions. Integrating predictive models into EMS frameworks has been shown to significantly improve the resilience and efficiency of microgrid operations, especially under variable weather and load conditions [39].

For example, in [40], a framework was proposed that utilizes the Internet of Things (IoT) to optimize microgrid operation. The proposed scheme focuses on balancing energy supply and demand by using IoT devices to monitor and control energy consumption in real time. The system incorporates smart appliances, sensors, and control algorithms to enable efficient energy usage, reduce peak loads, and enhance the integration of renewable energy sources such as solar and wind. Energy consumption is optimized based on cost, availability, and user preferences, improving both the efficiency and sustainability of the microgrid. This study highlights the potential of IoT in enhancing grid reliability and achieving energy savings.

Typically, three EMS control architectures are utilized: centralized, decentralized, and distributed. In a centralized approach, a single controller oversees all system components and optimizes their performance collectively. The decentralized approach allows each component to operate independently without interacting with others, while the distributed architecture involves controllers that share some information [41].

Moreover, energy management is generally structured hierarchically, consisting of three control levels. The primary and secondary control levels focus on maintaining voltage and frequency stability and restoring them after load variations. The tertiary control level plays a crucial role in control. This level considers economic, environmental, and technological factors from external sources (such as electricity prices and weather forecasts) as well as the dynamics of the microgrid components, such as energy storage and distributed generation. Tertiary control coordinates the microgrid operation with the distribution system to effectively tackle energy management challenges [42,43].

Various objective functions can guide a management system, such as reducing operational costs, CO_2 emissions, and import from the grid, and extending the lifetime of energy storage systems (ESSs) [44,45]. These objectives have to be met while respecting constraints that exist within the energy system, such as maintaining power balance, respecting generation and transmission capacity limits, and avoiding overcharging and overdischarging [44,46].

A diverse range of optimization and management models has been proposed for microgrids. Four main categories of techniques can be identified: traditional mathematical models, metaheuristic models, predictive control models, and AI-based approaches [47]. Traditional models include PI/PID controllers, mixed integer linear programming (MILP), linear quadratic regulators (LQRs), and robust control, among others. Recently, predictive control methods have gained traction for advanced system management. These controllers can anticipate future actions and decisions, relying on forecasted inputs such as power consumption or generation. Consequently, these forecasts are vital for making informed and proactive control decisions. Notably, model predictive control (MPC) and generalized predictive control (GPC) are well-known methodologies [48]. AI methods often use (un-)supervised learning, such as neural networks and reinforcement learning, with the latter offering adaptability in environments with frequently changing operating conditions [49].

Recent research focuses on integrating PV forecasting into data-driven control approaches, such as deep reinforcement learning (DRL). By incorporating future PV forecast data into the DRL algorithm, the observation space can be enriched with time series information [50]. This provides predictive insights to the algorithms and enables them to better handle generation uncertainties. An intermediate time-series processing layer receives the PV forecasts and generates an input vector for the DRL algorithm. This layer has been implemented using gated recurrent units (GRUs) in microgrid applications [51] and using RNNs and transformer architectures in building-integrated photovoltaic and battery (BIPVB) systems [52].

One approach to minimize reliance on nonrenewable energy sources is to focus on managing the energy demand rather than just the power generated, utilizing DSM strategies. DSM is a method used by electricity utilities to control demand by encouraging consumers to modify their electricity usage patterns to better align with the predicted power production profiles. This strategy is especially important when integrating nondispatchable renewable energy sources such as solar and wind [53].

The benefits of DSM include enhanced system efficiency, reduced operational costs, and lower environmental impact. However, DSM becomes increasingly complex as more variables and degrees of freedom are introduced. While traditionally driven by utility providers, DSM is now shifting toward the customer side, with end users playing a more active role through participation in energy programs and response schemes. Several studies highlight the impact of accurate forecasts in DSM optimization. De La Nieta *et al.* [54] showed that incorporating LSTM-based PV and EV forecasts into the microgrid EMS significantly improves scheduling decisions and reduces operating costs. Similarly, Zarei *et al.* [55] demonstrated the effectiveness of short-term load forecasting using deep learning in multiobjective DSM frameworks. Furthermore, real-time DSM algorithms [56] allow for probabilistic load forecasting for robust energy management in smart buildings.

Recent research emphasizes the importance of forecast-based models to enable more accurate and efficient demand control. For example, variable pricing models, such as time-of-use (TOU) rates or real-time pricing, are coupled with forecasting systems to better schedule energy use. Dai *et al.* [57] presented a dynamic pricing model for EV charging stations powered by PV systems. Using Stackelberg game theory, it adjusts prices in real time based on forecasts of solar generation and user behavior, improving both grid integration and user satisfaction. DR is a technique used in energy systems to temporarily adjust electricity consumption by end users in response to grid conditions, such as high demand or electricity prices; such programs increasingly rely on real-time forecasting to adjust energy loads dynamically. Huang *et al.* [58] described a stochastic optimization framework that accounts for uncertainties in PV generation and battery operation. Considering forecast data, the model adjusts DR actions and battery usage in real time to minimize operational costs. Forecast accuracy is also critical in load-shifting strategies. Instead of only incentivizing off-peak usage, modern systems aim to align energy-intensive activities with periods of expected high renewable generation.

10.3   Regulatory framework

Addressing regulatory and policy frameworks is essential for facilitating the integration of PV systems into energy networks and grid operations. PV forecasting plays a key role in this context, as it is often leveraged by policy measures to enhance the reliability of solar energy and to incentivize its production and integration into the grid.

Policies can encourage grid operators to adopt forecasting technologies by providing financial incentives or establishing performance standards that reward accurate predictions. Regulations should also promote data sharing and collaboration between meteorological services and energy providers, ensuring that reliable solar generation forecasts are accessible for operational planning. Additionally, policies that support the development of flexible grid architectures can enhance the ability to integrate renewable energy sources [59].

Regulatory frameworks and policies aim to incentivize the integration of PV forecasting to improve grid reliability and promote the use of renewable energy. Renewable portfolio standards (RPSs), for example, mandate that utilities source a specified percentage of their electricity from renewable sources [60]. In some US states, where the goal is 100% clean energy by 2045, accurate solar forecasts are critical to ensuring grid stability while meeting renewable energy targets [61]. Grid modernization policies, including the US Department of Energy's Smart Grid Investment Grant program, push for the integration of solar forecasting to improve energy distribution and optimize the grid's performance under fluctuating renewable generation [62]. Similarly, feed-in tariffs (FiTs) in countries such as Germany provide solar operators with fixed payments for the energy they feed into the grid, encouraging them to use accurate PV forecasts to maximize production during peak demand [63]. Japan's energy policy actively supports the integration of PV forecasting into grid operations as part of its broader vision for a smart society, known as "Society 5.0" [64]. Additionally, Japan's FiT program has encouraged large-scale PV deployment, further reinforcing the need for accurate forecasting to manage variability and maintain grid stability [65]. Also, Australia has an active approach to encourage the use of PV forecasts. The Australian Energy Market Operator (AEMO) requires aggregators and Virtual Power Plant (VPP) operators to submit real-time and forecast data to participate in energy markets and maintain grid security [66].

Additionally, performance-based regulation (PBR) frameworks, such as the UK Revenue = Incentives + Innovation + Outputs (RIIO) model [67], reward utilities for operational efficiency and innovation, creating a financial incentive to adopt advanced solar forecasting methods that improve grid performance. In ancillary services markets, PV operators can also be compensated for helping stabilize the grid. For instance, in Europe, grid codes from the European Network of Transmission System Operators (ENTSO-E) require accurate forecasts for PV operators to participate in ancillary services such as frequency regulation [68]. Beyond these incentives, carbon pricing mechanisms and emissions trading systems (ETS), such as the EU's cap-and-trade program [69], push renewable energy by making fossil fuel-based generation more expensive. Accurate PV forecasts become essential to integrating more solar energy into the grid and reducing reliance on carbon-intensive sources. Finally, the use of renewable energy certificates (RECs) in markets such as in the United States provides further financial motivation for solar operators to optimize output through forecasting, as each REC certifies that a megawatt-hour of electricity was generated from renewable sources. Collectively, these policies encourage the widespread adoption of PV forecasting technologies, facilitating a smoother and more reliable integration of renewable energy sources into energy systems [59].

10.4   Case studies

10.4.1   Case 1: robust unit commitment with characterized PV forecast

A day-ahead unit commitment decision is a trade-off between security and cost, in which deterministic methods minimize the cost by relying on the PV forecasts. The occurrence of even rare events may lead to loss of loads and significant cost implications due to a lack of enough ancillary services or exceeding network constraints in the system. In contrast, the commitment decision will be expensive and unrealistic if the solution is very conservative for the rare events.

The conservativeness of decisions directly depends on the modeling of the uncertainty set. Distributionally robust optimization (DRO) for power system operations has been proposed in recent literature to improve the modeling of the uncertainty. DRO uses a probability distribution function, unlike the classic uncertainty set-based and worst-case scenario methods. Incorporating an accurate model that can represent the variability and uncertainty of the PV generation can improve the reliability of the DRO approach.

The case study by Veysi Raygani [70] addresses a critical issue on the robust unit commitment (RUC), i.e., the construction of an accurate and reliable uncertainty set for solar PV systems. The uncertainty set for solar generation is constructed based on the type of day, levels of daily uncertainty index (DUI) and daily energy index (DEI), and the uncertainty level of solar ramps. Then, it formulates a two-stage RUC with solar generations for a solar day categorized as clear, overcast, or highly uncertain. The RUC models the loads and solar generations in the uncertainty sets. The results are demonstrated on the IEEE 118-bus test system [71].

10.4.1.1   Uncertainty set for PV generations

The uncertainty set for solar generations is composed of three parts. The first part defines the lower and upper bounds for the day-ahead solar forecast. The second part accounts for the variation of ramp rate for aggregated solar generation. The third part defines the level of DUI and DEI for each type of day. The standard deviation of ramps, DUI, and DEI can be determined by analyzing historical data.

The upper and lower bounds of the solar generation forecast are given as follows [70]:

$$pv_{j,t} \leq (1 + e^{up})PV_{j,t}^{up} \tag{10.4}$$

$$pv_{j,t} \geq (1 - e^{low})PV_{j,t}^{low} \tag{10.5}$$

where $pv_{j,t}$ is the realization of solar generation j at time t. e^{up} and e^{low} are the upper- and lower-bound forecast errors, respectively. For a highly uncertain day, as shown in Figure 10.1, PV^{up} is the solar forecast under clear sky and PV^{low} is the solar forecast under overcast and obscure conditions. On a purely clear-sky day, PV^{up} and PV^{low} are clear-sky solar generation forecasts, and on a completely overcast day, PV^{up} and PV^{low} are overcast generation forecast.

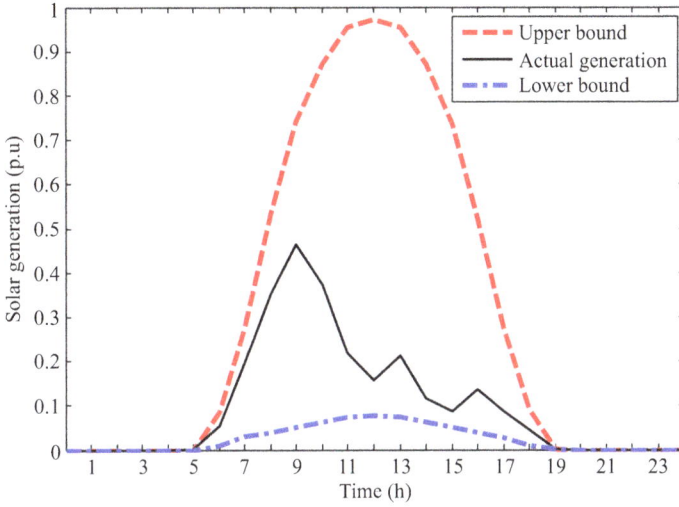

Figure 10.1   PV generation on a highly uncertain day

The uncertainty set for solar generation's ramps is defined as follows [70]:

$$\left| \sum_{j \in \mathrm{NPV}} (\mathrm{pv}_{j,t} - \mathrm{pv}_{j,t-1}) \right| \leq k\sigma_{\mathrm{NPV}}. \tag{10.6}$$

$$\text{where } \sigma_{\mathrm{NPV}} = \sqrt{\sum_{j \in \mathrm{NPV}} \sum_{j' \in \mathrm{NPV}} \sigma_j \sigma_{j'} \rho_{j,j'}} \quad \forall j, j' \in \mathrm{NPV} \tag{10.7}$$

where $\mathrm{pv}_{j,t} - \mathrm{pv}_{j,t-1}$ represents the ramp of solar generation j in two consecutive time intervals and NPV is the total number of solar generation units. σ_j and $\sigma_{j'}$ represent the standard deviation of the ramps of the solar generations j and j', respectively, and $\tilde{\rho}_{j,j'}$ is their correlation coefficient. Note that the solar generation's ramps are fitted to a normal distribution function with the mean value of zero and the standard deviation, σ. The term $k\sigma_{\mathrm{NPV}}$ defines the budget of uncertainty, where σ_{NPV} is the standard deviation of ramps of aggregated solar generations, and the parameter k is the width of the confidence interval.

To control the conservatives of the model, the DUI and DEU are defined as follows [70]:

$$\mathrm{DEI}_j = \frac{\sum_{t \in T} \mathrm{pv}_{j,t}}{\sum_{t \in T} \hat{\mathrm{pv}}_{j,t}} \tag{10.8}$$

where $\mathrm{pv}_{j,t}$ is the realization of solar generation of unit j at time t and $\mathrm{pv}_{j,t} \in [\hat{\mathrm{pv}}_{j,t} - \overline{\mathrm{pv}}_{j,t}, \hat{\mathrm{pv}}_{j,t} + \underline{\mathrm{pv}}_{j,t}]$. $\hat{\mathrm{pv}}_{j,t}$, $\overline{\mathrm{pv}}_{j,t}$, and $\underline{\mathrm{pv}}_{j,t}$ are the solar generation forecast, the upper bound, and the lower bound of forecast errors of unit j at time t, respectively. The DEI for a solar unit j is the ratio of the realization of solar energy to the forecast of solar energy. The DUI represents the ratio of the length of realized solar generation over the solar generation forecast, defined as follows:

$$\text{DUI}_j = \frac{\sum_{t\in T} |pv_{j,t} - pv_{j,t-1}|}{\sum_{t\in T} |\widehat{pv}_{j,t} - \widehat{pv}_{j,t-1}|} \tag{10.9}$$

The maximum value of DUI on a clear-sky day is when the solar generation is always above the clear-sky forecast. The minimum value of the DUI on all types of days is when the solar generation is always below the solar forecast.

10.4.1.2   Two-stage RUC model

The general form of the traditional UC for a fixed forecast of demand and solar generation is as follows [70]:

$$\min_{x,y}\ \ (c^T x + d^T y) \tag{10.10}$$

$$\text{s.t.}\ \ Ax \geq b, \quad x\ \text{binary} \tag{10.11}$$

$$Ex + Fy \geq h \tag{10.12}$$

$$Gy + Hp_{pv} = Iu \tag{10.13}$$

The cost function is composed of two parts: the first part, $c^T x$, represents the unit's start-up and shut-down costs, and the second part, $d^T y$, represents the dispatch cost. The x and y represent the vectors of binary and dispatch-related variables, respectively. The binary decision variables are shut-down, start-up, and commitment states, and the dispatch-related decision variables are the operating level of generation units. Equation (10.11) refers to shut-down/start-up and minimum on/off time constraints. Equation (10.12) represents the general form of all dispatch-related constraints. It includes coupling constraints of the dispatch and commitment decisions, as well as ramp and reserve capabilities. Equation (10.13) represents the generation and demand energy balance for solar generation and demand forecasts.

Considering the PV forecast uncertainty, the general form of the two-stage RUC is as follows [70]:

$$\min_{x}\ \left(c^T x + \max_{pv\in PV, u\in U}\ \min_{y\in\omega(u,pv,x)}\ d^T y\right) \tag{10.14}$$

$$\text{s.t.}\ \ \omega(u,pv,x) = \{y : Ex + Fy \geq h, Gy + Hp_{pv} = Iu\} \tag{10.15}$$

$$Ax \geq b, x\ \text{binary} \tag{10.16}$$

where $\omega(u,pv,x)$ defines the set of feasible dispatch solutions under uncertainties for a fixed commitment decision x and the realization of solar generations. The second-stage (dispatch) cost represented by '$\max_{pv\in PV, u\in U} \min_{y\in\omega(u,pv,x)} d^T y$' defines the minimum dispatch cost for a fixed commitment decision under worst-case uncertainty sets of U and PV.

This case study uses Bender's decomposition technique based on the two-level hierarchy of the RUC. The first-stage master problem (MP) is a mixed integer program (MIP) and the second-stage subproblem finds the cost of dispatch under the worst-case scenarios of solar and load lying within the defined uncertainty sets. The modified IEEE 118-bus test system is used to analyze and test the RUC. There are 54 generating units, 12 large-scale solar systems, and 91 loads, where the peak load is 5400 MW. The CPLEX solver is selected and the convergence parameters are $\epsilon = 1 \times 10^{-3}$ and $\Delta = 5 \times 10^{-5}$.

The optimization results show the impact of PV generation uncertainty on RUC performance across three distinct solar conditions: clear, overcast, and uncertain days. In the clear-day scenario (Figure 10.2), the PV forecast follows a smooth and predictable curve with narrow uncertainty bounds, resulting in relatively small differences between day-ahead and worst-case unit commitment costs.

However, under overcast (Figure 10.3) and uncertain conditions (Figure 10.4), the PV forecast becomes significantly less reliable, exhibiting erratic patterns and broader uncertainty bounds. For these two conditions, the results are compared not to the day-ahead forecast but to a classical robust unit commitment (CRUC) approach. The comparison shows that the proposed data-driven RUC (DRUC) method achieves

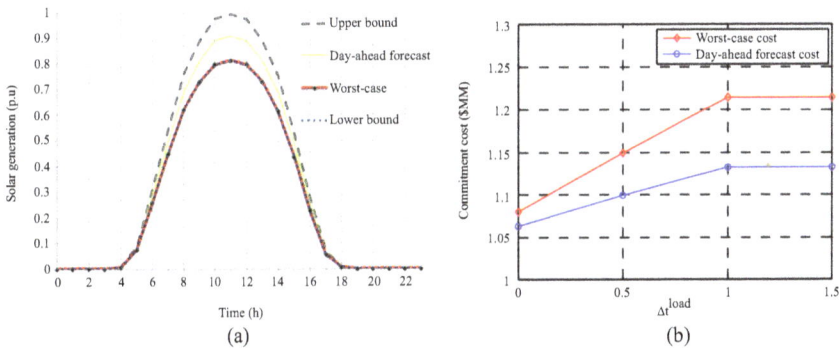

Figure 10.2   Forecast and unit commitment cost for a clear day [70]. (a) PV forecast uncertainty. (b) Unit commitment cost.

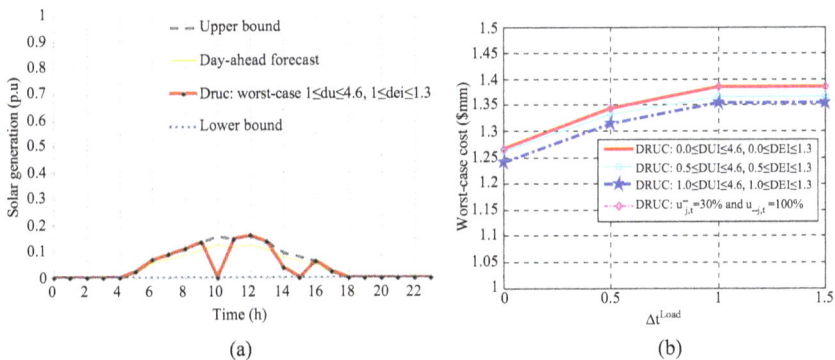

Figure 10.3   Forecast and unit commitment cost for an overcast day [70]. (a) PV forecast uncertainty. (b) Unit commitment cost.

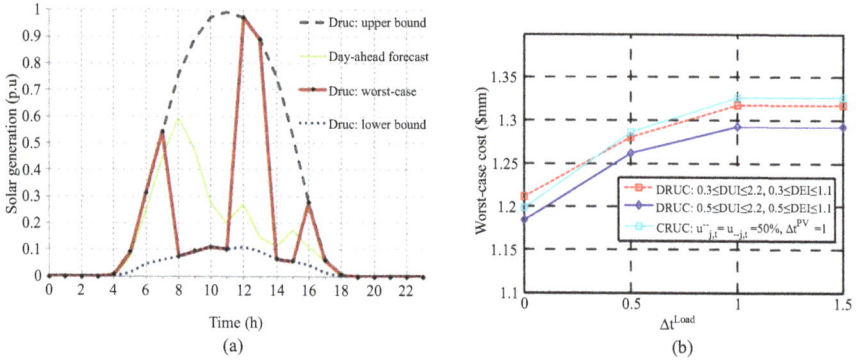

Figure 10.4   Forecast and unit commitment cost for an uncertain day [70]. (a) PV
             forecast uncertainty. (b) Unit commitment cost.

lower worst-case costs under various uncertainty set configurations, especially as load deviation increases. This demonstrates the value of more adaptive and scenario-tailored uncertainty modeling, which outperforms traditional CRUC methods in handling high variability and uncertainty in renewable generation.

This case study presented a reliable approach for the two-stage RUC that relies on the clear-sky and overcast solar forecasts. The solar generation forecast has been categorized into a clear, overcast, or highly uncertain day. The uncertainty set for solar PV generations is constructed based on the level of DUI, DEI, and the uncertainty level of solar ramps. The numerical results on the IEEE 118-bus test systems verified that the utilized approach is substantially efficient in cost management for all types of days.

10.4.2   Case 2: overvoltage control in distribution networks using PV forecast

Overvoltage is a critical issue that limits the integration of PV systems into the grid. In distribution grids, voltage levels tend to rise during periods of peak solar generation, which can lead to reverse power flow and compromise grid stability. To maintain reliable operation, advanced voltage control methods are required.

This case study presents a coordinated voltage control approach to address local overvoltage issues using local devices such as on-load tap changers (OLTCs), batteries, and the RPC capability of PV inverters [33]. The control method employs a stochastic two-level optimization approach. In the first level, the goal is to determine the optimal schedule for the OLTC at the reference bus and the batteries over 24 h. This step is based on day-ahead forecasts and considers PV generation uncertainty. In this control level, batteries are scheduled to reduce the impact of PV power on voltage rise. The optimal decisions from this stage serve as inputs to the second level. In the second level, the focus is on keeping the voltage within an acceptable range by using the RPC capability of PV inverters during 1-h intervals.

In this study, each PV-battery bus consists of a PV panel and a battery. The bus is connected to the distribution grid through a three-phase inverter, as shown in Figure 10.5. As illustrated, the typical bus includes a PV system, a battery, and a local load as the main connected components [33].

On the DC side of each PV-battery bus, the PV system is connected to a battery. On the AC side, the inverter converts overall power to AC and generates active power $P_{g_{i,\omega,t}}$ and reactive power $Q_{g_{i,\omega,t}}$. Depending on operating conditions, the active and reactive power can flow in both directions. In a stochastic problem, the net active and reactive power at each bus is calculated as follows:

$$P_{i,\omega,t} = P_{g_{i,\omega,t}} - P_{l_{i,t}} \qquad \forall i, \forall \omega, \forall t \tag{10.17}$$

$$Q_{i,\omega,t} = \pm Q_{g_{i,\omega t}} - Q_{l_{i,t}} \qquad \forall i, \forall \omega, \forall t \tag{10.18}$$

Here, $P_{i,\omega,t}$ and $Q_{i,\omega,t}$ represent the net active and reactive power at bus i, scenario ω, and time period t, respectively. Subsequently, the net power is used to support voltage control.

The control objective is to improve the voltage profile in distribution grids while maximizing the power injected by PV systems, ensuring that voltage remains within a safe operating range. The objective function in both the first and second levels of the voltage control strategy is to minimize voltage deviation from the nominal value of one per unit [33]:

$$\min \left(\sum_{i=1}^{N_I} \sum_{\omega=1}^{N_\Omega} \sum_{t=1}^{N_T} \pi_\omega \cdot |V_{i,\omega,t} - 1| \right) \tag{10.19}$$

where π_ω is the probability of occurrence for each scenario, and $V_{i,\omega,t}$ is the voltage magnitude (in per unit) at bus i, scenario ω, and time interval t. N_I, N_ω, and N_T represent the number of PQ buses, scenarios, and time steps, respectively. In the following, a detailed explanation is provided of each level in the proposed voltage control strategy.

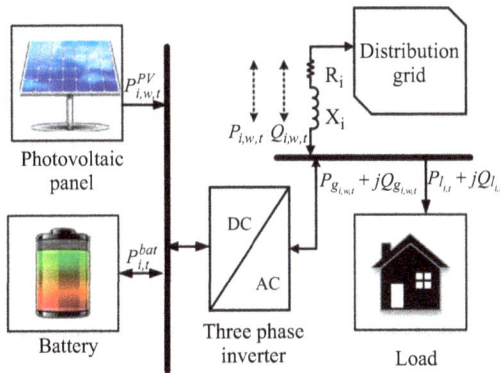

Figure 10.5 PV-battery bus system [33].

10.4.2.1   First control level: daily scheduling

One of the key advantages of batteries is their ability to control the rate of active power injection into the grid. By storing energy and releasing it at appropriate times, batteries help maintain a more stable voltage profile. For instance, on a typical sunny day, high PV generation combined with low load demand can cause the voltage to rise. If the voltage exceeds the allowable range, some or all of the PV output may need to be curtailed. In this case, the battery can act as a buffer, storing energy during peak solar generation and discharging it during periods such as nighttime.

In the first level of the voltage control strategy, the goal is to schedule the tap positions of the OLTC at the reference bus and determine the optimal battery operation over the next 24 time intervals to support voltage regulation. The objective function for this level is given by (10.19), maintaining the voltage profile close to 1 per unit. This optimization problem includes the following sets of constraints.

The OLTC constraint determines the voltage at the reference bus based on the tap position at each time interval:

$$V_{\text{ref},t} = 1 + \left(tp_t \times \frac{V^{\max} - V^{\min}}{tp^{\max} - tp^{\min}} \right) \quad \forall t \tag{10.20}$$

Battery dynamics describe the state of charge (SoC) for each battery regarding the active power exchanged with the grid:

$$SoC_{i,t} = SoC_{i,t-1} + P_{i,t}^{\text{ch}}\eta_c - \frac{P_{i,t}^{\text{dch}}}{\eta_d} \quad \forall i, \forall t \tag{10.21}$$

Power flow equations represent the active and reactive power flow at each bus, reflecting the electrical behavior of the distribution grid:

$$P_{i,\omega,t} = \sum_{\substack{j=1 \\ j \neq i}}^{N_I} V_{i,\omega,t} V_{j,\omega,t} Y_{ij} \cos(\delta_{i,\omega,t} - \delta_{j,\omega,t} - \theta_{ij}) \quad \forall i, \forall \omega, \forall t \tag{10.22}$$

$$Q_{i,\omega,t} = \sum_{\substack{j=1 \\ j \neq i}}^{N_I} V_{i,\omega,t} V_{j,\omega,t} Y_{ij} \sin(\delta_{i,\omega,t} - \delta_{j,\omega,t} - \theta_{ij}) \quad \forall i, \forall \omega, \forall t \tag{10.23}$$

Under normal conditions, the OLTC at the reference bus helps maintain the voltage profile within an acceptable range. Optimal battery scheduling can support voltage regulation and improve the overall voltage profile during peak hours. However, the excess PV generation may still cause voltage levels to exceed the acceptable range, even with battery support. The second level of voltage control addresses these scenarios by employing the RPC capability of PV inverters.

10.4.2.2   Second control level: reactive power compensation

In the second level, the RPC capability of PV inverters is used to adjust the voltage profile over the next hour. According to [72], the distribution grid is first partitioned into multiple microgrids to enable more effective control and operation. After apply-ing the network partitioning method, the distribution system is divided into *N* separate partitions, as illustrated in Figure 10.6. Since the partitions are optimally defined, the voltage control strategy can be applied independently to each one. Within each partition, two types of buses are identified: PV buses, which have voltage control capability, and normal buses, which do not. Therefore, the injection or absorption of reactive power at the PV buses is regulated in this control level.

To implement the second-level voltage control, the process begins with a load flow analysis. Based on the results, the partition that includes the bus with the high-est overvoltage is identified as the most critical. Then, a zonal voltage control strategy is applied, where only the PV buses within this critical partition are allowed to absorb reactive power. After this adjustment, another load flow analysis is conducted. If overvoltage persists, the process is repeated for the next most critical partition. This iterative approach continues until either the voltage at all buses is within the accept-able range or the reactive power absorption capacity of all PV inverters in all partitions is fully utilized. The optimization problem in the second stage must also satisfy the following voltage (10.24) and inverter capability (10.25) constraints:

$$V^{\min} \leq V_{i,\omega,t} \leq V^{\max} \quad \forall i, \forall \omega, \forall t \tag{10.24}$$

$$S_{\text{inv}_{i,\omega,t}} = \sqrt{P_{g_{i,\omega,t}}^2 + Q_{g_{i,\omega,t}}^2} \quad \forall i, \forall \omega, \forall t \tag{10.25}$$

Figure 10.7a presents the voltage profiles of buses under different scenarios, including the base case, installation of PV units at candidate buses, voltage control using only the first level, and the two-level voltage control strategy. Moreover, Fig-ure 10.7b shows the results for bus 25, identified as the most critical node in the system, over the next 24 time periods. As can be seen, the two-level control strat-egy with operation scheduling in the first level and reactive power absorption in the second level can effectively maintain the voltage within the acceptable range.

Table 10.1 shows an evaluative analysis of the voltage control method for mul-tiple buses from different partitions of the system. The results highlight the effect of

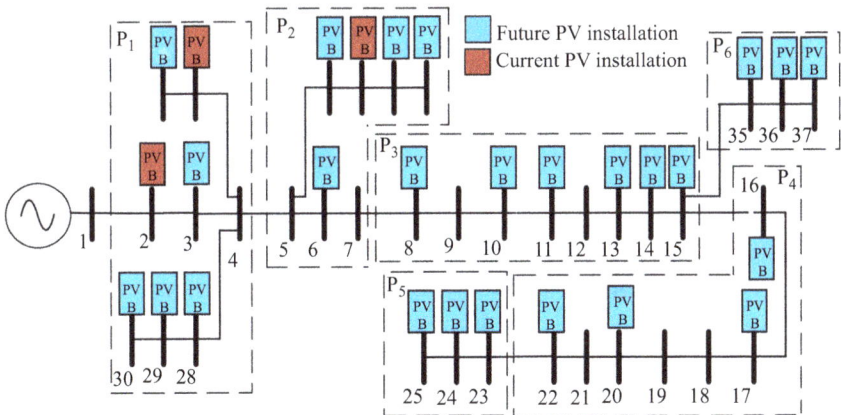

Figure 10.6   37-Bus distribution feeder

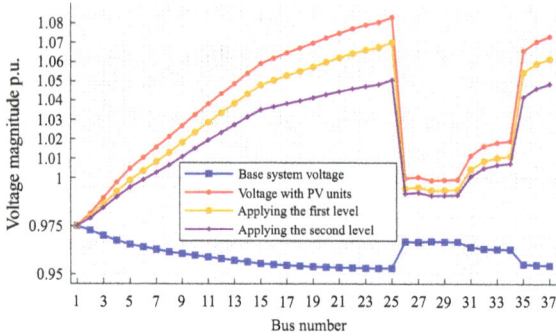

(a)

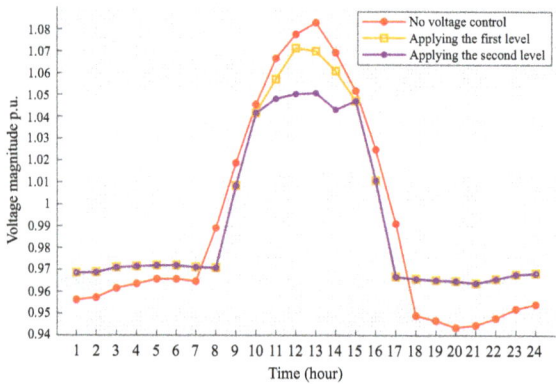

(b)

Figure 10.7   *Two-level voltage control using PV inverters. (a) Voltage profile for*
different modes in the time period with the highest solar penetration.
(b) Voltage profile of the bus 25 for the next 24 time periods.

battery scheduling in the first level and RPC in the second level of the control. The data corresponds to hour 13, which represents the time with the highest solar generation. During this hour, the OLTC tap position is set to level 4, and the corresponding voltage at the reference bus is 0.975 p.u. For example, the voltage magnitude at bus 25 before applying any control strategy is 1.083 p.u., which exceeds the acceptable limit. After applying the first level of control, the voltage improves to 1.071 p.u., but it is still outside the permissible range. The second level is then applied, which further reduces the voltage to 1.05 p.u. by absorbing 429 kVar. These results confirm that during periods of high PV generation, the presented two-level control strategy can maintain voltage stability of the distribution grid.

This case study highlights the role of PV inverters in grid voltage control and emphasizes the critical role of accurate PV forecasting. It demonstrates how long-term, day-ahead forecasts can be used for operational scheduling, while

Table 10.1 Buses characteristics before and after applying voltage control.

Bus No.	No control V (p.u.)	After first level (Scheduling)		After second level (RPC)	
		Battery charging (kW)	V (p.u.)	Q absorption (kVar)	V (p.u.)
2	0.981	40	0.978	0	0.978
8	1.021	10	1.013	0	1.006
20	1.072	20	1.060	165	1.043
25	1.083	52	1.071	429	1.050
32	1.016	78	1.008	0	1.005
37	1.073	38	1.061	0	1.048

short-term forecasts support real-time control through the reactive power capabilities of PV inverters.

10.4.3 Case 3: EV charging in a microgrid and PV forecast accuracy

This section presents a case study that analyzes the effect of forecast accuracy on the overall electricity costs at an EV charging station, which is integrated with solar PV power and stationary battery energy storage. The EMS utilizes a rolling horizon (RH) strategy, implemented with a MILP. The study is extensively described in [73]. The EMS takes inputs from EV load forecasts generated by LSTM neural networks and persistence models, as well as PV production forecasts using a physical hybrid artificial neural network (PHANN) [5]. The accuracy of the EMS was previously validated on a microgrid, demonstrating less than 10% variation in state variables compared to offline simulations. Simulations conducted across different seasons revealed that forecast accuracy significantly influences operational costs, with a 10% increase in cost attributed to inaccuracies in both PV and EV load forecasts. Additionally, combining PV and EV forecasts plays a more crucial role in reducing errors than focusing on each individually.

The testing facility is the MG2Lab microgrid (Figure 10.8), which is composed of several key components selected for this study [74]. The grid connection is limited to 100 kW and enables both the import and export of electricity to and from the distribution network. Two PV arrays, with installed capacities of 23 kWp and 25 kWp, respectively, provide nondispatchable renewable energy. Energy storage is managed through two lithium-ion battery energy storage systems (BESS1 and BESS2), offering capacities of 70 kWh and 67.5 kWh, respectively. Both systems operate with a C-rate of 1 and maintain a minimum SOC of 35% and a maximum of 85%, resulting in a combined effective storage capacity of 68.75 kWh. The system load is emulated using a back-to-back (B2B) converter, which functions as a controllable load that follows setpoints derived from an EV load dataset, scaled to a peak power of 25 kW. The microgrid employs a two-tier EMS that operates on an RH framework. The first layer (see Section 10.2.2), executed every 15 min, focuses on optimizing the BESS

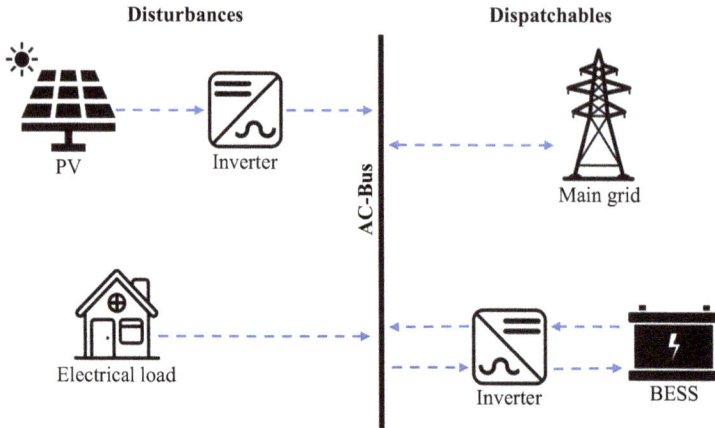

Figure 10.8   *Simple scheme of a system comprising PV generation, battery storage, an electrical domestic load, and a connection to the main grid. The scheme also shows the PV inverter that converts PV-generated DC current into AC and the battery bidirectional inverter.*

dispatch over the next 24-h period. This layer uses forecasts for both EV load and PV generation as inputs to minimize the system's operational costs while respecting technical constraints. Due to the inevitable errors in these forecasts, power imbalances may occur, which are then corrected in real time by the second EMS layer.

The first layer uses a MILP model that accounts for various power flows within the microgrid, such as PV production, BESS charge/discharge, and grid import/export. The objective of the MILP is to minimize overall operational costs across the planning horizon. The EMS uses forecasted PV production as an input, with the option to curtail excess generation when needed. The BESSs are modeled based on energy capacity, factoring in self-discharge rates and charge/discharge efficiencies. Additionally, the system balances the SOC between multiple BESS units. Interaction with the grid is regulated by limiting the maximum power that can be imported or exported, ensuring grid stability.

The second layer employs a heuristic strategy that takes inputs from the BESS and the grid setpoints established in the first layer. It also incorporates discrepancies between the forecasts for PV generation and EV loads, compared to the actual measurements. Depending on the direction and magnitude of these discrepancies, the microgrid may experience a "power excess," where incoming power exceeds outgoing power, or a "power deficit." The approach is to maintain the setpoint for power exchange with the grid as determined by the first layer.

If the first layer indicates that both unmet demand and PV curtailment are at zero, the BESS is activated to help balance the power. Should the BESS be unable to rectify the imbalances, adjustments to the power-exchange setpoint with the grid will be made.

This study employs a predict-then-optimize framework that combines forecasting models with optimization methods to improve decision-making in uncertain conditions while ensuring computational efficiency for application in an experimental setup. In this framework, forecasts for EV loads and PV generation serve as inputs for the optimization process.

10.4.3.1   PV forecasting with PHANN

The solar PV forecasting method employs a hybrid approach that combines physical models with regression techniques. This method was developed and validated at the MG2Lab before [5]. It generates a time series of solar production forecasts for the following day, with data points at 15-min intervals. The PHANN model (Figure 10.9) inputs include a numerical weather forecast specific to the site, specifications of the PV modules, and the latest available solar production time series. It predicts solar output based on numerical weather forecasts for the upcoming week at the specified locations and the clear-sky radiation (CSR) incident on the solar array. The weather data utilized is sourced from the ECMWF [3] and is provided by a private entity. The forecasting tool incorporates several factors: GHI, air temperature, wind speed and direction, precipitation, humidity, and atmospheric pressure. The CSR model relies on established evaluations of solar transmittance, specifically using the Reference Evaluation of Solar Transmittance, 2 bands (REST2) methodology [75], which incorporates more variables compared to other atmospheric models [76]. Additionally, the GHI must be adjusted to align with the solar plant's orientation based on its azimuth and tilt angles. An ANN is trained daily using the measured data from the previous day in a supervised learning manner, allowing for the optimization of weights and biases to produce a series of outputs accurately based on prior inputs. This ANN undergoes retraining each day to adapt to short-term trends, such as consistent underestimations in temperature forecasts.

10.4.3.2   Operational scheduling

The case study represents an EV charging station powered by the grid, PV, and BESS, see Figure 10.10. The EV load is based on the records from the Jet Propulsion Laboratory (JPL) database from the adaptive charging network (ACN) [77]. This site has 52 EV supply equipment (EVSE) stations and is only accessible by employees, making it a good representation of a typical workplace schedule. The power curve of all 52 EVSEs is aggregated to reconstruct the whole charging station demand. The

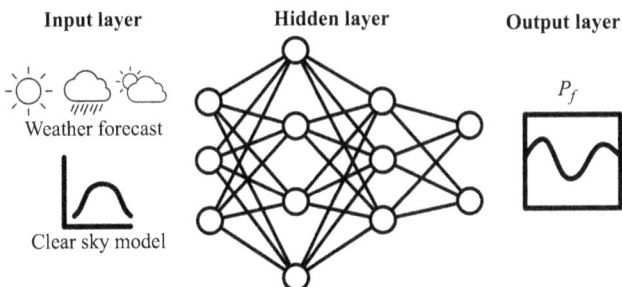

Figure 10.9   Physical hybrid artificial neural network architecture

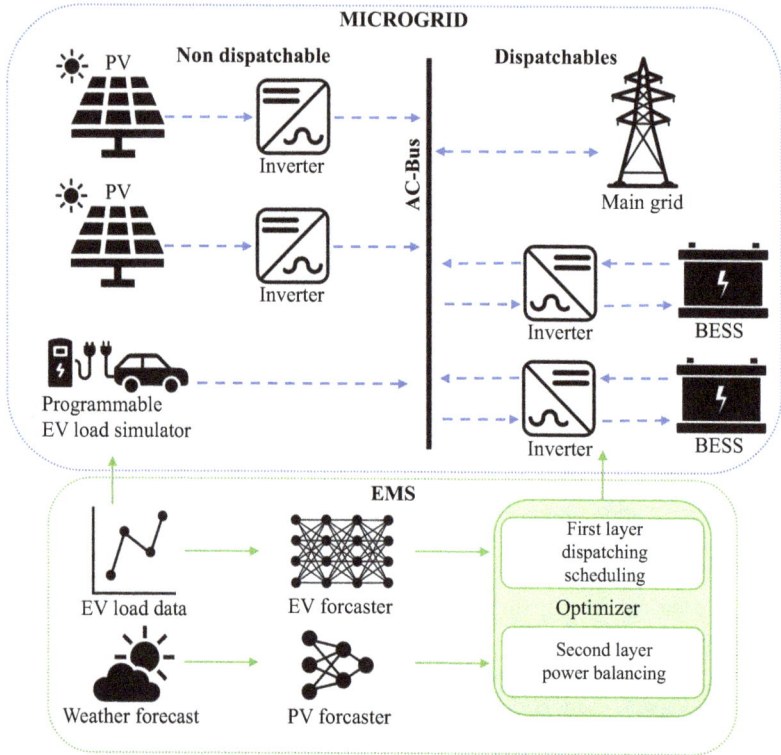

Figure 10.10   Control architecture and dispatching scheme

considered simulations and experiments cover the sessions from December 2018, 3 to January 2020, 4. To accommodate the charging station demand within the MG2Lab capabilities, the aggregated load power is scaled to a maximum of 25 kW. The cost of purchasing the electricity from the grid is assumed to be twice the Italian national unified price (Prezzo Unico Nazionale, PUN) to account for transmission and other costs, while the revenue from selling electricity to the national grid is set equal to the zonal price (Prezzo Zonale, PZ) of the Italian North region. These costs are taken from the Italian electricity market [78] for the year 2023.

Both online and offline simulations were conducted to validate the methods developed. The goal of the offline simulations is to assess the impact of each forecasting method in the overall optimization and the impact of combining PV and EV forecasting. Online simulations involve implementing the EMS and forecasting models in the MG2Lab experimental facility under real-world conditions, whereas offline simulations are computer-based, using real data inputs. The purpose of the online simulations is to demonstrate that the developed models, including the PV and EV forecasting methods, can be applied in actual system operations. These simulations were conducted over a 2-week period on the MG2Lab microgrid, utilizing either the persistence model or the LSTM model for EV load prediction. For consistency, the same EV load was programmed during both weeks, while the PV production and forecast were based on actual irradiance data at MG2Lab. The data gathered from the online experiments were subsequently used to run offline simulations with both forecasting models for comparison and validation of the approach.

Offline simulations highlight the impact of PV and EV load forecasts under varying conditions by comparing different forecasting methods over 8 weeks and 2 weeks per season of the year. These simulations integrate both PV and EV load forecasts, utilizing the JPL EV dataset and actual measurements from the MG2Lab PV field, corresponding to the specific season. The offline simulations, for each season, incorporate various EV forecasting techniques, including persistence and LSTM with attention [79,80]. Additionally, a benchmark scenario is introduced, called the hourly mean lookahead (HML), which assumes perfect forecasts for both PV and EV. The HML forecast predicts the exact hourly average of PV and EV load measurements over 15-min intervals. Since the error metrics are based on forecasted data (averaged over 15 min) and measured data (collected at 1-min intervals), statistical error metrics will not be zero, even for HML, unless all measurements within a 15-min window match their average exactly. The HML case establishes a lower bound for the simulations. This approach allows the offline simulations to assess not only the accuracy of individual forecasting models but also the combined statistical accuracy of PV and EV forecasts, the operational costs of the EV station, and the effect of forecast accuracy in comparison to the ideal scenario.

Since the EMS operates using a rolling horizon strategy, SOC of the BESS cannot be predetermined before the simulation. As various combinations of PV and load forecasts can result in different final SOC levels, any discrepancy between the initial and final storage content is treated as equivalent to energy imported from the grid. This energy is then valued at the lowest import price observed during the simulation period.

The results demonstrate to be consistent between the EMS operating in the microgrid with all the physical components and the simulations. Therefore, simulation results performed offline can be considered representative of the real operation of a microgrid. Minor discrepancies between online and offline are attributed to parasitic losses from the battery power supply systems and other equipment within the microgrid laboratory, primarily occurring during nighttime. Regarding the dispatching behavior, it can be noted that on weekdays when the electrical load is high, the microgrid imports electric energy at night when the electricity price is lower to charge the batteries, which are then discharged during the day in the absence of PV production. After validating the MG2Lab approach, offline simulations were performed to evaluate the effects of the selected PV and EV forecasting methods over an extended period. We only present the results from the spring weeks in detail for brevity. PV and EV were assessed both separately and combined as summarized in Table 10.2, in which MAE, symmetric mean absolute percentage error (SMAPE), and RMSE are used as error metrics. It can be seen that the LSTM model surpasses the persistence method across all statistical performance metrics. However, when combined with the PHANN day-ahead PV forecast, it produces a higher net error in electrical load forecasting compared to the other predictors. This initial finding highlights the importance of evaluating accuracy based on the overall forecast rather than

Table 10.2   Statistical performance of PV and EV forecasters (single and combined)

	Forecast	MAE	SMAPE	RMSE
PV	PHANN	4.10	71.91	7.49
	HML	0.72	6.85	2.13
EV	Persistence	1.58	44.89	3.23
	LSTM	1.06	44.66	1.87
	HML	0.16	7.33	0.35
PV+EV	PV HML + EV Persistence	1.94	−69.67	3.87
	PV HML + EV LSTM	1.44	−30.41	2.82
	PV HML + EV HML	0.76	0.29	2.15
	PV PHANN + EV Persistence	2.88	−7.73	5.27
	PV PHANN + EV LSTM	4.30	−388.93	7.63
	PV PHANN + EV HML	2.37	35.51	4.60

focusing on individual forecasters in isolation. Likewise, the development and calibration of forecasting tools should consider the entire system's performance, rather than concentrating solely on the accuracy of individual predictions.

Figure 10.11 presents the offline simulation results for one of the spring weeks analyzed. The operational scheduling for both forecasting methods is fairly consistent throughout the week, with only slight differences in the amount of electricity imported from the grid, which is marginally lower in the persistence model. A significant variation occurs on April 7 under the persistence model, and the EMS imports grid power only in the early hours of the day to meet demand, followed by discharging the BESSs. In contrast, the LSTM model scheduling imports electricity from the grid throughout the entire day without utilizing the BESSs, leading to higher operational costs, as detailed in Table 10.3.

In summary, the most favorable scenario for the two spring weeks under consideration happens when LSTM is paired with the HML PV forecast, whereas the persistence method performs better when combined with the PHANN PV forecast. This outcome highlights the importance of incorporating system management costs into the error metrics of forecasting algorithms, aiming to minimize errors related to costly balancing efforts.

The overall results (among all seasons) are summarized in Table 10.4 and Table 10.5. The overall net electrical errors show a strong correlation with the economic outcomes. Specifically, as the precision of the combined forecast errors improves, the operational costs tend to rise. The LSTM model shows better performance when paired with the HML PV forecast but underperforms compared to the persistence method when the PHANN PV forecast is utilized. Furthermore, the effects of EV and PV forecasts on economic results are quite comparable when assessed against the ideal scenario; in both instances, the variation in operating costs between perfect and actual forecasts is approximately 5%.

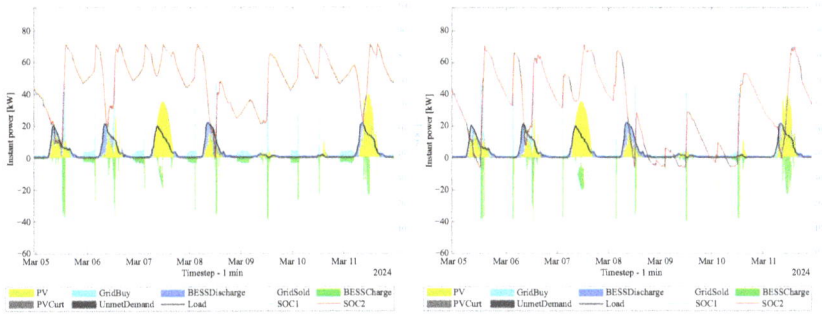

Figure 10.11 Operational scheduling for a considered week using LSTM forecast: experimental results (left) and offline results with the simulator (right)

Table 10.3 Comparison of forecasting methods for PV and load

Load Forecast	PV forecast: HML			PV forecast: PHANN		
	Persist.	LSTM	HML	Persist.	LSTM	HML
EV load [kWh]	1665.24	1665.24	1665.24	1665.24	1665.24	1665.24
PV production [kWh]	3328.39	3328.39	3328.39	3328.39	3327.34	3328.39
Electricity purchased [kWh]	214.59	214.88	144.99	184.22	176.48	150.00
Electricity sold [kWh]	1830.40	1833.37	1771.80	1774.63	1740.01	1757.46
Total BESS losses [kWh]	57.22	53.86	46.44	75.56	86.79	65.65
Initial SOC [kWh/%]	17.5/50.0	17.5/50.0	17.5/50.0	17.5/50.0	17.5/50.0	17.5/50.0
Final SOC [kWh/%]	12.6/35.9	12.9/36.9	12.5/35.6	16.1/46.0	23.4/66.8	12.5/35.8
Unmet demand [kWh]	0.00	0.00	0.00	0.00	0.00	0.00
Curtailment [kWh]	0.00	1.05	0.00	0.00	1.05	0.00
Purchased electricity [€]	50.61	48.34	29.14	47.42	51.47	35.07
Sold electricity [€]	278.30	279.24	267.79	265.40	253.35	261.38
Unmet demand [€]	0.00	0.00	0.00	0.00	0.00	0.00
Curtailment [€]	0.00	0.00	0.00	0.00	0.00	0.00
BESS residual [€]	0.40	0.37	0.40	0.11	-0.47	0.40
Total [€]	−227	−231	−238	−218	−202	−226

Table 10.4 Net load forecast error (RMSE)

Season	PV forecast [kW]	HML [kW]	PHANN [kW]	EV Forecast [kW]	Persistence [kW]	LSTM [kW]
Spring	3.87	2.82	2.15	5.27	7.63	4.60
Summer	3.39	2.22	1.38	4.89	5.05	3.78
Autumn	3.64	2.25	1.25	5.16	5.71	3.80
Winter	2.87	4.40	0.93	4.68	4.48	4.23

Table 10.5   Economic result

Season	PV forecast [€]	HML [€]	PHANN [€]	EV Forecast [€]	Persistence [€]	LSTM [€]
Spring	−227.29	−230.54	−238.35	−217.87	−202.35	−225.91
Summer	−197.85	−199.21	−201.44	−187.02	−181.94	−191.97
Autumn	46.50	44.86	38.92	48.49	49.68	43.90
Winter	218.31	216.65	205.86	218.19	218.15	215.59

In summary, the integration of PV and EV forecasts can result in discrepancies of up to 10% compared to the ideal case. Overall, these findings indicate that the effectiveness of an individual forecasting model diminishes when used alongside others. Additionally, forecasting models with higher accuracy may be outperformed in scenarios where the timing of the predictions is crucial, particularly in energy systems with fluctuating electricity prices. Thus, it is essential to accurately predict load or PV generation during peak electricity price periods, and this factor should be considered when developing forecasting methods.

10.5   Conclusions

The integration of PV systems into modern power grids represents a necessary step toward decarbonization. However, the variable and weather-dependent nature of solar generation introduces operational challenges. PV forecasting has emerged as critical tool for managing these challenges. This chapter presents the benefits of the integration of PV forecasting in energy systems from several points of view. One of the central challenges in PV integration is ensuring power quality and system reliability despite the inherent intermittency of solar resources. As demonstrated, short-term forecasting directly helps maintain voltage and frequency stability and reduce reliance on fossil-based reserves. We highlighted the growing importance of PV forecasting in real-time grid operations, particularly through techniques such as APC, RPC, and advanced inverter coordination. Accurate short-term forecasts enhance these control mechanisms, minimizing energy losses, curtailments, and disruptions. At the systems level, PV forecasting is necessary in the design and operation of EMSs, especially in hybrid systems combining renewables, storage, and flexible loads. In this case, medium-term PV forecasters, typically merging ML and statistical methods, provide essential input to EMS, which typically works in a classical predict-then-optimize approach. Indeed, forecast-informed EMS architectures can make predictive, economically optimal decisions regarding energy storage scheduling, DR participation, and grid interaction. The integration of AI-driven techniques such as reinforcement learning further enhances the potential of solar PV forecasting. Long-term PV forecasting provides information during the planning stage, when climate models and scenario analyses are used as a decision-making tools for investment definition.

To better foster PV integration with accurate forecasting, a supportive regulatory and incentive framework is needed. In fact, the advantage of operating a PV system with accurate forecasting not only affects the system owner but also generally contributes to broader social welfare. Increasing incentives on PV integration, especially when coupled with accurate forecasting and energy storage, in energy systems creates a virtuous cycle that will reduce the reliance on carbon-intensive sources while addressing the challenges of an intermittent energy source.

The presented case studies showcase the versatility and effectiveness of PV forecasting integration, improving the performance of energy systems when dealing with unit commitment, voltage control in distribution networks, and optimal management of an electric vehicle charging station.

References

[1]   Nwaigwe KN, Mutabilwa P, and Dintwa E. An overview of solar power (PV systems) integration into electricity grids. *Materials Science for Energy Technologies*. 2019;2(3):629–633.

[2]   Mohamad Radzi PNL, Akhter MN, Mekhilef S, *et al.* Review on the application of photovoltaic forecasting using machine learning for very short- to long-term forecasting. *Sustainability*. 2023;15(4).

[3]   Forecast charts and data. ECMWF; 2025 [Assessed 2025 Aug 1]. Available from: https://www.ecmwf.int/en/forecasts.

[4]   Mellit A, Massi Pavan A, Ogliari E, *et al.* Advanced methods for photovoltaic output power forecasting: A review. *Applied Sciences*. 2020;10(2).

[5]   Nespoli A, Ogliari E, Leva S, *et al.* Day-ahead photovoltaic forecasting: A comparison of the most effective techniques. *Energies*. 2019;12(9):1621.

[6]   Mosavi A, Salimi M, Faizollahzadeh Ardabili S, *et al.* State of the art of machine learning models in energy systems, a systematic review. *Energies*. 2019;12(7):1301.

[7]   Voyant C, Notton G, Kalogirou S, *et al.* Machine learning methods for solar radiation forecasting: A review. *Renewable Energy*. 2017;105:569–582.

[8]   Yao X, Fu X, and Zong C. Short-term load forecasting method based on feature preference strategy and LightGBM-XGboost. *IEEE Access*. 2022;10:75257–75268.

[9]   Abdel-Nasser M and Mahmoud K. Accurate photovoltaic power forecasting models using deep LSTM-RNN. *Neural Computing and Applications*. 2019;31(7):2727–2740.

[10] Sardarabadi A, Heydarian Ardakani A, Matrone S, *et al.* Multi-temporal PV power prediction using long short-term memory and wavelet packet decomposition. *Energy and AI*. 2025;21:100540.

[11] Badwawi RA, Abusara M, and Mallick T. A review of hybrid solar PV and wind energy system. *Smart Science*. 2015;3(3):127–138.

[12] Frank C, Fiedler S, and Crewell S. Balancing potential of natural variability and extremes in photovoltaic and wind energy production for European countries. *Renewable Energy*. 2021;163:674–684.

[13]   Santos-Alamillos FJ, Pozo-Vázquez D, Ruiz-Arias JA, *et al.* Combining wind farms with concentrating solar plants to provide stable renewable power. *Renewable Energy.* 2015;76:539–550.

[14]   Lindberg O, Arnqvist J, Munkhammar J, *et al.* Review on power-production modeling of hybrid wind and PV power parks. *Journal of Renewable and Sustainable Energy.* 2021;13(4):042702.

[15]   Kakoulaki G, Gonzalez Sanchez R, Gracia Amillo A, *et al.* Benefits of pairing floating solar photovoltaics with hydropower reservoirs in Europe. *Renewable and Sustainable Energy Reviews.* 2023;171:112989.

[16]   Liu Z and Du Y. Evolution towards dispatchable PV using forecasting, storage, and curtailment: A review. *Electric Power Systems Research.* 2023;223:109554.

[17]   Pellow MA, Emmott CJM, Barnhart CJ, *et al.* Hydrogen or batteries for grid storage? A net energy analysis. *Energy & Environmental Science.* 2015;8(7):1938–1952.

[18]   Mason J, Fthenakis V, Zweibel K, *et al.* Coupling PV and CAES power plants to transform intermittent PV electricity into a dispatchable electricity source. *Progress in Photovoltaics: Research and Applications.* 2008;16(8):649–668.

[19]   Angenendt G, Zurmühlen S, Axelsen H, *et al.* Comparison of different operation strategies for PV battery home storage systems including forecast-based operation strategies. *Applied Energy.* 2018;229:884–899.

[20]   Sorour A, Fazeli M, Monfared M, *et al.* Forecast-based energy management for domestic PV-battery systems: A U.K. case study. *IEEE Access.* 2021;9:58953–58965.

[21]   Yan Q, Zhang B, and Kezunovic M. Optimized operational cost reduction for an EV charging station integrated with battery energy storage and PV generation. *IEEE Transactions on Smart Grid.* 2019;10(2):2096–2106.

[22]   Al-Waeli AHA, Kazem HA, Chaichan MT, *et al.* The impact of climatic conditions on PV/PVT outcomes. In: Al-Waeli AHA, Kazem HA, Chaichan MT, *et al.*, editors. *Photovoltaic/Thermal (PV/T) Systems: Principles, Design, and Applications.* Cham: Springer; 2019. pp. 173–222.

[23]   Zahedi A. Maximizing solar PV energy penetration using energy storage technology. *Renewable and Sustainable Energy Reviews.* 2011;15(1):866–870.

[24]   Milano F, Dörfler F, Hug G, *et al.* Foundations and Challenges of Low-Inertia Systems (Invited Paper). In: *2018 Power Systems Computation Conference (PSCC)*; 2018. pp. 1–25.

[25]   Groß D, Bolognani S, Poolla BK, *et al.* Increasing the resilience of low-inertia power systems by virtual inertia and damping. In: *Proceedings of IREP'2017 Symposium. International Institute of Research and Education in Power System Dynamics (IREP)*; 2017. p. 64.

[26]   Poolla BK, Groß D, and Dörfler F. Placement and implementation of grid-forming and grid-following virtual inertia and fast frequency response. *IEEE Transactions on Power Systems.* 2019;34(4):3035–3046.

[27]   Bevrani H, Ise T, and Miura Y. Virtual synchronous generators: A survey and new perspectives. *International Journal of Electrical Power & Energy Systems.* 2014;54:244–254.

[28] Badrudeen TU, Nwulu NI, and Gbadamosi SL. Low-inertia control of a large-scale renewable energy penetration in power grids: A systematic review with taxonomy and bibliometric analysis. *Energy Strategy Reviews.* 2024;52:101337.

[29] Liu Z and Du Y. Evolution towards dispatchable PV using forecasting, storage, and curtailment: A review. *Electric Power Systems Research.* 2023;223:109554.

[30] Gui Y, Nainar K, Bendtsen JD, *et al.* Voltage support with PV inverters in low-voltage distribution networks: An overview. IEEE Journal of Emerging and Selected Topics in Power Electronics. 2024;12(2):1503–1522.

[31] Wang J, Xu W, Gu Y, *et al.* Multi-agent reinforcement learning for active voltage control on power distribution networks. *Advances in Neural Information Processing Systems.* 2021;34:3271–3284.

[32] Ahmed EEE, Demirci A, Poyrazoglu G, *et al.* An equitable active power curtailment framework for overvoltage mitigation in PV-rich active distribution networks. *IEEE Transactions on Sustainable Energy.* 2024;15(4):2745–2757.

[33] Emarati M, Barani M, Farahmand H, *et al.* A two-level over-voltage control strategy in distribution networks with high PV penetration. *International Journal of Electrical Power & Energy Systems.* 2021;130:106763.

[34] Zheng W, Wu W, Zhang B, *et al.* A fully distributed reactive power optimization and control method for active distribution networks. *IEEE Transactions on Smart Grid.* 2016;7(2):1021–1033.

[35] Xu X, Li J, Xu Z, *et al.* Enhancing photovoltaic hosting capacity—A stochastic approach to optimal planning of static var compensator devices in distribution networks. *Applied Energy.* 2019;238:952–962.

[36] Jianguo Z, Qiuye S, Huaguang Z, *et al.* Load balancing and reactive power compensation based on capacitor banks shunt compensation in low voltage distribution networks. In: *Proceedings of the 31st Chinese Control Conference*; 2012. p. 6681–6686.

[37] Cordova S, Rudnick H, Lorca Á, et al. An efficient forecasting-optimization scheme for the intraday unit commitment process under significant wind and solar power. *IEEE Transactions on Sustainable Energy.* 2018;9(4): 1899–1909.

[38] Furukakoi M, Adewuyi OB, Matayoshi H, *et al.* Multi-objective unit commitment with voltage stability and PV uncertainty. *Applied Energy.* 2018;228:618–623.

[39] Wang R, Xiong J, Fan He M, *et al.* Multi-objective optimal design of hybrid renewable energy system under multiple scenarios. *Renewable Energy.* 2020;151:226–237.

[40] Sedhom BE, El-Saadawi MM, El Moursi MS, *et al.* IoT-based optimal demand side management and control scheme for smart microgrid. *International Journal of Electrical Power & Energy Systems.* 2021;127:106674.

[41] Elmouatamid A, Ouladsine R, Bakhouya M, *et al.* Review of control and energy management approaches in micro-grid systems. *Energies.* 2021;14(1):168.

[42] Vasilakis A, Zafeiratou I, Lagos DT, *et al.* The evolution of research in microgrids control. *IEEE Open Access Journal of Power and Energy.* 2020;7: 331–343.

[43] Fathima AH and Palanisamy K. Optimization in microgrids with hybrid energy systems – A review. *Renewable and Sustainable Energy Reviews.* 2015;45:431–446.

[44] Gao K, Wang T, Han C, *et al.* A review of optimization of microgrid operation. *Energies.* 2021;14(10):2842.

[45] García Vera YE, Dufo-López R, and Bernal-Agustín JL. Energy management in microgrids with renewable energy sources: A literature review. *Applied Sciences.* 2019;9(18):3854.

[46] Ahmad Khan A, Naeem M, Iqbal M, *et al.* A compendium of optimization objectives, constraints, tools and algorithms for energy management in microgrids. *Renewable and Sustainable Energy Reviews.* 2016;58:1664–1683.

[47] Thirunavukkarasu GS, Seyedmahmoudian M, Jamei E, *et al.* Role of optimization techniques in microgrid energy management systems—A review. *Energy Strategy Reviews.* 2022;43:100899.

[48] Camacho EF and Bordons C. *Model Predictive Control.* 2nd ed. London: Springer; 2013.

[49] Mahmoud MS, Alyazidi NM, and Abouheaf MI. Adaptive intelligent techniques for microgrid control systems: A survey. *International Journal of Electrical Power & Energy Systems.* 2017;90:292–305.

[50] Heydarian Ardakani A, Hurink J, Gibson I, *et al.* Embedding temporal awareness in reinforcement learning models for energy system control. In: *Proceedings of the 16th ACM International Conference on Future and Sustainable Energy Systems*; 2025. pp. 1002–1004.

[51] Zhang Y, Ren Y, Liu Z, *et al.* Federated deep reinforcement learning for varying-scale multi-energy microgrids energy management considering comprehensive security. *Applied Energy.* 2025;380:125072.

[52] Gao Y, Hu Z, Yamate S, *et al.* Unlocking predictive insights and interpretability in deep reinforcement learning for Building-Integrated Photovoltaic and Battery (BIPVB) systems. *Applied Energy.* 2025;384:125387.

[53] Palensky P and Dietrich D. Demand side management: Demand response, intelligent energy systems, and smart loads. *IEEE Transactions on Industrial Informatics.* 2011;7(3):381–388.

[54] De La Nieta A, Pérez-Ruiz D, Hernández-Díaz AG, *et al.* Impact of PV and EV forecasting in the operation of a microgrid. *Inventions.* 2021;6(3):32.

[55] Zarei M, Chen J, Li Y, et al. Forecast-based demand side management for microgrids using deep learning. *Frontiers in Energy Research.* 2025;13:1567808.

[56] Han Y, Wu J, Yu W, *et al.* Real-time demand side management algorithm using stochastic optimization. *Energies.* 2018;11(5):1166.

[57] Dai Y, Qi Y, Li L, *et al.* A dynamic pricing scheme for electric vehicle in photovoltaic charging station based on Stackelberg game considering user satisfaction. *Computers & Industrial Engineering.* 2021;154:107117.

[58]   Huang C, Zhang H, Song Y, *et al.* Demand response for industrial micro-grid considering photovoltaic power uncertainty and battery operational cost. *IEEE Transactions on Smart Grid.* 2021;12(4):3043–3055.

[59]   Yang D, Li W, Yagli GM, *et al.* Operational solar forecasting for grid integration: Standards, challenges, and outlook. *Solar Energy.* 2021;224:930–937.

[60]   Carley S. State renewable energy electricity policies: An empirical evaluation of effectiveness. *Energy Policy.* 2009;37(8):3071–3081.

[61]   Lukanov BR and Krieger EM. Distributed solar and environmental justice: Exploring the demographic and socio-economic trends of residential PV adoption in California. *Energy Policy.* 2019;134:110935.

[62]   U S Department of Energy. *Smart Grid Investment Grant Program – Final Report.* Office of Electricity Delivery and Energy Reliability; 2016. Accessed: 2025-07-14. Available from: https://www.smartgrid.gov/document/smart-grid-investment-grant-program-final-report.html.

[63]   Lukanov BR and Krieger EM. The impact of the German feed-in tariff scheme on innovation: Evidence based on patent filings in renewable energy technologies. Energy Economics. 2017;67:545–553.

[64]   Ministry of Economy, Trade and Industry (METI), Japan. *Society 5.0: Co-creating the Future*; 2019. Accessed: 2025-07-14. https://www.meti.go.jp/english/policy/society5_0/index.html.

[65]   Alliance IS. *Forecasting and Grid Integration: Insights from Japan's Society 5.0 Energy Vision*; 2022. https://isolaralliance.org/uploads/docs/a85d84a7a402f9209262f855b273da.pdf.

[66]   The Australian. *Energy regulator tweaks rules to boost household earnings and improve forecasting*; 2024. https://www.theaustralian.com.au/business/renewable-energy-economy/energy-regulator-tweaks-rules-to-boost-household-earnings-and-improve-forecasting/news-story/0d35c9e57352a949178ebe86e0981f10.

[67]   Jamasb T. Incentive regulation of electricity and gas networks in the UK: From RIIO-1 to RIIO-2. *Economics of Energy & Environmental Policy.* 2020.

[68]   ENTSO-E. *Connection Network Codes – Requirements for Generators*; 2019. Accessed: 2025-07-14. https://www.entsoe.eu/network_codes/cnc/.

[69]   European Commission. *EU Emissions Trading System (EU ETS)*; 2023. Accessed: 2025-07-14. https://climate.ec.europa.eu/eu-action/eu-emissions-trading-system-eu-ets_en.

[70]   Veysi Raygani S. Robust unit commitment with characterised solar PV systems. *IET Renewable Power Generation.* 2019;13(6):867–876.

[71]   Peña I, Martinez-Anido CB, and Hodge BM. An Extended IEEE 118-Bus Test System With High Renewable Penetration. *IEEE Transactions on Power Systems.* 2018;33(1):281–289.

[72]   IEEE Guide for Design, Operation, and Integration of Distributed Resource Island Systems with Electric Power Systems. IEEE Std 15474-2011. 2011; pp. 1–54.

[73]   Manzolini G, Fusco A, Gioffrè D, *et al.* Impact of PV and EV forecasting in the operation of a microgrid. *Forecasting.* 2024;6(3):591–615.

[74] Polimeni S, Moretti L, Manzolini G, *et al.* Numerical and experimental testing of predictive EMS algorithms for PV-BESS residential microgrid. In: *2019 IEEE Milan PowerTech*; 2019. pp. 1–6.

[75] Gueymard CA. REST2: High-performance solar radiation model for cloudless-sky irradiance, illuminance, and photosynthetically active radiation – Validation with a benchmark dataset. *Solar Energy*. 2008;82(3): 272–285.

[76] Stein JS, Hansen CW, and Reno MJ. *Global Horizontal Irradiance Clear Sky Models: Implementation and Analysis*. Livermore, CA: Sandia National Laboratories (SNL); 2012.

[77] Lee ZJ, Li T, and Low SH. ACN-data: Analysis and applications of an open EV charging dataset. In: *Proceedings of the Tenth ACM International Conference on Future Energy Systems*; 2019. pp. 139–149.

[78] Gestore dei Mercati Energetici (GME), editor. *Prezzo Unico Nazionale (PUN) – Italian Day-Ahead Market Price*. Gestore dei Mercati Energetici (GME); 2024. Accessed: 2025-05-19. Available from: https://www.mercatoe lettrico.org/It/Default.aspx.

[79] Matrone S, Ogliari E, Nespoli A, *et al.* Electric vehicle supply equipment day-ahead power forecast based on deep learning and the attention mechanism. *IEEE Transactions on Intelligent Transportation Systems*. 2024;25(8): 9563–9571.

[80] Matrone S, Ardakani AH, Ogliari E, *et al.* Probabilistic forecast of EV charging demand using quantile regression and LSTM with attention mechanism. In: *Proceedings of the 16th ACM International Conference on Future and Sustainable Energy Systems*; 2025. pp. 1005–1007.

Index